INFRARED AND MILLIMETER WAVES

VOLUME 2 INSTRUMENTATION

CONTRIBUTORS

B. L. BEAN

J. R. BIRCH

N. C. LUHMANN, JR.

WALLACE M. MANHEIMER

T. J. PARKER

S. PERKOWITZ

D. VÉRON

INFRARED AND MILLIMETER WAVES

VOLUME 2   INSTRUMENTATION

Edited by   *KENNETH J. BUTTON*

NATIONAL MAGNET LABORATORY
MASSACHUSETTS INSTITUTE OF TECHNOLOGY
CAMBRIDGE, MASSACHUSETTS

1979

ACADEMIC PRESS
A Subsidiary of Harcourt Brace Jovanovich, Publishers

New York   London   Toronto   Sydney   San Francisco

ACADEMIC PRESS, INC.
111 Fifth Avenue, New York, New York 10003

United Kingdom Edition published by
ACADEMIC PRESS, INC. (LONDON) LTD.
24/28 Oval Road, London NW1   7DX

Library of Congress Cataloging in Publication Data
Main entry under title:

Infrared and millimeter waves.

    Includes bibliographies and index.
    CONTENTS: [etc.] −−v. 2. Instrumentation.
    1. Infrared apparatus and appliances. 2. Milli−
meter wave devices. I. Button, Kenneth J.
TA1570.I52   621.36'2   79−6949
ISBN 0−12−147702−9 (v. 2)

PRINTED IN THE UNITED STATES OF AMERICA

79 80 81 82   9 8 7 6 5 4 3 2 1

D
621·362
INF

CONTENTS

Chapter 4   Far Infrared Submillimeter Spectroscopy with an Optically Pumped Laser
B. L. Bean and S. Perkowitz

Chapter 5   Electron Cyclotron Heating of Tokamaks
Wallace M. Manheimer

LIST OF CONTRIBUTORS

Numbers in parentheses indicate the pages on which the authors' contributions begin.

B. L. BEAN* (273), *Physics Department, Emory University, Atlanta, Georgia 30322*

J. R. BIRCH (137), *Division of Electrical Sciences, National Physical Laboratory, Teddington, England*

N. C. LUHMANN, JR. (1), *Department of Electrical Sciences, University of California, Los Angeles, California 90024*

WALLACE M. MANHEIMER (299), *Naval Research Laboratory, Washington, D.C. 20375*

T. J. PARKER (137), *Department of Physics, Westfield College, University of London, London, England*

S. PERKOWITZ (273), *Physics Department, Emory University, Atlanta, Georgia 30322*

D. VÉRON (67), *Association Euratom—CEA sur la Fusion, Département de Physique du Plasma et de la Fusion Contrôlée, Centre d'Études Nucléaires, Fontenay-aux-Roses, France*

*Present address: Science Applications, Inc., White Sands Missile Range, New Mexico 88002.

PREFACE

The rapid development of this emerging technology has resulted in a large number of new instruments and techniques. We intend to include in this treatise a state-of-the-art description of all of these.

The previous volume (Volume 1: Sources of Radiation) dealt with the optically pumped laser, the gyrotron, solid state devices for generation and modulation of millimeter waves, backward wave oscillators, the ledatron, and relativistic electron beam devices. In this volume we proceed to the description of the newest instrumentation and methods, which can be used for a variety of applications. Neville Luhmann has provided a summary of techniques used in plasma diagnostics, and his chapter is backed up by D. Véron's thorough treatment of the use of lasers for precise interferometric measurements. Bean and Perkowitz show that this sort of instrumentation can generally be applied not only to plasma diagnostics but also to spectroscopic measurements in liquids and in semiconductor and superconductor reflection and transmission. Their chapter on submillimeter spectroscopy with an optically pumped laser not only describes the construction and operation of the laser spectrometer, including the CO_2 pumped laser, the theory and performance of the far infrared laser, and the detection system, but also the present and future applications.

At the first opportunity we included a chapter on an important new far infrared technique, dispersive Fourier transform spectroscopy, and we were fortunate to have J. R. Birch and T. J. Parker prepare it. Their chapter could almost stand as a monograph on this most effective of the far infrared spectroscopic methods. It covers the great improvements that have come to classical Fourier transform spectroscopy.

The plasma diagnostic theme of this volume would not be complete without Wallace Manheimer's chapter on electron cyclotron heating in tokamaks (magnetically confined plasmas). This is the "timely" chapter of Volume 2 for those entering the specialty of plasma research or, more broadly, fusion research. A statement that heating is the "name of the game" in plasma studies at this moment would be too strong because it would imply that the problem will soon be solved and that when solved we shall have fusion power. Not at all. The break-even fusion reaction in magnetically confined plasmas will be demonstrated in the laboratory, but one can hardly say how or when. Since fusion research is so well financed, we shall follow the development of its infrared and millimeter wave diagnostics in order to pick up the fall-out benefits for other areas of research and for military applications.

Volumes 3, 4, and 5 will contain several chapters on detectors and receivers (Volume 3), spectroscopic methods and the interaction of electromagnetic waves with matter (Volume 4), and millimeter wave components, techniques, and systems (Volume 5). These will be followed by a book on millimeter and submillimeter wave astronomy.

We wish to express our gratitude to the authors and to those industrial and academic institutions that have created the atmosphere in which scholarly research and reporting can thrive. We all realize, at least occasionally, that the sponsors of the arts and sciences have played an essential role during the past 200 years in bringing us to that point which we now call the state of the art.

CONTENTS OF OTHER VOLUMES

INFRARED AND MILLIMETER WAVES

VOLUME 2   INSTRUMENTATION

CHAPTER 1

Instrumentation and Techniques for Plasma Diagnostics: An Overview*

N. C. Luhmann, Jr.

I. Introduction

In recent years there has arisen considerable interest in the plasma state of matter. The rapidly expanding research areas include controlled thermonuclear fusion, isotope separation, high-energy collective effect particle accelerators, magnetospheric physics, and cosmic ray physics. Basically, a plasma is a collection of charged particles that interact with each other by means of the long-range Coulomb forces. This gives rise to collective modes of behavior not found in ordinary neutral gases. One characteristic plasma scale length is the Debye length

$$\lambda_D = [KT/(4\pi n_0 e^2)]^{1/2},$$

where K is Boltzmann's constant, T the electron temperature, n_0 the electron density, and e the electron charge. This is essentially the distance over which a stationary test charge will alter the electrostatic potential in a plasma. One can distinguish a plasma from the low-density charged particle beams, found in many particle accelerators and television cathode ray tubes, by the plasma parameter $g = 1/n\lambda_D^3$. When $g \ll 1$ there are many particles within a Debye

---

* Supported by U.S. Department of Energy, National Science Foundation, and U.S. Air Force Office of Scientific Research.

sphere and one has a plasma. There are also two characteristic frequencies, the plasma frequency and cyclotron frequency, that prove useful in our description of a plasma. The plasma frequency $\omega_p = [4\pi n_0 e^2/m]^{1/2}$ and is the frequency at which particles displaced from their equilibrium positions will oscillate due to the electrostatic Coulomb restoring force. For electrons, we have $f_{pe} = \omega_{pe}/(2\pi) = 8.87 \times 10^3 [n_e \,(\text{cm}^{-3}]^{1/2}$. The cyclotron frequency $\omega_c = qB/(mc)$ is the frequency at which charged particles gyrate about magnetic field lines of strength B. For electrons, we have $f_{ce} = \omega_{ce}/(2\pi) = 2.8B \,(\text{kG})\text{GHz}$.

Much of the effort aimed at an understanding of the plasma state has been oriented toward the achievement of controlled thermonuclear fusion. Here one obtains energy through the fusion of light nuclei. Two reactions often considered involve deuterium and tritium (Rose and Clark, 1961):

$$D + T \rightarrow (\text{He}^4 + 3.52 \text{ MeV}) + (n + 14.06 \text{ MeV})$$
$$D + D \rightarrow (T + 1.01 \text{ MeV}) + (p + 3.03 \text{ MeV})$$
$$\searrow (\text{He}^3 + 0.82 \text{ MeV}) + (n + 2.45 \text{ MeV}).$$

The collision cross section of the first reaction is larger so that it will be used in the first fusion reactors. However, the D–T fusion reaction cross section is still insignificant for energies below $\simeq 5$ keV so that one requires temperatures in excess of 10^6 K to penetrate the strong Coulomb repulsive barrier between colliding nuclei. To achieve fusion energy, one must prevent the hot plasma from contacting material walls until the fusion reaction has taken place and meaningful energy released. To obtain useful energy from the D–T reaction, one requires that $T > 10$ keV and that the Lawson criterion (Lawson, 1957) $n\tau > 10^{14}$ be satisfied where n is the plasma density (particles per cubic centimeter) and τ the confinement time in seconds. It is obvious that the confinement time restriction is relaxed as the plasma density increases. This has led to the so-called inertial confinement approach to fusion in which a pellet of D–T is compressed to $\simeq 10^4$ times solid density by means of an intense laser beam (Brueckner and Jorna, 1974) or charged particle beam. This scheme shall not be discussed further as there do not, at the present time, appear to be any applications for far infrared systems. The alternate approach is to employ magnetic fields to confine plasmas in the 10^{14}–10^{16} cm^{-3} density range.

There are many important applications of far infrared systems in the determination of the physical parameters of a magnetically confined fusion plasma. Ideally, one would hope to obtain the electron and ion distribution functions $f(\mathbf{r}, \mathbf{v}, t)$. Then by taking the appropriate velocity moment, one has the density, temperature, etc. However, one must usually settle for less than a complete description and be content with the temperatures, density, and lowest-order transport coefficients. As we shall see, submillimeter wave systems are essential for several of these measurements.

II. Interferometry

The electron density in laboratory and fusion plasmas can be determined indirectly from a measurement of the dielectric constant, which is density dependent. For a cold, unmagnetized plasma, the dielectric constant is given by

$$\varepsilon(r) = 1 - [\omega_{pe(r)}^2/\omega^2] = 1 - [n_{e(r)}/n_{crit}]$$

where ω_{pe} is the electron plasma frequency, n_e the electron density, ω the probing frequency, and n_{crit} the so-called critical density where $\omega^2 = \omega_{pe}^2$. For $\omega < \omega_{pe}$, the index of refraction becomes purely imaginary and the wave is evanescent. The dielectric constant becomes anisotropic and wave number dependent for a warm, magnetized plasma. The interested reader is referred to the excellent reference books available (Krall and Trivelpiece, 1973; Stix, 1962; Montgomery and Tidman, 1964).

The phase shift that an electromagnetic wave undergoes in traversing a plasma slab of thickness L is given by

$$\phi = k_0 L - \int_0^L k_p(x) \, dx$$

$$= (\omega_0 L/c)\left\{1 - (1/L) \int_0^L [1 - (n_{e(x)}/n_{crit})]^{1/2} \, dx\right\}.$$

When the probe frequency is much higher than the cutoff frequency, this expression simplifies considerably for a constant density plasma:

$$\phi = (L\omega_0/2c)(n_e/n_{crit}) = (\pi L/\lambda_0)(n_e/n_{crit}).$$

Figure 1 shows the phase shift per unit path length as a function of electron density for various probe frequencies including those appropriate for available lasers. The graphs are only continued to approximately $n_{crit}/4$ beyond which strong refractive effects dominate. To obtain several fringe shifts in the next generation of fusion devices ($r_p \simeq 1\,\text{m}$, $n_e \simeq 10^{14}\,\text{cm}^{-3}$) requires a far infrared laser.

The standard interferometer employed in the microwave region is shown schematically in Fig. 2. The rf signal traversing the plasma chamber undergoes a phase change as indicated in the preceding expressions. The signal is then combined with a reference signal on a detector whose output is proportional to $\cos\phi$. The variable phase shifter permits the signal to be nulled in the absence of a plasma. It is obviously convenient to have an interferometer

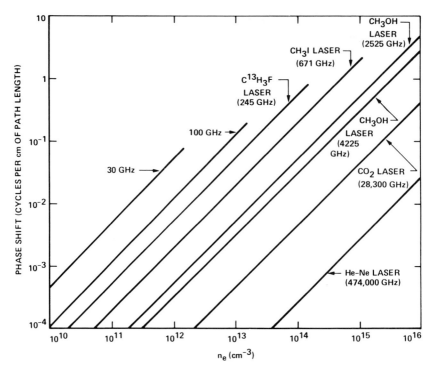

FIG. 1. Phase shift per unit path length as a function of electron density for several presently available lasers.

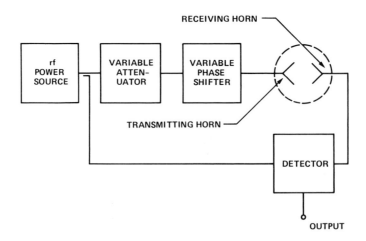

FIG. 2. Standard microwave interferometer.

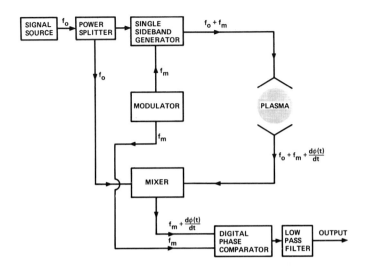

FIG. 3. Schematic of improved microwave interferometer. The single-sideband modulator and digital phase comparator provide an output directly proportional to plasma density.

system whose output signal is directly proportional to the phase shift and hence the plasma density. A realization of such a system in the microwave region is shown in Fig. 3. The single-sideband generator shifts the signal frequency f_0 to $f_0 + f_m$. This signal is further shifted in frequency (phase modulated) as it traverses the plasma column due to the aforementioned phase shift. The signal is then down-converted to $f_m + (d\phi/dt)$ and enters the digital phase comparator together with a reference signal of f_m. The digital phase comparator (Ernst, 1971; Meddas and Taylor, 1974) basically consists of two digital counters; one counting upward and the other downward. The outputs are processed by a digital adder, fed to a D/A converter which provides a signal proportional to the phase shift instead of cos ϕ as in conventional systems (Jahoda and Sawyer, 1971). The system response in the microwave region is obviously limited by the phase comparator circuitry rather than the maximum attainable modulation frequencies. Present systems have frequency responses of $\simeq 1$ MHz. However, it is conceivable that 100-MHz response can be achieved with premium digital integrated circuits.

Although simple to implement in the microwave region, the single-sideband generator requires more ingenuity in the submillimeter wave region. Presently, there are only two workable schemes. The first, an HCN laser system devised by Véron (1974), makes use of a rotating grating to provide the frequency offset indicated in Fig. 3. As shown in Figs. 4 and 5, the FIR laser beam undergoes a Doppler shift as it is reflected from the rotating grating.

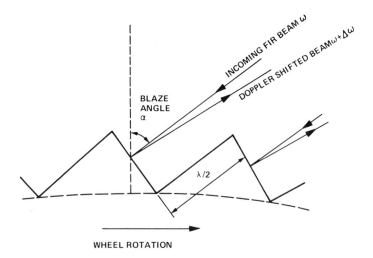

FIG. 4. Rotating grating used to provide the frequency shift in a far infrared interferometer (after D. Véron, 1974).

Although in typical operation only $\simeq 10$ kHz frequency shifts are obtained, it appears that with multiple reflections and high-speed operation of the rotating grating frequency shifts in the megahertz range can be achieved. More recently, Wolfe and his co-workers (1976) at the MIT National Magnet Laboratory have made use of two optically pumped 118.8-μm CH_3OH lasers (see Fig. 6) which are set to a frequency difference of 1 MHz $\pm$ 2% by

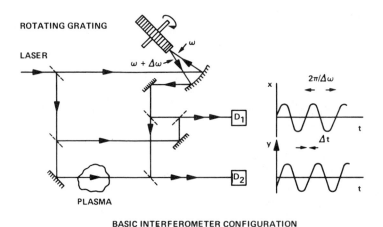

FIG. 5. Basic far infrared laser interferometer (after D. Véron, 1974).

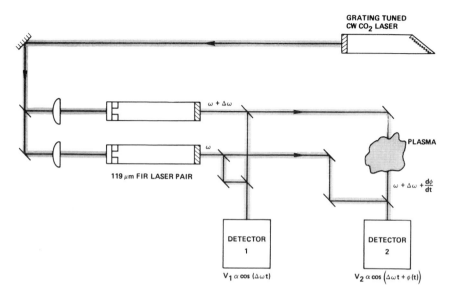

FIG. 6. Dual beam modulated far infrared interferometer (after S. Wolfe *et al.*, 1976).

a feedback system. Although both systems have proven to work well, it would still be desirable to have a solid state modulator.

Nothing has yet been said about source power requirements. This is extremely important as the available power is quite restricted above $\simeq 100$ GHz as compared to the more familiar centimeter wave region. The long path lengths and beam transport distances and a desirability for multi-channel operation and fast time response impose a burden on the source. Figure 7 contains a comparison of coherent cw source technologies in the region between 50 and 10 mm. The cw CO_2 laser is indicated for comparison purposes. At frequencies below $\simeq 300$ GHz, backward wave oscillators will provide $\gtrsim 30$ mW of output power ($\gtrsim 1$ W for narrowband tubes). While IMPATT diodes at the 10-mW level exist below 200 GHz, for frequencies above $\simeq 300$ GHz, one must employ far infrared lasers. It is, therefore, useful to review the present cw FIR laser technology as listed in Table I. With the exception of the electrically excited HCN (Belland and Véron, 1973) and DCN lasers (Véron *et al.*, 1978), all of the preceding results were obtained with optimal pumping of the indicated molecular gases using a dc CO_2 laser (Hodges *et al.*, 1976; Danielewicz and Weiss, 1978; Galantowicz *et al.*, 1978).

The optical pumping process is relatively simple to understand and is illustrated schematically in Fig. 8 for cw operation. The near coincidence of a vibrational rotation absorption line of a polar molecule and an infrared laser emission line results in the selective population of a particular rotational

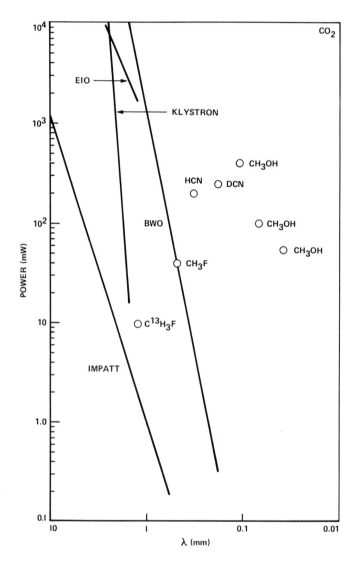

Fig. 7. Comparison of cw FIR optically pumped laser performance with conventional millimeter sources.

TABLE I

CW FIR Laser Technology

Molecule	Wavelength (μm)	FIR power (mW)
CH_3OH	42	55
CH_3OH	71	100
CH_3OH	97	25
CH_3OH	119	200
CH_2F_2	185	33
DCN	190	250
CH_2F_2	215	24
HCN	337	200
HCOOH	393	30
HCOOH	420	30
CH_3I	447	40
CH_3F	496	11
CH_3OH	570	8
CH_2CF_2	890	3.5
$C^{13}H_3F$	1222	10

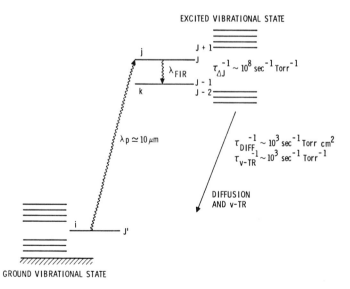

FIG. 8.   Energy level representation of optical pumping of polar molecules.

sublevel in an excited vibrational state. The far infrared lasing transition occurs between adjacent rotational states J and $J - 1$ in the upper vibrational state. Collisions occurring at a rate τ_{Λ}^{-1} tend to thermalize the rotational levels. In cw operation molecular diffusion to the walls maintains the population inversion against these thermalizing collisions. Although optical pumping is an inherently inefficient process the high power available from tunable cw CO_2 lasers has resulted in the discovery of hundreds of FIR lasing transitions, many of which produce power outputs sufficiently high for plasma diagnostics application.

Several choices of detector are available for use in FIR plasma interferometry systems. The MIT National Magnet Laboratory CH_3OH 118.8-μm interferometer system employs a cooled Ge:Ga detector, while the single-channel, 337-μm, HCN system developed by Véron utilized a cooled InSb detector. Both detectors possess an NEP in the neighborhood of 10^{-11} to 10^{-12} W/Hz$^{1/2}$ when utilized in the video mode. The high power available at 337 μm together with an efficient optical system has permitted the use of less sensitive room-temperature pyroelectric detectors in the eight-channel HCN interferometer operating on the TFR tokamak (Véron *et al.*, 1977). Finally, the recent advances in room-temperature Schottky barrier diodes makes them attractive candidates as detectors for FIR interferometry applications (McColl, 1977; McColl *et al.*, 1977; Fetterman *et al.*, 1974).

It should be clear from the preceding discussion that one measures the chord-averaged line density and not the density directly in plasma interferometry. Therefore, one must unfold the interferometric information to determine $n_e(r)$. The process is illustrated schematically in Fig. 9. Consider a beam passing through an inhomogeneous cylindrical plasma of radius R. The phase shift is given by

$$\phi(y_1) = (2x_1\omega_0/c)\left\{1 - (1/(2x_1)) \int_{-x_1}^{x_1} [1 - (n_e(x, y_1)/n_{crit})]^{1/2} \, dx\right\}$$

$$= (2x_1\omega_0/c)\{1 - [I(y_1)/x_1]\}.$$

If we assume axisymmetry, $n_e(x, y) = n_e(r)$, and this can be rewritten as

$$I(y_1) = x_1 - [\lambda_0 \phi(y_1)/(4\pi)] = \int_{y_1}^{R_1} \{1 - [n_e(r)/n_{crit}]\}^{1/2} r \, dr/(r^2 - y^2)^{1/2}.$$

The density $n_e(r)$ is then obtained by inverting this integral equation. We then obtain

$$\{1 - [n_e(r)/n_{crit}]\}^{1/2} = -(1/\pi) \int_r^R I'(y)/(y^2 - r^2)^{1/2},$$

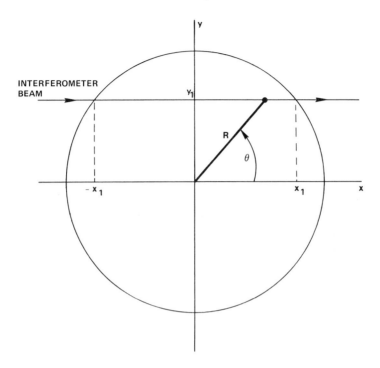

FIG. 9.   Schematic of FIR interferometer geometry.

where the prime denotes differentiation with respect to y. One then must have a sufficient number of interferometer channels to evaluate the preceding integral. In the limit $n_e \ll n_{crit}$, the two preceding expressions become, respectively,

$$\phi(y_1) \simeq 2\pi/\lambda_0 n_{crit} \int_{y_1}^{R_1} rn_e(r) \, dr/(r^2 - y^2)^{1/2}$$

and

$$n_e(r) \simeq (-\lambda_0 n_{crit}/2\pi^2) \int_r^R \phi'(y) \, dy/(y^2 - r^2)^{1/2}.$$

In the absence of the above assumed azimuthal symmetry, the determination of the density profile becomes more difficult. However, in this case two-dimensional image reconstruction similar to that employed in tomographic x-ray scans and astronomy may be used. This method has, for example, been used in both visible plasma spectroscopy (Myers and Levine, 1978) and in the x-ray studies of tokamak disruptive instabilities (Sauthoff et al., 1978).

The presence of a transverse density gradient, and therefore index of refraction gradient, gives rise to a bending of the laser beam as it traverses the plasma. Since the plasma index of refraction $n_p = \varepsilon^{1/2} = [1 - (n_e/n_{crit})]^{1/2}$, we see that an interferometer beam will experience a larger index of refraction in low-density regions of the plasma. The phase delay is therefore larger in these regions, and the beam will be deflected from the high-density regions. The angular deflection α in traversing the plasma of thickness L is

$$\alpha = \int_0^L (1/n_p)(dn_p/dx)\, dz.$$

In terms of the plasma dielectric constant this becomes

$$\alpha = (1/2n_c) \int_0^L (dn_e/dx)\, dz/[1 - (n_e/n_{crit})]$$

$$= [-e^2\lambda_0^2/(2\pi mc^2)] \int_0^L (dn_e/dx)\, dz/[1 - (n_e/n_{crit})].$$

For a tokamak plasma of 50-cm, semiminor radius and 10^{14}-cm^3 peak density, the deflection angle $\alpha \simeq 3°$ at 447 μm assuming a parabolic density profile.

It is obvious that the presence of an angular beam deflection greatly complicates the interferometry problem. The most serious effect occurs for densities $\gtrsim 0.1 n_{crit}$ and for large transverse gradients. In this case the departure of the ray trajectories from the straight line case illustrated in Fig. 9 becomes significant and must be accounted for in the inversion process. The neglect of the correction results in a density profile that is enhanced near the surface of the plasma and depressed at the center (Kahl and Wedemeyer, 1964). In addition, the refraction may actually result in insufficient power falling on the detectors to maintain a satisfactory signal-to-noise ratio. Hosea and Jobes (1975) have made an extensive study of the problems of long-wavelength, multichannel interferometry for large tokamaks such as PLT.

III.   Faraday Rotation

Since a plasma is diamagnetic, the local magnetic field is often quite different from the vacuum magnetic field and determined by the plasma velocity distribution. For example, in the Lawrence Livermore Laboratories 2XIIB mirror device there is presently a serious need for a measurement of the local magnetic field to see if field reversal has been achieved (Simonen et al., 1978). One approach to the magnetic field measurement is to make use of the Faraday rotation of an electromagnetic wave which propagates along the magnetic field through the plasma region.

If one launches a plane wave along the magnetic field, there is a rotation of the plane of polarization since the wave numbers are not equal for the right- and left-hand circularly polarized components of the wave. The wave numbers are given by (Krall and Trivelpiece, 1973)

$$k_R = (\omega/c)\{1 - (\omega_{pe}^2/\omega^2)/[1 - (\omega_{ce}/\omega)]\}^{1/2}$$

and

$$k_L = (\omega/c)\{1 - (\omega_{pe}^2/\omega^2)/[1 + (\omega_{ce}/\omega)]\}^{1/2}.$$

The ratio of the electric field components for a wave launched along the magnetic field axis at $z = 0$ is

$$E_x/E_y = \cot[\tfrac{1}{2}(k_L - k_R)z].$$

For both ω_{pe}, $\omega_{ce} \ll \omega$ this becomes

$$E_x/E_y = \cot\{[2\pi n e^3 B_0/(m^2 c^2 \omega^2)]^z\}$$

or numerically

$$\theta \text{ (degrees/cm-path length)} = 1.5 \times 10^{-17} \lambda^2 \text{ (mm)} n_e(\text{cm}^{-3}) B_0(\text{Gauss}).$$

If we assume a fusion plasma with $n_e = 10^{14}$ cm^{-3} and $B_0 = 50$ kG we see that large rotation angles are obtained in the FIR: 15°/cm for 447 μm (CH$_3$I laser) and 1.1°/cm for 118.8 μm (CH$_3$OH laser).

The Livermore group (Stallard et al., 1978) has employed an HCN laser (337 μm) for their measurements in the 2XIIB magnetic mirror device ($n_e \simeq 10^{14}$ cm^{-3}, $B_0 \simeq 6$ kG) with an expected rotation angle of $\simeq 1 - 30°$. The laser and modulation system were provided by Véron. Presently a modulation frequency of 100 kHz is employed. Here one of the greatest problems is beam propagation and machine access. The laser beam path is illustrated schematically in Figs. 10 and 11. Device complexities require the propagation of the laser beam $\simeq 20$ m to the beam splitter table where the beam undergoes an approximately 12-m roundtrip in order to enter the machine and traverse the 29–50-cm-long plasma. In order to reduce diffraction losses, they have employed 10-cm-diameter glass pipe as a dielectric waveguide in the EH$_{11}$ mode (Degnan, 1973). Approximately 10–15 mW of FIR are presently available at the beam splitter table.

The detection system presently uses Schottky diodes and off-axis parabolic reflection optics. Typical sensitivities of $\simeq 5$–10 V/W are obtained. The system has already been utilized in the interferometer mode and gives a density in agreement within 10 to 20% of the values obtained using their standard 2-mm microwave interferometer and neutral beam attenuation diagnostics. As indicated in Fig. 12 for Faraday rotation measurements, the linearly polarized signal beam ($\gtrsim 100 \pm 1$ extinction ratio) passes through

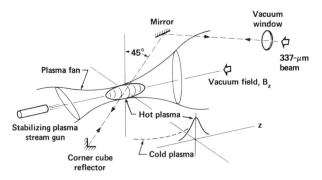

FIG. 10.   HCN laser Faraday rotation 337-μm beam path in the Lawrence Livermore Laboratory 2X11B magnetic mirror device (after Stallard *et al.*, 1973).

the plasma where its plane of polarization is rotated. The beam is then combined with the modulated reference beam, split in a polarization insensitive
quartz beam splitter, and the two resulting beams passed through wire
polarizers ($+30°$ and $-30°$). The resulting Faraday rotation phase shift is
then obtained in a differential manner. The present minimum sensitivity of
$\pm 2°$ is limited not by the interferometer but from the vibration of the mirrors
mounted within the plasma chamber.

   The use of Faraday rotation in the FIR has also been proposed as a means
of determining the poloidal field distribution in tokamaks. A proposed FIR
polarimeter for the simultaneous measurement of $\int n_e \, dl$ and $\int n_e B \, dl$ with
the same probe beam (Dodel and Kunz, 1978) is shown schematically in
Fig. 13. This system is essentially a modification of the FIR laser interferometer discussed in the preceding section. Here one introduces a probe beam
consisting of an R-wave at frequency ω and an L-wave of frequency $\omega + \omega_m$
($\omega_m \ll \omega$) into the plasma. In the absence of the plasma, the plane of polarization would slowly rotate at the frequency ω_m. In the presence of the plasma,

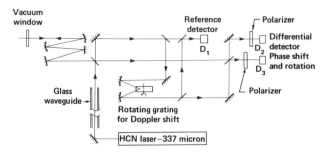

FIG. 11.   HCN laser Faraday rotation apparatus (after Stallard *et al.*, 1978).

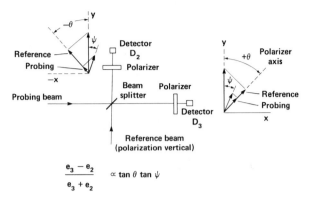

$$\frac{e_3 - e_2}{e_3 + e_2} \propto \tan \theta \tan \psi$$

FIG. 12.   Differential detector arrangement for Faraday rotation measurements (after Stallard *et al.*, 1978).

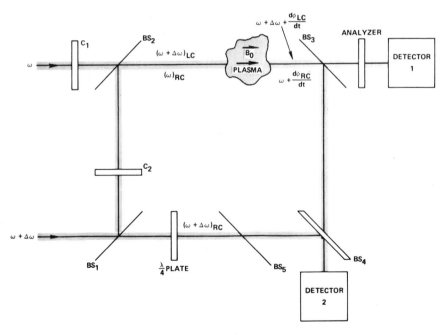

FIG. 13.   Far infrared polari-interferometer for the simultaneous measurement of $\int n_e \, dl$ and $\int n_e \, B \, dl$ (after Dodel and Kunz, 1978).

the plane of polarization undergoes an additional rotation due to the dielectric properties of the plasma. The phase information is obtained from the zero crossings as before. Dodel and Kunz (1978) estimate a sensitivity of $\simeq 0.5°$ for the polarimeter. This system was recently operated successfully on the TFR tokamak using one of the channels of the Véron interferometer described in the preceding subsection.

Hutchinson *et al.* (1978) at Oak Ridge National Laboratory have constructed a similar polarimeter operating at 393 μm using a cw optically pumped HCOOH laser. In tests simulating the plasma, a sensitivity of $\simeq 1$ mrad was obtained. The system polarization sensitivity was $\simeq 26$ mW/mR at a modulation angle of 149 mrad and is expected to increase significantly as the laser power has been increased from 6 to 50 mW. The frequency response of the system was limited to $\simeq 10$ kHz by the ferrite modulator and driver employed. However, this is to be replaced with a lithium tantalate modulator with an expected frequency response of $\simeq 2$ MHz. Present plans are to install the device on the RPI tokamak for initial testing.

IV.   Synchrotron Radiation Measurements

A useful plasma diagnostic technique has been the observation of the electromagnetic wave emission from a plasma. Since the plasma is immersed in a magnetic field, the electrons radiate energy at their local cyclotron frequency and its harmonics. The radiated power at the nth harmonic, per unit solid angle per frequency interval, for a single electron in a uniform magnetic field is given by the Schott–Trubnikov expression (Trubnikov, 1958; Bekefi, 1966; Jackson, 1962):

$$\frac{d^2 P_n(\omega, \theta)}{d\Omega \, d\omega} = \frac{e^2 \omega^2}{8\pi^2 \varepsilon_0 c} \left\{ \left(\frac{\cos \theta - \beta_\parallel}{\sin \theta} \right)^2 J_n^2 \left(n\beta_\perp \frac{\sin \theta}{1 - \beta_\parallel \cos \theta} \right) \right.$$

$$\left. + \beta_\perp^2 J_n'^2 \left(n\beta_\perp \frac{\sin \theta}{1 - \beta_\parallel \cos \theta} \right) \right\} \delta(n\omega_{ce} - \omega).$$

In the above expression, θ is the angle between the magnetic field and the direction of observation, $\beta_\parallel$ is the normalized velocity ($v_\parallel/c$) parallel to the magnetic field, $\beta_\perp$ is the perpendicular velocity component, and ω_{ce} is given by

$$\omega_{ce} = eB/[\gamma m_0 (1 - \beta_\parallel \cos \theta)]$$

with

$$\gamma = (1 - \beta_\parallel - \beta_\perp^2)^{-1/2} = (1 - \beta^2)^{-1/2}.$$

The spectral distribution of a collection of *uncorrelated* electrons is then obtained by performing the ensemble average of the single-particle spectrum over the electron distribution, i.e.,

$$\frac{d^2 P}{d\omega\, d\Omega} = \int \sum_{n=1}^{\infty} \frac{d^2 P_n}{d\omega\, d\Omega}\, f(\mathbf{r}, \mathbf{v}, t)\, d\mathbf{v}\, d\mathbf{r}$$

where $f(\mathbf{r}, \mathbf{v}, t)$ is the particle distribution function. In the above, reabsorption was ignored and it was assumed that $\omega^2 \gg \omega_{pe}^2$ so that the influence of plasma dielectric effects can be ignored. The possible uses of synchrotron radiation as a plasma diagnostic tool have been discussed by Engelmann and Curatoto (1973). First we note that for a 50-kG fusion device, the rest mass cyclotron frequency is $\simeq 140$ GHz or the wavelength $\simeq 2$ mm. Therefore the bulk of the radiation will be in the submillimeter wave region for $T_e > 10$ keV. For these short wavelengths one has good spatial resolution.

By measuring the synchrotron radiation spectrum one can determine the spatially resolved electron temperature and density. The first experimental measurements were reported by Lichtenberg *et al.* (1964) on a hot electron plasma confined in a magnetic mirror. Their results are reproduced in Fig. 14 where the synchrotron radiation spectrum from a mirror-confined hot electron plasma was measured using a cryogenically cooled InSb detector and grating spectrometer. The electron temperature is then determined by fitting the energy-dependent distribution function to the observed radiation spectrum. The temperature obtained in this manner was in good

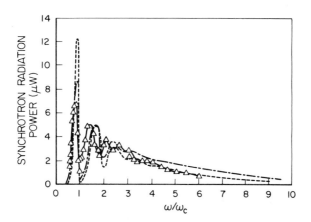

FIG. 14.   Spectral power distribution of synchrotron radiation emitted by a hot electron plasma confined in a magnetic mirror field of kG. The spectrum ($\triangle$) is compared with theoretical spectra for a two-dimensional Maxwellian distribution at 75 keV (– – –) and 100 keV (——) (after Lichtenberg *et al.*, 1964).

agreement with that obtained from conventional x-ray *bremsstrahlung* measurements.

More recently, there has been considerable interest in the use of synchrotron radiation measurements for tokamak diagnostics. For example, as pointed out by Engelmann and Curatolo (1973) the temperature profile of a thermal plasma can be obtained using synchrotron emission. Measurements of the emission in the extraordinary mode of polarization at the lower harmonics where the tokamak radiates as a blackbody directly give the temperature through (Engelmann and Curatolo, 1973; Clifton *et al.*, 1977; Celata and Boyd, 1977)

$$I_m^{(xo)}(\omega) = \omega^2 K T_e(R)/(8\pi^3 c^2).$$

At the higher harmonics where the plasma is optically thin the ratio of the intensities of the ordinary and extraordinary modes yields the temperature profile

$$I_m^{(O)}(\omega)/I_m^{(xo)}(\omega) \simeq K T_e(R)/(m_0 c^2).$$

Finally, another method described by Celata and Boyd (1977) and Clifton *et al.* (1977) permits the determination of the temperature profile through the measurement of the ratio of synchrotron emission intensities at successive harmonics in the optically thin region. The temperature is then given by

$$K T_e(R) \simeq M_m I_{m+1}[\omega_{m+1}(R)]/\{I_m[\omega_m(R)]\} m_0 c^2$$

where

$$M_m = 2em^{2m+1}/(m+1)^{2m+1}(2m+1/2m+3)^{m+1}.$$

In addition to the temperature profile, one may also obtain the density profile from the synchrotron emission spectrum of a thermal plasma. The intensity of the emission at an optically thin harmonic is then directly proportional to density (Engelmann and Curatolo, 1973; Stauffer and Boyd, 1978; Clifton *et al.*, 1977):

$$I_m(\omega) = [m^{2m-1}/(m-1)!]\omega_{pe}^2(R)[\omega K T_e(R)/(4\pi c^3)][K T_e(R)/(2m_0 c^2)]^{m-1} R.$$

In the above, the observation direction was assumed to be along a major radius. To understand this technique, we follow the discussion of Stauffer and Boyd (1978). Here they begin with the result of radiation transport (Bekefi, 1966) that the specific intensity of cyclotron radiation at frequency ω is

$$I(\omega) = I_{BB}(1 - e^{-\tau})/(1 - \rho e^{-\tau})$$

where I_{BB} is the blackbody intensity, τ the frequency-dependent optical depth, and ρ the power reflectivity of the vacuum vessel walls. For $\hbar\omega/KT \ll 1$ we have $I_{BB} \propto \omega^2 T_e$. Then for high harmonics such as $m = 3$ we have

$\tau \ll 1$. Furthermore for $\rho\tau \ll 1 - \rho$ we have $I(3\omega_{ce}) \simeq I_{BB}[\tau/(1 - \rho)]$. Using the fact that $\tau(\omega) \propto n_e T_e^{m-1}/(m\omega_{ce})^2$ we have $I(3\omega_{ce}) \propto n_e T_e^3/(1 - \rho)$. If the second harmonic is optically thick we have $I(2\omega_{ce}) \propto (2\omega_{ce})T_e(r)$. Therefore the ratio of the third harmonic specific intensity with respect to its value at some reference time t_R is

$$I_3(t)/I_3(t_R) = [n_e(t)/n_e(t_R)]|T_e(t)/T_e(t_R)|^3 = [n_e(t)/n_e(t_R)]|I_2(t)/I_2(t_R)|^3.$$

Therefore

$$n_e(t) = n_e(t_R)[I_3(t)/I_3(t_R)]|I_2(t_R)/I_2(t)|^3.$$

It should be stressed that it is difficult to obtain an *absolute* density profile from this method due to complications arising from wall reflections and antenna coupling losses. Therefore, the density profile data shown in Fig. 15 (Boyd *et al.*, 1979) were normalized to the ruby laser derived data at one point.

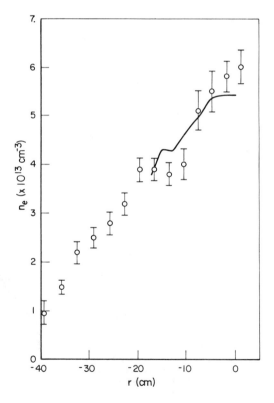

FIG. 15. Comparison of $n_e(r)$ from cyclotron radiation (solid line) with $\bar{n}_e(r)$ from ruby laser Thomson scattering (circles) in the PLT tokamak (after Boyd *et al.*, 1979).

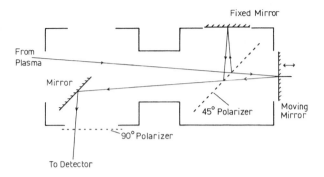

Fixed Mirror

FIG. 16. Modified Michelson spectropolarimeter for synchrotron radiation polarization (Boyd *et al.*, 1979).

It has also been proposed to use the polarization properties of the synchrotron radiation as a means of determining the poloidal field profile and hence the toroidal current distribution. Basically, this makes use of the fact that $\theta \simeq \pi/2$ nearly all of the emitted radiation is in the extraordinary mode. It should be noted that wall reflections can rotate the plane of polarization of the received radiation and even result in unity polarization ratio between the ordinary and extraordinary modes (Costley and TFR Group, 1977). However, measurements (Hutchinson and Komm, 1977) on the MIT Alcator tokamak have indicated a ratio of extraordinary to ordinary mode intensities of approximately 2. More recently, Hutchinson (1978) has employed both a polarizing Michelson interferometer and a Fabry–Perot and rotating polarizer combination to obtain frequency resolved measurements of the polarization parameters in the Alcator device. Figure 16 depicts schematically the modified Michelson spectropolarimeter. The output intensity I_o is proportional to

$$\tfrac{1}{2}\{|E_x|^2 + |E_y|^2 + (|E_y|^2 - |E_x|^2)\cos k\delta + i(E_y E_x^* - E_y E_x^*)\sin k\delta\}$$

where E_x and E_y are the complex field amplitudes and δ is the retardation produced by the moving mirror. One can thereby obtain the difference in polarization intensities as a function of frequency by performing the cosine transform of the resultant interferogram. In addition, the ellipticity is obtained from the sine transform. Typical cosine transform data are displayed in Fig. 17.

It should be noted that the abovementioned synchroton radiation measurements can be performed with excellent spatial resolution. The toroidal magnetic field in a tokamak and hence the electron cyclotron frequency vary inversely as the major radius R. Therefore, radiation of a given harmonic number m and frequency mf' is emitted from a definite surface of major radius R'. Using a frequency selective receiver to obtain spectral resolution

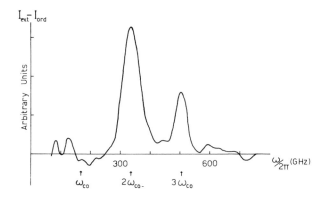

FIG. 17. Typical cosine transform data from the Alacator tokamak using the spectro-polarimeter (after Hutchinson, 1978).

in combination with spatially resolved viewing optics one can collect the radiation emitted from a small localized volume element in the plasma. This is illustrated in Fig. 18.

The results quoted above all assumed a thermal plasma. The spectrum can be drastically modified by the presence of a small polulation of suprathermal or runaway electrons. As can be seen in Fig. 19, the emission is both broadened and significantly increased at the higher harmonics. The emission is also a sensitive function of the electron pitch angle as can be seen in Fig. 20.

The first tokamak synchrotron radiation measurements in the FIR were performed by Costley *et al.* (1974) on the CLEO and TFR tokamaks and on the ATC, Alcator, and PLT tokamaks by Boyd *et al.* (1976). The apparatus of Costley *et al.* (1974) employed a Putley InSb cooled detector. Since this is a broadband detector, a frequency selective element was required to resolve the synchrotron radiation spectrum. Their system employed a scanning Michelson interferometer which, combined with the InSb detector, formed a Fourier transform spectrometer. This system had a frequency resolution $\Delta f/f \simeq 0.1$ at 1.7 mm and a time response of $\simeq 10$ msec. Much better time resolution [$\simeq 10$ μsec was achieved by Boyd *et al.* (1976, 1977a)], who performed the measurements over the interval 0.67–8.6 mm using a five-channel detector system with fixed bandwidth wire mesh filters in each channel ($\Delta f/f \simeq 0.5$). More recently, the multichannel bandpass filter system has been replaced by a multichannel grating polychromator (Boyd *et al.*, 1977b; Rutgers and Boyd, 1977). This combines fast time resolution ($\simeq 1$ μsec) together with good spectral resolving power ($f/\Delta f \simeq 20$). In addition, the Maryland group now also employ a Fourier transform spectrometer for spectral analysis of the synchrotron radiation emitted from the PLT tokamak. Figure 21

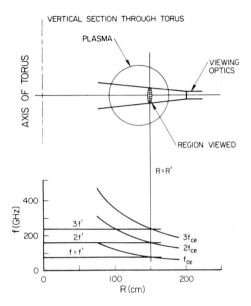

FIG. 18. Schematic of spatial resolution of synchrotron radiation detection system, for $B_t = 32$ kG.

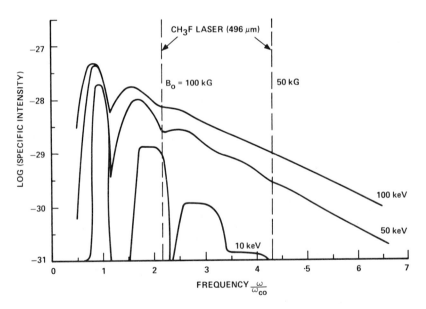

FIG. 19. Synchrotron radiation intensity as a function of energy for a 45° pitch angle (after Celata and Boyd, 1977).

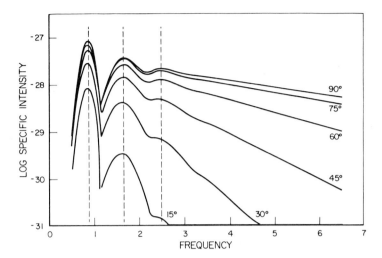

FIG. 20.   Synchrotron radiation intensity as a function of pitch angle for 75-keV electrons with $n_e = 1/m^3$ (after Celata and Boyd, 1977).

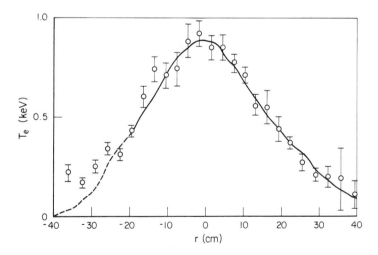

FIG. 21.   Comparison of PLT electron temperature profile obtained from $2\omega_{ce}$ emission with that obtained from ruby laser Thomson scattering (after Boyd *et al.*, 1979). Dashed line indicates region from which $2\omega_{ce}$ radiation has same frequency as $3\omega_{ce}$ radiation from $r = 6 - 40$ cm.

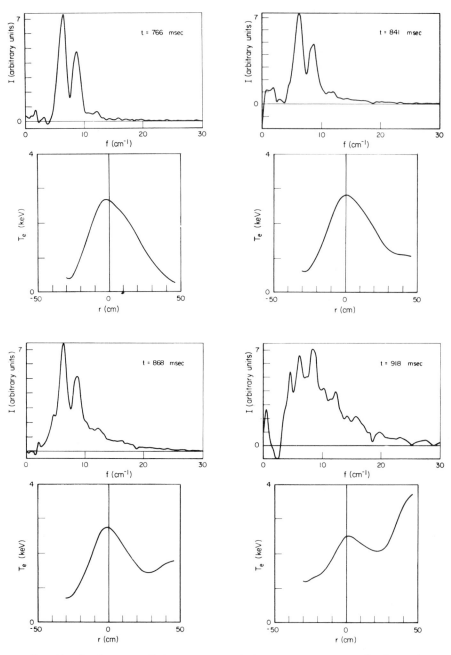

FIG. 22.   Synchrotron radiation spectrum and electron temperature profiles for various times in a PLT tokamak discharge (Boyd *et al.*, 1979).

displays a PLT radial temperature profile obtained from measurement of the $2\omega_{ce}$-emission using this system. The circles represent the electron temperature obtained from ruby laser. Thomson scattering. The electron temperature values obtained using this technique are found to be in excellent agreement with those obtained using conventional ruby laser Thomson scattering. The advantage of the synchrotron radiation measurement is that one can obtain the temperature profile almost as a continuous function of time. On the other hand, ruby laser systems provide at best only the profile at a few isolated times in a single plasma discharge.

The wide spectral range and moderate scan time of the Fourier transform spectrometer permits the study of the evolution of the electron distribution function. Figure 22 displays results obtained by the Maryland group (Boyd *et al.*, 1979) on the PLT tokamak. Both the synchrotron radiation spectrum and electron temperature profile are displayed for various times in the discharge. The increase in the high harmonic emission is due to the appearance of a significant runaway electron population. These cyclotron radiation measurements provide important information concerning tokamak supplementary heating techniques. Figure 23 displays the use of these measurements in determining the time evolution of the PLT electron temperature during neutral beam heating.

In addition to the high-frequency InSb-type systems described above, there has been considerable effort concerned with synchrotron radiation measurements using conventional heterodyne millimeter-wave receivers. The first measurements were made on the French TFR tokamak using two

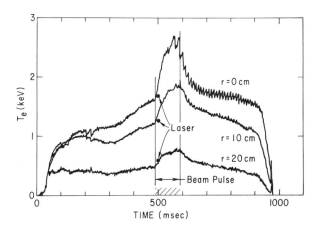

FIG. 23.   Temporal evolution of PLT electron temperature during neutral beam heating (Boyd *et al.*, 1979). 1.2 mW $H^0 \rightarrow D^+$; $\bar{n}_e = 4.3 \times 10^{13}$.

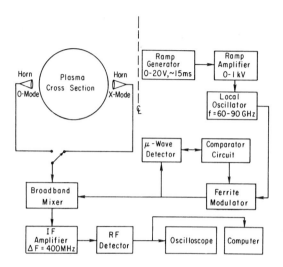

FIG. 24.   Millimeter wave scanning heterodyne receiver system for tokamak synchrotron radiation measurements (Efthimion *et al.*, 1978).

local oscillators to cover the second harmonic from 120 to 160 GHz (TFR Group, 1975). The radial electron temperature profile was obtained by changing the local oscillator frequency on a shot-to-shot basis. In addition, the time evolution of the electron temperature at a fixed radial position was obtained for a single tokamak discharge.

This heterodyne receiver technique has been utilized on the Princeton Large Torus with great success (Hosea, *et al.*, 1977; Arunaslam, *et al.*, 1977) for measurements of f_{ce} and $2f_{ce}$. The apparatus is shown schematically in Fig. 24 and is seen to consist of a standard heterodyne receiver with a sweepable BWO for the local oscillator. Measurements of both the ordinary and extraordinary mode emission are made with the latter performed with a horn located at the inner (accessible) side of the torus. The receiver is capable of scanning over the frequency range 60–90 GHz every 15 msec, which results in about 70 radial temperature profiles per tokamak discharge. Figure 25 (Efthimion *et al.*, 1978) shows the temperature profile obtained from their measurements at f_{ce}. As indicated in Fig. 26, the agreement with the profile obtained with the ruby laser Thomson scattering system is seen to be excellent. This same apparatus was employed to investigate turbulent temperature fluctuations in the PLT device (Arunasalam *et al.*, 1977). Such measurements provide important insight into the particle and energy confinement properties of fusion plasmas.

For measurements such as growth rates in lower hybrid heating studies one wishes better temporal response than can be obtained with the InSb

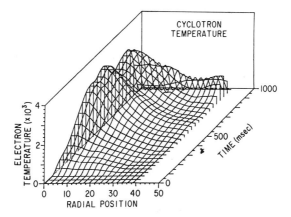

FIG. 25.   Time-resolved PLT radial electron temperature profile obtained using the scanning millimeter wave receiver (Efthimion *et al.*, 1978).

detection systems. Conventional millimeter-wave receivers possess the necessary temporal response but are limited to rather low field tokamaks. The latter statement is especially true if one wishes to examine the development of the higher harmonic radiation as a monitor of the suprathermal electron production. Even at the relatively low magnetic field strengths of present tokamaks such as PLT the radiation at the fourth and higher harmonics occurs in the far infrared. Unfortunately, present millimeter-wave single-mode waveguide mixer technology cannot be simply scaled down in size for

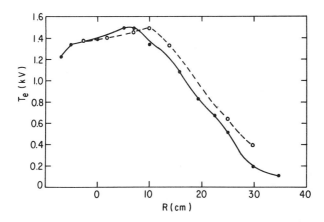

FIG. 26.   Comparison of PLT electron profiles obtained from ordering mode fundamental cyclotron emission (○) and from ruby laser Thomson scattering (●) (Efthimion *et al.*, 1978).

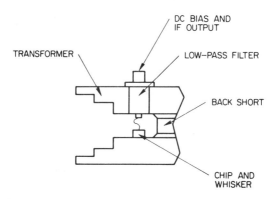

FIG. 27.   A typical waveguide mixer geometry.

operation at higher frequencies. A typical millimeter-wave single-mode waveguide mixer is shown schematically in Fig. 27. The input horn and waveguide impedance is matched to the diode impedance through a series of steps or tapers in the waveguide. At these frequencies the mixing element typically consists of a semiconductor substrate upon which has been deposited a large number of metallic dots. The Schottky diode consists of these metal contacts and the semiconductor surface. The diode is contacted by a whisker as shown in Fig. 27. The interested reader is referred to the excellent mixer review by McColl (1977) together with the diode fabrication article by Clifton *et al.* (1971). Complete detail concerning the actual techniques for whisker preparation and diode contacting can be found in several excellent technical reports (Kerr *et al.*, 1978; Hagström and Lidholm, 1977; and Lidholm, 1977). The diode and chip are then resonated by a movable backshort. The diode dc bias and i.f. signals are connected through the whisker to the i.f. port. Appropriate filters are required to prevent signal and local oscillator leakage. Recently, such a system has been scaled down for operation at 890 μm (345 GHz) by Erickson at the University of California, Berkeley, for radio astronomy applications (Erickson, 1978). It is interesting to note that the waveguide dimensions even at this relatively low frequency are only 0.026 × 0.008 in. Clearly, the scaling of this system to frequencies $\gtrsim 1$ THz is a formidable undertaking. A similar fundamental waveguide mount has been developed and successfully tested for operation to 400 GHz by Lidholm and de Grauuw (1979) for the European Space Agency millimeter-wave radio astronomy program.

During the past two years, a quasi-optical Schottky barrier diode receiver developed by Gustincic for the NASA–JPL atmospheric radiometry and radio astronomy programs has been tested to frequencies near 1 THz for

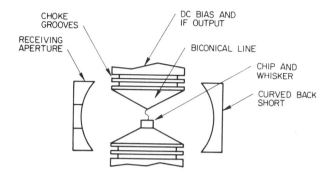

CHOKE
GROOVES

RECEIVING
APERTURE

DC BIAS AND
IF OUTPUT

BICONICAL LINE

CHIP AND
WHISKER

CURVED BACK
SHORT

FIG. 28. The biconical quasi-optical mount (after Gustincic, 1977a).

plasma diagnostics applications. The reader interested in the complete details of the receiver is referred to the references (Gustincic, 1976, 1977a,b; Gustincic et al., 1977). Here we will only briefly describe the quasi-optical mixer mount shown in Fig. 28. In this configuration, the chip and whisker are located at the apex of two terminated biconical transmission lines. The choke grooves which prevent rf leakage are made after the transmission lines flare out to a large diameter compared to a wavelength. The bicone is surrounded by a curved backshort which tunes the structure. Incoming radiation is focussed through a circular aperture into the mount. It is of interest to note that the entire mixer assembly can be machined on a standard lathe as opposed to the complicated fabrication techniques required for the conventional mount. This receiver has been successfully operated on the UCLA Microtor tokamak devices for synchrotron radiation measurements into the submillimeter region. These radiometers possessed an NEP $> 3 \times 10^{-19}$ W/Hz and excellent frequency resolution ($\Delta f/f > 10^{-2}$) and temporal response ($\tau < 20$ nsec).

As mentioned earlier, one can obtain spatially and temporally resolved synchrotron radiation measurements using sweepable local oscillators. However, at frequencies above 600 GHz, even carcinotrons produce less than 1 mW (Golant et al., 1969). In order to extend the operating frequency range of the local oscillator, one immediately thinks of second-harmonic or higher-harmonic mixing. This scheme makes use of the nonlinear nature of the mixing element to produce harmonics of the local oscillator frequency which in turn beat with the desired signal. This process is inherently inefficient and the increased conversion loss manifests itself in a higher system NEP. A more desirable alternative is the two-diode subharmonically pumped mixer (Carlson et al., 1978). In this two-diode mixer, the LO frequency is approximately half signal frequency. As long as the diodes are spaced a distance small compared to a wavelength, the conductance waveform contains only even harmonics of the LO frequency. This mixer is inherently as efficient as the single-diode fundamental mixer. Tests at 98 GHz (Carlson et al., 1978) have varied this efficiency with a measured single-sideband noise temperature of 400 K obtained for their subharmonically pumped waveguide mixer. Preliminary measurements by Gustincic (1978) have also demonstrated the feasibility of employing the quasi-optical biconical geometry in a subharmonically pumped mixer.

V. Thomson Scattering

A. REVIEW

Consider a plane electromagnetic wave $E = xE_0 \cos(k \cdot r - \omega_0 t)$ incident on a free electron initially at rest. For moderate field strengths $(eE_0/m\omega_0 c \ll 1)$, the motion is nonrelativistic and the electron oscillates at the wave frequency ω_0 and moves in the direction of the wave electric field. The electron then reradiates at the wave frequency ω_0 and the total cross section for this scattering process is given by the Thomson scattering cross section

$$\sigma_{th} = (8\pi/3)r_e^2 = 0.665 \times 10^{-24} \quad cm^2,$$

where r_e is the classical electron radius.

If the electron is initially moving, it sees the wave at a Doppler-shifted frequency

$$\omega' = \omega_0 - k_0 \cdot v_e.$$

The scattered radiation of wave number k_s is then observed at some angle θ with respect to k_0. The electron therefore reradiates with a further Doppler-shifted frequency corresponding to its velocity component along the scattered wave direction. We then have

$$\omega_s = \omega' + k_s \cdot v_e = \omega_0 + (k_s - k_0) \cdot v_e$$

or

$$\omega_s - \omega_0 = (k_s - k_0) \cdot v_e = k \cdot v_e.$$

For most practical cases $|k_0| \simeq |k_s|$, so that the law of cosines give us

$$k^2 = k_0^2 + k_s^2 - 2k_0 k_s \cos \theta \simeq 2k_0^2(1 - \cos \theta) = 4k_0^2 \sin^2(\theta/2).$$

Thus we have $\theta \simeq 2 \sin^{-1}[k/(2k_0)]$, which resembles the familiar Bragg equation for scattering from a crystal lattice.

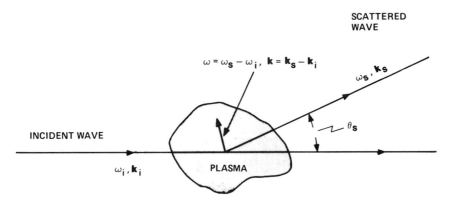

FIG. 29. Schematic of Thomson scattering process.

If one now has a collection of electrons that are distributed completely uniformly, the superposition of the scattered radiation from all electrons interferes destructively, and there is no net scattering. However, deviations in density occur due to random thermal fluctuations and also due to coherent collective motions in a plasma. These departures from uniformity give rise to a scattered intensity proportional to $\tilde{n}$ for random fluctuations and to $(\tilde{n})^2$ for coherent fluctuations. The scattering is shown schematically in Fig. 29.

The differential scattering cross section is then given by $d^2\sigma/d\omega\, d\Omega = (d\sigma_{th}/(d\Omega))\, S(k, \omega)$, where $S(k, \omega)$ is a form factor obtained from the Fourier transform of the electron density pair correlation function. It can be shown (Kunze, 1968) that $S(k, \omega) = S_e(k, \omega) + S_i(k, \omega)$, where S_e refers to the contribution from free moving electrons and S_i to the contribution from electrons that are correlated with the ion motion. As discussed earlier, potentials in a plasma are screened out in distances of the order of the Debye length λ_D. This arises since due to the Coulomb forces each particle is surrounded by a cloud of shielding charge. On a distance small compared to a Debye length ($k\lambda_D \gg 1$) the particle positions are uncorrelated and the scattering is simply from individual electrons. On the other hand, for $k\lambda_D \ll 1$, all of the scatterers within a Debye sphere will produce approximately the same phase factor $(\mathbf{k} \cdot \mathbf{r})$. In this case one obtains information about both plasma collective effects and the ion distribution (via scattering from their electron shielding clouds).

It is useful to define a characteristic parameter that is essentially the ratio of the effective scattering wavelength to the Debye length:

$$\alpha = 1/k\lambda_D \simeq 1/[2k_0 \sin(\tfrac{1}{2}\theta)\, \lambda_D] = \lambda_0/[4\pi\lambda_D \sin(\tfrac{1}{2}\theta)].$$

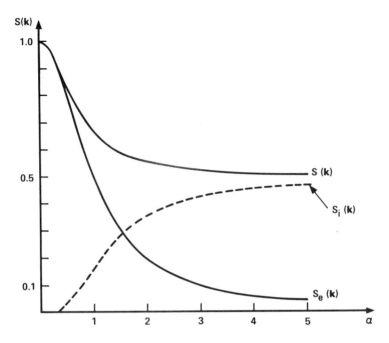

FIG. 30.   Form factors for an equal electron and ion temperature plasma.

For $\alpha \ll 1$, the scattering is from individual (uncorrelated) electrons, and the scattered radiation has a frequency spread determined by the electron temperature. One may then obtain the electron temperature from the scattered spectrum and also the density (with absolute calibration). For typical toka-mak and mirror temperatures and densities, a ruby laser ($\lambda \simeq 6943$ Å) yields $\alpha \ll 1$, and one obtains information about the electron collective motion. The dependence of S, S_e, and S_i on α is shown in Fig. 30.

In the large α limit, one may obtain information about ion fluctuations and, therefore, the ion temperature in a thermal plasma through scattering from the electrons correlated with the ions. The shape of the spectrum (Kunze, 1968) is shown in Fig. 31 as a function of α. The ion temperature T_i is determined from the central portion of the scattered spectrum in the large α limit. The determination by laser scattering of ion temperature in the next generation of fusion plasmas is of crucial importance as the present methods will not work in these large devices. For example, in present-generation fusion-plasma devices the most common method of ion-temperature deter-mination is via energy analysis of the charge–exchange, neutral atoms that escape from the plasma. However, even in these devices the mean free path of the neutral atom can be smaller than the plasma dimensions. Therefore, as

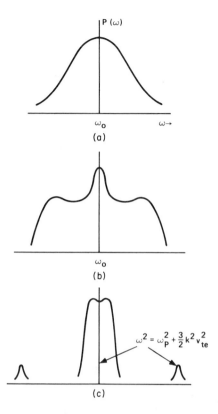

FIG. 31.   Dependence of the shape of the spectral power distribution of scattered radiation as a function of $\alpha = 1/k\lambda_D \simeq \lambda_0/4\pi\lambda_D \sin \frac{1}{2}\theta$; (a) $\alpha \ll 1$; (b) $\alpha \simeq 1$; (c) $\alpha \gg 1$.

devices continue to increase in size, one's knowledge of the ion temperature becomes restricted to the plasma surface.

The problem with laser Thomson scattering determination of ion temperature is that to obtain $\alpha \gg 1$, one requires long wavelengths or extremely small scattering angles. This is illustrated in Fig. 32, where the $\alpha = 1$ lines are plotted for various wavelength lasers as a function of plasma density and temperature. As you can see, for the proposed next-generation mirror and tokamak devices (MFTF and TFTR), one requires a near forward angle scattering ($\theta \ll 1°$) for a 10.6-μm CO_2 laser. On the other hand, a far infrared laser (see Fig. 33) operating near 0.5 mm relaxes this restriction so that a scattering angle at more nearly 30° is required. This problem has served to motivate the significant improvement in the state-of-the-art for both the pulsed FIR laser and the detection apparatus which has occurred over the past few years. The problems associated with both areas shall be discussed.

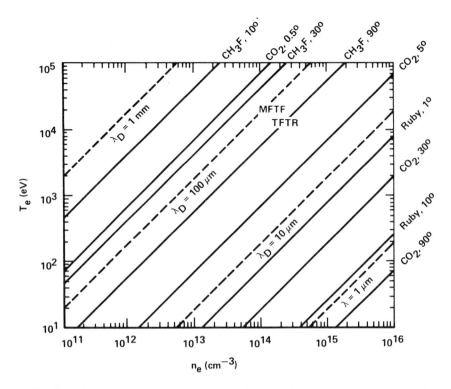

FIG. 32. The $\alpha = 1$ lines (solid) as plotted as a function of plasma temperature and density for various lasers and scattering angles. The constant plasma Debye length regions are indicated by the dashed lines.

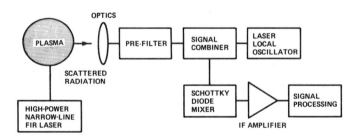

FIG. 33. Schematic of FIR laser Thomson scattering diagnostic for ion temperature measurement.

One need not be restricted to a thermal scattering experiment for ion-temperature measurements. An alternative approach is to excite a plasma wave whose dispersion relation depends strongly on the ion temperature. Since the scattered power is proportional to $(\tilde{n})^2$ in this case, one may contemplate the use of relatively low-power cw FIR lasers. An example of an appropriate wave is the electrostatic, ion-cyclotron wave which has been successfully employed in laboratory plasma devices for ion-temperature measurements (Ault and Ikezi, 1970; Ono et al., 1977). In addition, a recent 75-GHz scattering experiment on the TFR tokamak has provided confirmation of the existence of the electrostatic ion cyclotron (ESIC) instability in a tokamak plasma (TFR Group, 1978). More recently, the ESIC instability has been identified in the PLT tokamak using coherent scattering of 2-mm microwaves (Yamada et al., 1978).

B. PULSED FIR LASER ION THOMSON SCATTERING DIAGNOSTIC

The Thomson scattering determination of the ion temperature in a fusion plasma places severe constraints on the required pulsed laser. The wavelengths presently under consideration range between 200 and 500 μm and are dictated by the plasma density and temperatures, as we previously saw. Due to the small size of the Thomson cross section, the scattered power is quite small. Therefore, one requires a sufficiently intense laser for the scattered power to exceed the noise background. The principal noise contributions from the plasma arise from continuum radiation (*bremsstrahlung*) and synchrotron radiation at the electron cyclotron frequency and harmonics. An upper bound to the latter is easily obtained by assuming the plasma to radiate as a blackbody. On this long wavelength limit, the Rayleigh–Jeans approximation can be made, and we have (Robinson, 1973)

$$dP_{BB} = KT \, d\omega \quad \text{W.}$$

We then find that the synchrotron radiation contribution will be $\simeq 1.5 \times 10^{-12}$ W/Hz assuming a 10-keV plasma. The plasma is only optically thick for the first few harmonics so that we can assume that for a 50-kG plasma this expression will be valid up to $\simeq 300$ GHz. This is unfortunate since at 245 GHz there exist both strong laser sources ($C^{13}H_3F$) and also low noise (~ 2000 K) receivers. One could also argue that high-density ($n_e > 10^{14}$ cm^{-3}) mirror or tokamak operation would also preclude the use of such a source due to strong refractive effects. However, for present mirror devices this may prove to be a quite acceptable region of operation due to the low synchrotron radiation background ($T_e \ll T_i$ and absence of runaways). In the neighborhood of 0.5 mm, Boyd et al. (1977c) predict that for the proposed TFTR tokamak the synchrotron radiation background will be in the neighborhood of 2.5×10^{-18} W/Hz. We shall shortly make use of this number in determining the required laser power. The *bremmstrahlung* contribution when $\omega_p/\omega \ll 1$ and $\hbar\omega/(KT) \ll 1$ is given for one polarization by (Heald and Wharton, 1965)

$$\zeta\omega \, d\omega = 3.2 \times 10^{-41}\bar{g}\{Z[n(\text{cm}^{-3})]^2/(KT_{\text{ev}})^{1/2}\} \, df(\text{MHz}) \quad \text{W}/\text{cm}^3,$$

where $\bar{g}$ is the Gaunt factor. This yields for a 0.5-mm wavelength laser and 10-keV hydrogen plasma a noise power of about 2×10^{-21} W/Hz.

The scattered power assuming thermal fluctuations is given by

$$P_s = P_0 n_e(d\sigma_{\text{th}}/d\Omega) \, L \, \Delta\Omega \, S_i(k),$$

where L is the length of the scattering volume along the direction of the incident wavevector and $\Delta\Omega$ the solid angle subtended by the detector. We can obtain a conservative estimate of the required FIR laser power by requiring that we obtain a signal-to-noise ratio of unity. Assuming a 1-GHz bandwidth, we find that the synchrotron and *bremsstrahlung* radiation contributions are 2.5 nW and 2×10^{-12} W, respectively. Then for $T_e \simeq T_i$, $\alpha \simeq 3$, $\Delta\Omega \simeq 10^{-3}$ sr, $S_i \simeq 0.4$, and $\theta = 30°$, we find that $P_0 \simeq 1$ MW.

In the preceding discussion we naively assumed that the scattered power was uniformly distributed over the bandwidth of interest. We realize, of course, that the spectral shape is quite different and that we must resolve it beyond the e^{-1} point to determine adequately the ion temperature. Then, folding in a safety margin, we see that we require an FIR laser power of $\simeq 2$ to 4 MW.

The spectral purity of the laser is of great importance. One requires a laser spectral linewidth narrow compared to the scattered spectrum. For a 10-keV fusion plasma with $T_i \simeq T_e$, the Doppler broadening for a Maxwellian plasma is given by (Kunze, 1968)

$$\Delta\omega = 4\omega_0 \sin(\theta/2)[2KT/(mc^2)] \ln 2$$

which yields $\Delta f \simeq 1.7$ GHz for a 10-keV deuterium plasma. If one had perfectly absorbing viewing and incident beam dumps, the preceding scattered spectrum could be adequately resolved with a laser bandwidth as large as several hundred megahertz. However, stray scattered light together with the small Thomson scattering cross section ($P_{\text{scat}} \simeq 10^{-12} P_{\text{inc}}$) impose much more stringent demands on the laser bandwidth. Unfortunately, with the exception of the ORNL work (Staats *et al.*, 1978), little attention has been paid to the design of FIR beam dumps. Using their measured rejection ratio of 10^4, we can estimate that an FIR laser with bandwidth less than 40 MHz is dictated.

The actual choice of laser wavelength is dictated by a number of competing effects. Strong refractive effects and the increasingly large synchrotron radiation background (see Fig. 19) at long wavelengths necessitate laser

TABLE II

NARROW-LINE FIR OUTPUT

Molecule	Wavelength (μm)	Narrow-line output (kW)
CH_3F	193	7
HCN	337	1
D_2O	385	250–1000
CH_3I	447	10
CH_3F	496	750
HCOOH	512	2
$C^{13}H_3F$	1222	150

wavelengths shorter than 600 μm. We also wish to employ a reasonably large scattering angle such as $\theta > 10°$. This latter statement, of course, assumes adequate machine access for other than forward angle scattering. Then for $\alpha = 3$ and a 10-keV, 10^{14} cm^{-3} plasma we find that $(\lambda_0)_{min} = 200$ μm. At present there are only a few high-power laser sources in this spectral region as can be seen from Table II. To date, most of the effort at obtaining high-pulsed power has been devoted to the 385-μm line of D_2O and the 496-μm line of CH_3F. Megawatts of power have been achieved with superradiant D_2O and CH_3F systems, but with too large a bandwidth. There are presently, therefore, various attempts underway to achieve narrow-line, megawatt-level FIR output using oscillator–amplifier schemes.

One of the approaches (Semet and Luhmann, 1976) under investigation at UCLA to obtain high-power 496-μm output is shown schematically in Fig. 34. The design of the system is based on the observation that the CH_3F molecule, when excited by a linearly polarized pump, has a tendency to lase with its electric field vector polarized normal to the pump. The output of a grating-tuned TEA CO_2 laser oscillator–amplifier configuration is divided by a beam splitter with part of the output used to pump the FIR oscillator

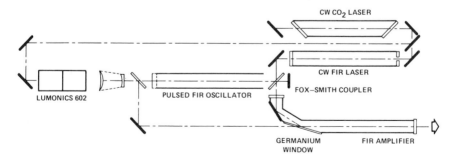

FIG. 34.   Schematic of UCLA pulsed CH_3F 496-μm FIR laser system.

and the remainder used to pump the FIR amplifier. The cw CO_2 laser is employed for injection locking of the pulsed CO_2 laser and results in a temporally smooth output pulse. The output from the FIR oscillator has a polarization perpendicular to that of the CO_2 pump and is reflected by the Brewster angle Ge flat into the FIR amplifier. The CO_2 pump beam is also incident on the Ge at the Brewster angle but with the opposite polarization and, therefore, enters the FIR amplifier. The amplified 496-μm radiation exits through a TPX or Teflon window at the opposite end. This system has produced up to about 750 kW at 0.1 % conversion efficiency from 9.55- to 496-μm power.

In this system a Fox–Smith interferometer (Smith, 1965) was used in the oscillator to ensure operation on a single longitudinal mode. This method of mode control was first employed in an optically pumped FIR laser by the Culham group (Evans *et al.*, 1977). The Culham oscillator configuration is shown schematically in Figs. 35 and 36. A 2.43-m long cavity is formed by the Cu-mesh mirror and the Au-coated rear mirror. The 70-cm separation Fox–Smith interferometer was formed by the Cu-mesh output mirror and a 100 % reflecting Cu-mirror. The linewidth of the FIR oscillator was determined to be substantially less than 27 MHz at the 0.6-kW level, as can be seen from the video pulse signal shown in Fig. 37 and by the Fabry–Perot trace shown in

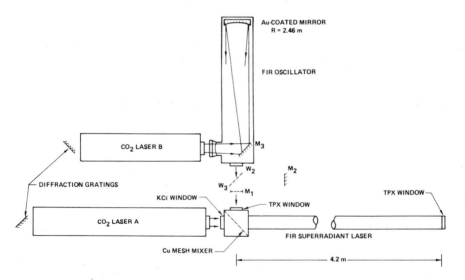

FIG. 35.   Schematic of the Culham Laboratory injection laser assembly. Important features include FIR oscillator, superradiance tube with its independent CO_2 laser pump and Cu mesh mixer, by means of which single-mode FIR radiation from the oscillator is directed into the superradiance tube (after Evans *et al.*, 1977).

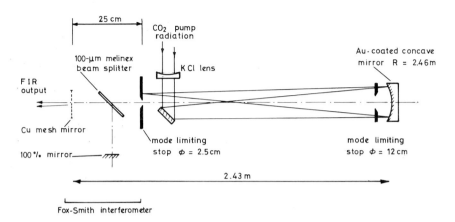

FIG. 36.   Schematic of Culham FIR oscillator employing a Fox–Smith interferometer for mode control (after Evans *et al.*, 1977).

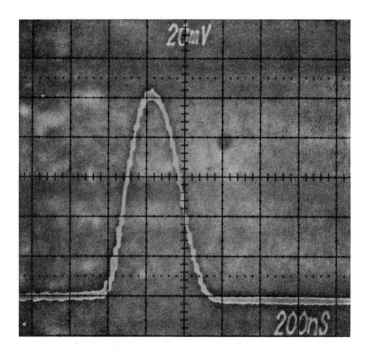

FIG. 37.   Video pulse shape of Culham FIR oscillator with Fox–Smith interferometer (after Evans *et al.*, 1977).

Fig. 38. The narrow-line output of this oscillator was then injected into the 4.2-m-long superradiant laser (Evans *et al.*, 1976). The bandwidth of the superadiant laser was found to narrow from $\simeq 300$ MHz (see Fig. 39) to a value between 50 and 60 MHz (see Fig. 40) with an output power of $\simeq 250$ kW.

The MIT National Magnet Laboratory group has done considerable work on pulsed FIR lasers (Drozdowicz *et al.*, 1976, 1977; Woskoboinikow *et al.*, 1978, 1979; Cohn *et al.*, 1975). Recently they have concentrated on the study of the off-resonance, pumped, D_2O laser (Woskoboinikow *et al.*, 1979). Here they have taken advantage of the fact that the FIR emission for off-resonance pumping is via a Raman, two-photon process. The lasing linewidth is then strongly determined by the spectral qualities of the CO_2 pump laser. Using an etalon-tuned CO_2 laser they demonstrated: (1) the

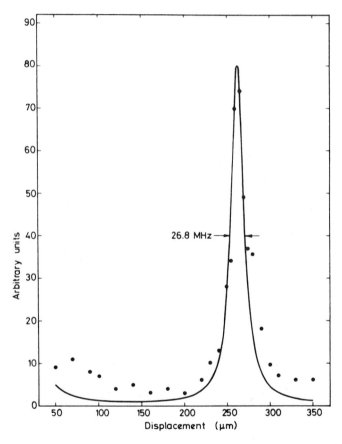

FIG. 38.   Fabry–Perot scan of output of Culham FIR oscillator (after Evans *et al.*, 1977).

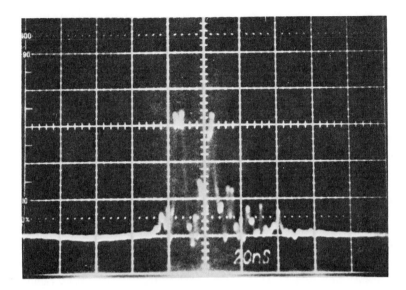

FIG. 39.   Video pulse shape of output from Culham superradiant FIR Laser (after Evans *et al.*, 1977).

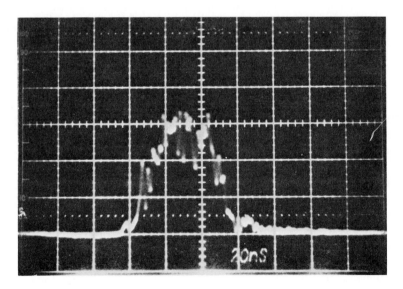

FIG. 40.   Video pulse slope of output from Culham FIR laser with narrowband FIR injection (after Evans *et al.*, 1977).

tunability of the D_2O output near 385 μm, (2) a maximum energy conversion efficiency at frequencies below the D_2O line center as expected for the Raman process, and (3) an absolute conversion efficiency of 0.3 %. The latter number is approximately a factor of three better than that obtainable at 496 μm and argues strongly for the use of D_2O in a scattering experiment.

Hutchinson and Vander Sluis at Oak Ridge National Laboratory have concentrated on the development of compact, large-volume FIR systems (Vander Sluis *et al.*, 1978). Their FIR oscillator employs an unstable resonator cavity and pumps an amplifier with an active volume of 250 liters and a length of $\simeq 1$ m. Output power levels of 250 kW ($\tau_p \simeq 150$–200 nsec) have been obtained at 385 μm with D_2O. The spectral qualities of the laser are determined by performing a fast Fourier transform of the video signal from a photon drag detector. Using an injection-smoothed CO_2 pump pulse they have measured FIR linewidths comparable to their instrumental resolution ($\simeq 5$ MHz). Here the reader contemplating similar measurements is cautioned to first measure the time response of the photon drag detector, as studies at UCLA have indicated time responses significantly slower than Schottky diodes for available photon drag detectors. The oscillator output has been amplified to the 1-MW level with no apparent increase in linewidth using a 5-m-long amplifier. They are presently concentrating on the reduction in size of this system prior to an ion scattering measurement.

The techniques necessary for spectral linewidth measurements of pulsed FIR lasers deserve brief mention. One can, of course, employ high-finesse Fabry–Perot interferometers to obtain the spectral characteristics of the laser by averaging over many pulse repetitions. However, to eliminate effects of pulse-to-pulse variation of this low-repetition rate system ($\simeq 1$–10 sec) one desires a single-shot spectrum. As mentioned previously, the ORNL group obtains such information by employing a transient digitizer to provide a fast Fourier transform of the video output of a photon drag detector. Similar measurements are made routinely at UCLA with a high-speed, Schottky diode detector in place of the photon drag detector. In addition, heterodyne linewidth measurements are performed using a tunable, carcinotron, local oscillator and harmonic mixer. The i.f. output is processed using a filter bank similar to conventional radio astronomy systems. The filter bank output is analyzed with a computer after passing through multiple gated sample-and-hold circuits, multiplexer, and digitizer. Fetterman *et al.* (1979) have recently made clever use of a surface acoustic wave (SAW) dispersive delay line to obtain single-shot spectra of their pulsed FIR lasers. In their system, shown in Fig. 41, the output of a 385-μm D_2O laser was first mixed with the 11th harmonic of a tunable klystron in a Schottky diode mixer resulting in a 420-MHz i.f. signal which was fed into the SAW device. As indicated in Fig. 41, an input transducer converts the $\simeq 400$-MHz i.f. signal into an acoustic

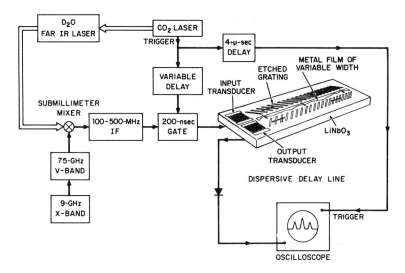

FIG. 41.   Determination of pulsed FIR laser delay bandwidth using a SAW dispersive delay line (after Fetterman *et al.*, 1978a).

wave that is dispersed by the etched grating, and the various frequency components are then sent to the output transducer after an appropriate delay. The output delay time is an increasing function of frequency and results in a single-shot spectral display as shown in Fig. 42. In part (b), the FIR laser was operating simultaneously on three longitudinal modes.

Assuming the successful development of a high-power, pulsed, FIR laser, one still requires a detection system for the scattered radiation. The low-level, scattered power ($\simeq 2.4 \times 10^{-8}$ W) distributed over a large bandwidth ($\simeq 1.7$ GHz) necessitates the use of a heterodyne receiver for the detection system. For a signal-to-noise ratio of unity, the maximum receiver noise equivalent power will then be $(\text{NEP})_{max} = 1.4 \times 10^{-17}$ W/Hz or equivalently the maximum noise temperature $(KT)_{max} = 970,000$ K. Therefore, to facilitate detection and account for system losses, one actually requires a receiver with $(\text{NEP})_R < 10^{-18}$ W/Hz or $(KT)_R < 70,000$ K. This, together with the wide bandwidth requirements, have led to the consideration of the Schottky barrier diodes or Josephson junctions as the mixer element.

The approaches under investigation for the receiver package are quite varied. As mentioned earlier, Gustincic (1976, 1977a,b) has developed a novel receiver employing quasi-optical techniques for operation at frequencies in excess of 1 THz. Presently, the best performance using Schottky diodes in the 0.4–0.5-mm region has been a receiver noise temperature of $\simeq 20,000$ K, (Gustincic *et al.*, 1977). This figure was obtained using Schottky diodes

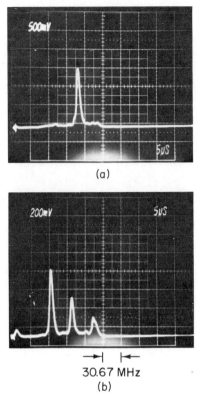

FiG. 42.   Spectral power distribution of pulsed D_2O laser using SAW dispersive delay line;
(a) 500 mV; (b) 200 mV (after Fetterman *et al.*, 1979).

developed for much lower frequency operation ($\simeq 100$ GHz). Mixer scaling
studies from 150 to 800 GHz indicate that the high-frequency performance
degradation is due to the diodes, and significant noise improvement is
anticipated when higher-frequency diodes are employed. The interested
reader is referred to the excellent review article by Wrixon and Kelly (1978)
for further details on low-noise Schottky diode mixers. Several groups
(Fetterman *et al.*, 1978; Sauter *et al.*, 1978) have recently reported extremely
low-temperature, quasi-optical, Schottky diode mixers using the corner-
reflector configuration previously examined by Kräutle *et al.* (1977, 1978).
In this configuration, shown in Fig. 43, the diode is contacted by a whisker
(long wire antenna) mounted in a 90° corner reflector. This improves the

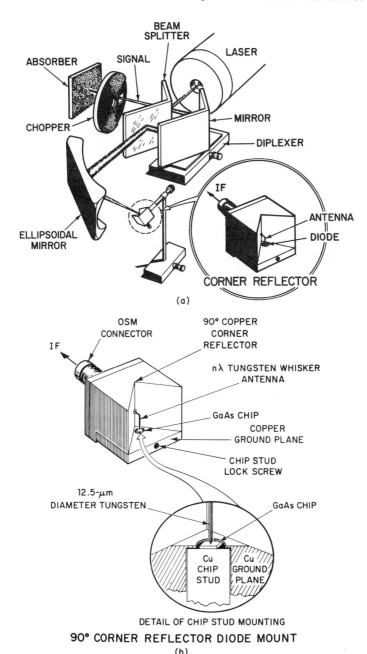

FIG. 43.   Quasi-optical FIR mixer employing a corner cube reflector (after Fetterman *et al.*, 1978).

coupling efficiency by $\simeq 12$ dB, compared to a whisker mounted without reflector. Basically, the conical antenna pattern of the long wire is converted to a beam-type pattern with an attendant increase in efficiency. The Lincoln Laboratory group (Fetterman *et al.*, 1978) has reported receiver system temperatures (DSB) of $\simeq 10,000$ K at frequencies as high as 762 GHz with a conversion loss (DSB) of $\simeq 13.6$ dB. Sauter *et al.* (1978) have measured a single-sideband conversion loss of 17.8 dB at 762 GHz, which should result in performance at least comparable to the Lincoln results. Finally, the Aerospace Corp. group is investigating the use of scaled-down single-mode conventional waveguide mixer mounts.

We have so far not addressed the local oscillator problem. The receiver noise temperature is given by

$$T_R = L_i L_c (T_D + T_{i.f.}) + T_{i.f.}(L_i - 1),$$

where L_c is the conversion loss. For typical, submillimeter wave, Schottky diode receivers the required local oscillator power is $\simeq 10$–100 mW. This means that at frequencies above $\simeq 400$ GHz one must employ cw, optically pumped, molecular lasers. However, despite the approximately 2000 lasing transitions which have been discovered in the 50–2000-μm region, there exist only a few lines suitable for local oscillator purposes. This situation has been ameliorated somewhat by the recent use of SCR pulsers to provide 1-kW, 150-μs pulses from 50-W, cw, CO_2 lasers. This has resulted (Semet *et al.*, 1978; Grossman *et al.*, 1979) in the development of 1-W FIR lasers on the strongest transitions and the addition of a host of 10- to 100-mW lasing lines.

An interesting alternative to the fundamental mixer ($\omega_{LO} \simeq \omega_s$) is the subharmonically pumped mixer ($\omega_{LO} \simeq \frac{1}{2}\omega_s$) discussed earlier. In this scheme two diodes spaced a fraction of a wavelength apart are contacted with reverse polarities. They are therefore switched on and off at twice the local oscillator frequency. Recently, Gustincic has managed to adapt his quasi-optical mixer to such a configuration for a NASA–JPL balloon-borne observational program. The system works well at 280 GHz ($f_{LO} = 140$ GHz) and offers the promise of extension to $\simeq 1$ THz. This would allow the use of presently available local oscillators such as IMPATT diodes and carcinotrons. The former source is particularly attractive due to its small size and low power consumption. Recently, cw operation of IMPATT diodes has been reported at frequencies as high as 420 GHz (Ohmori *et al.*, 1977; Ishibashi *et al.*, 1977) with power outputs of 50 mW at 200 GHz (Chao *et al.*, 1977). The major problem associated with the use of IMPATT diodes for submillimeter wave, local oscillators is their broad spectral characteristics. To solve this problem, Gustincic has spectrally filtered the diode output using tandem-folded Fabry–Perot filters. Alternatively, Mizuno *et al.* (1978) have frequency-stabilized IMPATT diodes in the submillimeter region by phase-locking to a stable, low-noise reference operating at low frequency.

Although the preceding discussion stressed the use of Schottky diodes as mixing elements, the reader should not assume that Josephson junctions have been ruled out. On the contrary, the performance figures of FIR, point-contact, Josephson mixers have exceeded the aforementioned requirements. For example, Blaney et al. (1978) have reported a receiver noise temperature $T_{DSB} \simeq 2100$ K at 660 μm using a CH_2CF_2 optically pumped laser local oscillator. They further project ultimate noise temperatures of $\simeq 300$ K ($\simeq 4 \times 10^{-21}$ W/Hz). It should be further mentioned that in contrast to the Schottky diode, the Josephson junction requires quite modest local oscillator power levels ($\simeq 5$–100 nW). Therefore, many of the hundreds of weak FIR lasing transitions ($P < 1$ mW) in the 50–200-μm region can be employed as local oscillators. It should be noted, however, that the Josephson mixer is not without problems. In particular, the required operation at liquid He temperatures and contact reliability under recycling are definite drawbacks. However, despite these problems, the Culham Laboratory group plans to employ a Josephson mixer for the ion scattering measurements.

The conclusion to be reached from this is that an FIR ion temperature diagnostic is a difficult but not intractable problem. It appears reasonable to expect an actual ion temperature measurement to be completed within the year due to the rapidly developing FIR technology.

C. CW WAVE SCATTERING

Thus far we have assumed that our FIR scattering was from fluctuations in a thermal plasma. However, in the case of an unstable plasma, enhanced scattering may result from coherent scattering from a density wave. In particular, if the plasma supports a density wave $\tilde{n} \exp(i\mathbf{k} \cdot \mathbf{r})$, which is coherent over the scattering volume, then as long as $k_0 > k/2$, coherent scattering with a maximum at $\mathbf{k} = \mathbf{k}_s - \mathbf{k}_0$ occurs. The scattered power integrated over the diffraction maximum is then

$$P_s = P_i L^2 (\tilde{n})^2 \sigma_{th} \lambda_0^2,$$

which can greatly exceed the incoherent scattered power. Here λ_0 is the incident wavelength, L the scattering length, P_i the incident power, σ_{th} the Thomson scattering cross section, and $\tilde{n}$ the density fluctuation amplitude. For fluctuation levels significantly above thermal, the scattering enhancement may easily exceed 10^6 so that one may even employ a cw FIR laser for the scattering source.

This collective scattering was first performed by Arunasalam and Brown (1965) in their investigation of the current driven ion acoustic instability. This has since become a powerful tool for the investigation of waves and instabilities in turbulent plasmas. The technique is well illustrated by a consideration of the recent study of drift-wave turbulence by Okabayashi and Arunasalam (1977) Fig. 44 depicts the experimental arrangement of the 8-mm wavelength microwave scattering from the Princeton FM-1 plasma device. The observed spectral power distribution is shown in Fig. 45, as obtained by both probe and by microwave scattering. The advantages of the unperturbing electromagnetic scattering are obvious.

Collective scattering from low-frequency microinstabilities has been successfully performed on tokamaks using both microwave systems (Mazzucatto, 1976, 1978) and CO_2 lasers (Surko and Slusher, 1976; Slusher and Surko, 1978). In both cases scattering was observed from density fluctuations ($\tilde{n}/n_0 \gtrsim 1\%$) in the frequency range for drift wave turbulence. Such measurements will hopefully eventually lead to an understanding of anomalous loss processes in tokamak plasmas.

More recently, this technique has been applied in the submillimeter wave region (Luhmann et al., 1978). The goal is to develop a wave-scattering diagnostic that will provide high-frequency response and spatial resolution for both fusion plasmas and laboratory nonlinear wave experiments.

In the development of this FIR wave-scattering diagnostic, it was decided to use initially a well-diagnosed, pilot plasma and to scatter from easily monitored, driven, plasma waves. A successful cw scattering measurement at both 800 and 447 μm from $\sim$50 to 150 kHz ion acoustic waves is described next. Present laboratory diagnostic efforts are concerned with scattering at

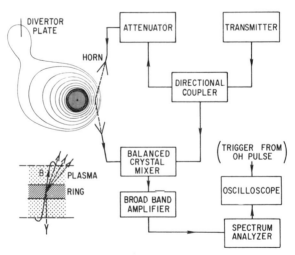

FIG. 44. Schematic of apparatus employed by Okabayashi and Arunasalam (1977) to observe 8-mm collective scattering from unstable drift waves in the Princeton FM-1 plasma device (after Okabayashi and Arunasalam, 1977).

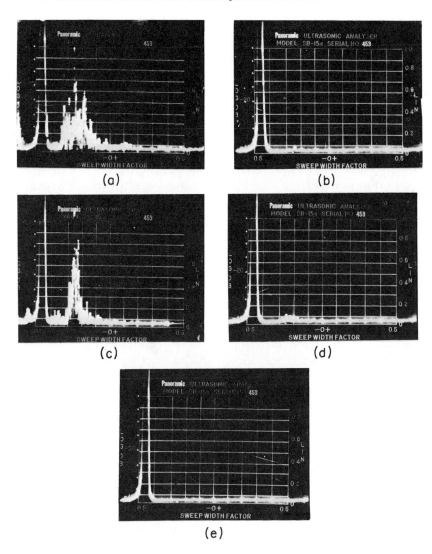

FIG. 45.   Observed spectral power distribution obtained by probe and by microwave scattering in high magnetic shear configuration in a helium plasma (after Okabayashi and Arunasalam, 1977). (a) Probe; (b) $k_\phi \rho_i = 0.25$ ($k_\phi = 2k_0 \cos 75°$); (c) $k_\phi \rho_i = 0.5$ ($k_\phi = 2k_0 \cos 60°$); (d) $k_\phi \rho_i = 0.71$ ($k_\phi = 2k_0 \cos 45°$); (e) $k_\psi \rho_i = 0.97$ ($k_\psi = 2k_0 \cos 30°$).

$\lambda_0 \lesssim 0.5$ mm from electron plasma waves (~ 1–3 GHz) using optically pumped FIR lasers. In addition, as will be discussed later, FIR scattering measurements on tokamak plasmas have begun.

The target plasma in both initial laboratory scattering experiments was a 33-cm-diameter, 100-cm-long, argon hot-cathode discharge plasma with $n_e \simeq 10^{10}$ cm^{-3} and $T_e \simeq 2.5$ eV ($T_e/T_i \sim 8$). The ion waves are launched using a gridded (20 lines/cm), single-ended exciter resulting in a typical fluctuation amplitude ($\tilde{n}/n \simeq 0.5\%$). Launching frequencies of 50 to 150 kHz produced ion acoustic wavelengths of $\simeq 1.5$ to 4.5 cm. The 375-GHz source was a Thomson CSF carcinotron with an output power of $\simeq 25$ mW and a beam divergence of $\simeq 5°$. For the 671-GHz measurements, a 38-mm i.d., 1.6-m-long CH_3I dielectric waveguide laser (Hodges et $al.$, 1976) producing about 30 mW of output power was used. The FIR laser was pumped by a 40-W grating-tuned CO_2 laser operating on the P(18), 10.6-μm line. The FIR cylindrical waveguide resonator consisted of a hollow, dielectric pipe located axially between two reflectors. The CO_2 pump was injected into the FIR resonator through a 4-mm diameter coupling hole using $f/50$–$f/200$ optics. The FIR silicon output mirror coupler was coated with a highly reflective (98 %) multilayer dielectric film for the 9.5- to 10.5-μm region. This substrate was then masked and overcoated with 5000 Å of gold, leaving a 22-mm-diameter coupling hole for the FIR output. The ion acoustic wavelength resulted in a scattering angle of $\simeq 3°$, and consequently, in these preliminary measurements, homodyne detection techniques were employed. The experimental arrangement is shown in Fig. 46.

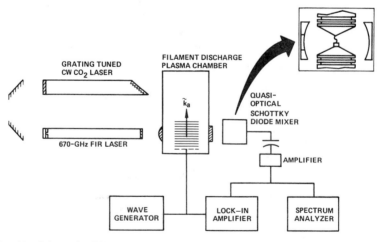

FIG. 46.   Schematic of FIR laser scattering from driven acoustic waves (after Luhmann et $al.$, 1978).

The scattered signal was combined with the transmitted input radiation in a quasi-optical, Schottky diode mixer (Gustincic, 1976; Gustincic, 1977, 1977a,b). The output from the mixer was then amplified and fed into the lock-in amplifier (FWHM bandwidth $\simeq 10$ kHz). A reference signal for the lock-in was taken directly from the wave generator.

The amplitude of the ion acoustic waves was monitored by directly observing the density fluctuations with a single Langmuir probe biased to the plasma potential. Figure 47 illustrates some typical signals. Separation of the direct pick up from the density fluctuations was achieved by gating the wave generator. In trace (a) the direct electromagnetic pick up appears first and is followed by the more slowly propagating ion acoustic waves. In this instance the waves are still reasonably sinusoidal. However, at higher launching voltages these waves degenerate into shock wave structures which have only a small Fourier component at the launching frequency.

As noted in the preceding discussion, the scattered power is proportional to the square of the fluctuation amplitude. Figures 48 and 49 display the observed dependence of scattered power upon the square of the wave amplitude. The open squares indicate the experimental points for the sinusoidal ion waves, and it can be seen that the expected linear dependence was obtained. The open circles represent data where shock structures were present. It

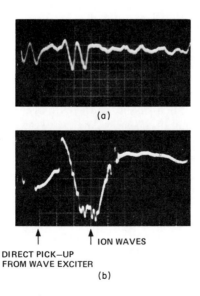

(a)

↑   ↑ ION WAVES
DIRECT PICK–UP
FROM WAVE EXCITER
(b)

Fig. 47.   Driven ion acoustic waves detected via Langmuir probe: (a) sinusoidal ion waves $\tilde{n}/n_0 \simeq 0.4\%/$div; (b) shock waves $\tilde{n}/n_0 \simeq 1\%/$div (after Luhmann et al., 1978).

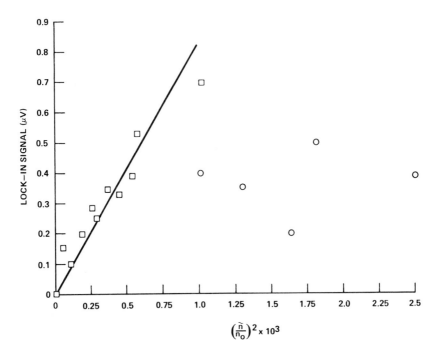

FIG. 48.   Scattered FIR radiation as a function of ion acoustic wave amplitude. A 375-GHz carcinotron was employed as a scattering source (after Luhmann *et al.*, 1978).

appears that for higher fluctuation amplitudes the scattered signal has fallen. The reason for this is that the lock-in amplifier selects a *particular* Fourier component of the scattered spectrum (i.e., the launching frequency). This component is generally much smaller in the case of the shock structures as compared to the lower-amplitude, sinusoidal waves.

The scattered signal was also found to exhibit a definite angular dependence. When the mixer was placed normal to the incident beam, resulting in maximum bias current and receiver sensitivity, the lock-in signal was significantly smaller than obtained with a $3°$ mixer orientation and the associated reduced sensitivity (factor of four).

In summarizing, a FIR scattering measurement from 50- to 150-kHz, ion acoustic waves has been achieved at 375 and 671 GHz. The scattered data possessed a linear dependence upon the square of the fluctuation amplitude and also exhibited the expected angular dependence. Fluctuation amplitudes as low as 10^7 cm^{-3} have been observed, and a calculation of the expected scattered signal ($\sim 10^{-13}$ W) was in reasonable agreement with the measured values, taking account of the amplifier gain and mixer conversion loss. This

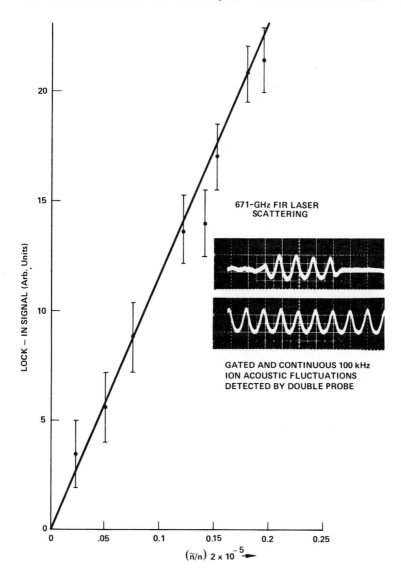

FIG. 49.   Amplitude of scattered 671-GHz radiation as a function of ion wave amplitude (after Luhmann *et al.*, 1978).

bodes well for tokamak and mirror-scattering measurements where fluctuation amplitudes of $\sim 10^{10}$ to 10^{13} cm^{-3} can be anticipated.

As discussed earlier, fusion plasma ion temperature measurements may be made using Thomson scattering from electrostatic ion cyclotron waves. The real part of the dispersion relation is given for $\omega \simeq \omega_{ci}$ by

$$\omega = \omega_{ci}\left[1 + \frac{T_e}{T_i + \varepsilon(b_i)T_e} I_1(b_i)e^{-b_i}\right],$$

where $b_i = \frac{1}{2}k_\perp^2\zeta_i^2$ with ζ_i the ion cyclotron radius. For $k_\perp\zeta_i \ll 1$ we have $(\omega - \omega_{ci})/\omega \simeq (T_e/T_i)I_1(b_i)e^{-b_i}$. Therefore a measurement of ω versus $k_\perp$ yields T_i, assuming T_e is known (i.e., from synchrotron radiation). The waves can be either spontaneously occurring, as in the recent TFR and PLT experiments (TFR Group, 1978; Yamada et al., 1978), or externally excited. This technique is particularly applicable in the case of either nonthermal plasmas or plasmas with large impurity levels.

Electrostatic, ion-cyclotron waves can be generated by electron currents such as occur in ohmically heated tokamaks. The critical drift velocity v_d is given by (Drummond and Rosenbluth, 1962)

$$\frac{v_d}{v_{thi}} \geq \left(\frac{T_i}{T_eI_1(b_i)e^{-b_i}} + 1\right)\left(\ln\frac{m_i}{m_e}\right)^{1/2} \simeq 14\frac{T_i}{T_e} + 3$$

for hydrogen. This condition is easily satisfied in the UCLA Microtor tokamak where an FIR, laser-scattering experiment at 1.22 and 0.447 mm is presently

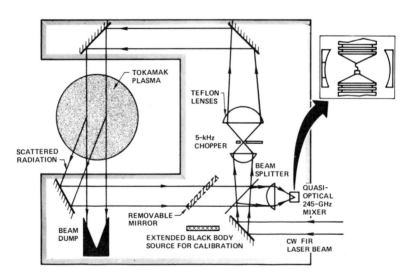

FIG. 50.   Schematic of tokamak cw FIR scattering apparatus.

under way. A homodyne scattering measurement is under way with the apparatus indicated in Fig. 50. The scattering and frequency-shifted FIR beam is distinguished from the synchrotron radiation background by means of a 5-kHz chopper and coherent detection.

Initial microtor scattering has been concerned with a study of low frequency microturbulence in the drift wave frequency range in order to make contact with the aforementioned microwave and CO_2 scattering. The radially resolved scattering form factor $S(\mathbf{k}, \omega) \propto \tilde{n}(\mathbf{k}, \omega)^2$ has been obtained in the range $5 < k_\perp < 20$ cm^{-1} and $f < 1$ mHz. Figure 51 displays $\int_{\omega/2\pi = 5\,\mathrm{kHz}}^{1\,\mathrm{mHz}} S(\mathbf{k}, \omega)\, d\omega$ as a function of $k_\perp$ and r. A further integration over $k_\perp$

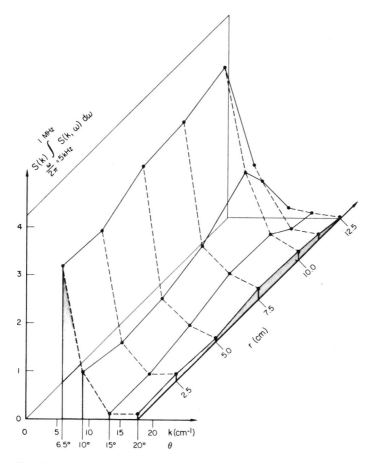

FIG. 51. Wave number spectrum of low frequency density fluctuations in the UCLA microtor tokamak obtained from cw FIR laser scattering.

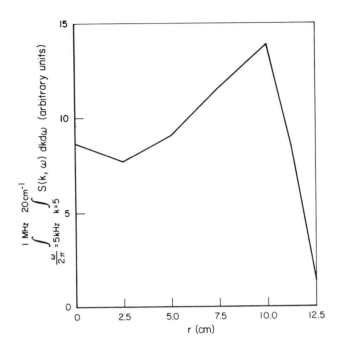

.Fig. 52. Radially resolved low frequency density fluctuations in the UCLA microtor tokamak plasma obtained by cw FIR laser scattering.

has been performed to provide the density fluctuation profile shown in Fig. 52.

VI. FIR Holographic Imaging of Plasmas

Plasmas may exhibit long scalelength collective behavior, such as convective cells, which can greatly affect the containment properties of magnetic confinement systems. A detailed knowledge of these instabilities is greatly needed in order to pursue fusion power successfully. However, present diagnostic techniques permit only a limited measurement of plasma parameters. Under study at the present time by the Aerospace Corp. and UCLA groups is the potential application of FIR holographic imaging to this problem. Figure 53 depicts schematically the imaging system under consideration. The image and reference beams will be obtained from cw optically pumped lasers. The beams are combined on an array of Schottky diode detectors with the holographic information stored in a digital computer for reconstruction and display at a later time. Eventually, it is hoped that three-dimensional,

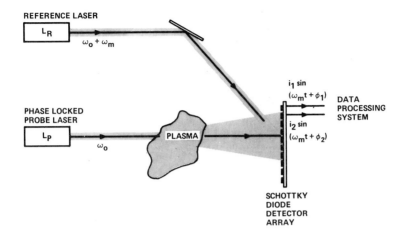

HOLOGRAPHIC PLASMA IMAGING SYSTEM

FIG 53.   Proposed FIR holographic imaging system for measurement of plasma transport properties.

time-resolved, plasma images can be obtained. Schottky diodes possess more than adequate time response (<1 nsec) and can be made in very regular fashion.

It is of interest to estimate the required receiver sensitivity for the proposed system. Let us assume that a 10-mW cw optically pumped FIR laser is used for the probe beam and that a 10×10 cm^2 plasma cross section is investigated. The resultant FIR beam intensity is therefore approximately 10^{-4} W/cm^2. One must then couple this radiation efficiently onto an array of diodes each of which has a geometrical area of only $\simeq 10^{-8}$ cm^2. Assuming that each element has an effective coupling area of $\simeq \lambda^2$ leads to a received power per element of $\simeq 2 \times 10^{-7}$ W. If we then further require a time resolution of 100 nsec, we arrive at a required sensitivity of 10^{-13} W/Hz for unity signal-to-noise ratio. As discussed in the preceding sections, quasi-optical, FIR receivers have recently obtained sensitivity of $\simeq 10^{-19}$ W/Hz. However, both of these receivers have utilized whisker antennas in some sort of resonant structure. This is obviously not practical for a large area where one requires $\gtrsim 10^4$ elements. A solution is to fabricate antenna structures directly on the diode substrate. A number of groups are presently addressing this problem (Murphy et al., 1977; Daiku et al., 1978; Mizuno et al., 1977). Figures 54 and 55 show a realization of such a configuration currently under investigation at the MIT Lincoln Laboratory (Murphy and Clifton, 1978). They find video responsitivity of $\simeq 1$ V/W at 337 μm. However, it should be noted that there remains

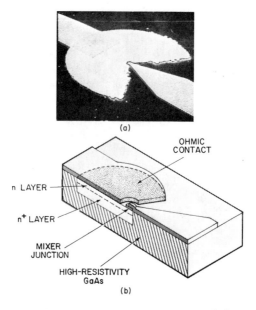

FIG. 54. (a) Scanning electron micrograph showing 2-μm diode (small dot in center), ohmic contact establishing connection to n$^+$-layer of diode, and two metal strip contacts. (b) Planar diode as fabricated by growth of n-n$^+$ epitaxial layers on a high-resistivity GaAs substrate.

much theoretical work before the mutual coupling between the antennas is understood in such a configuration. An alternative design is the monolithically integrated mixer chip.

Another problem to be surmounted in FIR holographic imaging is the fabrication of large arrays with identical detectors. However, as indicated in Fig. 57, it is now possible to fabricate large Schottky diode arrays using high-field pulsed plating so that the diodes are completely uniform. Also shown in Fig. 57 is an array fabricated using a dc plating process with the attendant diode nonuniformity. Much of the technology required for this proposed imaging has been developed. Active imaging experiments have previously been conducted by Hartwick *et al.* (1976) in the 0.3- to 1-mm region of the spectrum. Their apparatus is shown schematically in Fig. 58. In this work the concern was with the application of low-resolution, FIR imaging for nondestructive testing and law enforcement. In their test apparatus only a single detector was employed so that either the detector or object was required to be scanned. As shown in Fig. 59, reasonably detailed images were obtained.

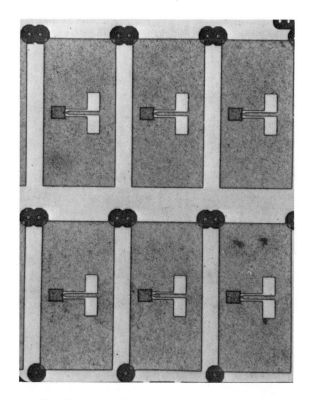

FIG. 55.   Array of planar diodes on a GaAs wafer.

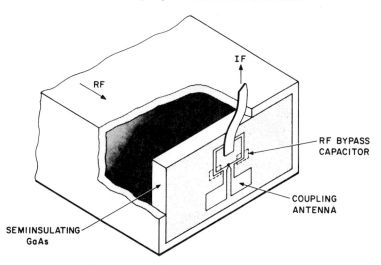

FIG. 56.   Monolithically integrated mixer chip.

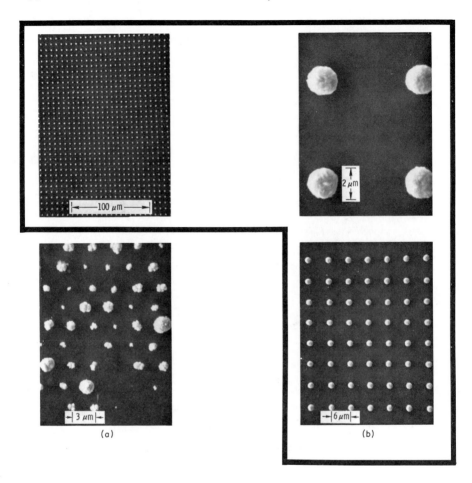

FIG. 57.   Comparison of Schottky diode array fabrication using (a) dc and (b) high-field pulse plating (after McColl, 1977).

More recently, there has been an attempt at microwave holographic imaging of large-diameter, low-density plasmas conducted at UCLA by K. Iizuka, G. Steiner and A. Y. Wong (Steiner, 1977). Here they employed the hologram matrix radar technique developed by Iizuka *et al.* (1976) for mapping ice layer thickness profiles. This radar employs the spatial distribution of scattered waves to determine distance as opposed to time delay as in coventional radars. Here both the illuminating source and the receiver are scanned to construct the hologram matrix. In this region both the phase and amplitude of the scattered field are recorded relative to some reference. These quantities

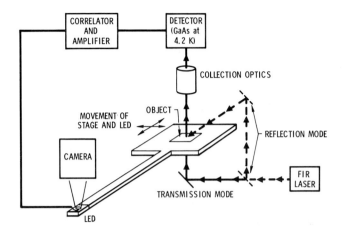

FIG. 58. Schematic of FIR imaging apparatus (after Hartwick *et al.*, 1976).

(a)   (b)   (c)

FIG. 59. FIR imaging of an ordinary key (after Hartwick *et al.*, 1976).

can be mapped in the form of a complex hologram. One can then reconstruct a unique, real or virtual image.

A preliminary test of the system at 12.4 GHz was made in a large-diameter (several meters), low-density ($\simeq 10^{11}$ cm^{-3}), plasma device. The system employed a one-dimensional array of 32 transmitting and 32 receiving antennas. By scanning every permutation of the transmitting and receiving antennas they form the 1024 complex elements of their hologram matrix. Encouraging results were obtained using several modes of operation. In one method, the plasma properties are deduced from the apparent shift of a small target relative to its vacuum position. In the other method, the reflected waves from the inhomogeneous plasma are directly employed.

VII. Summary

As can be seen from the preceding discussion, there are at present millimeter and FIR systems that can provide us with important information in both laboratory and fusion plasmas. The relative youth of the FIR field implies a large increase in submillimeter wave diagnostics in the coming years.

ACKNOWLEDGMENTS

I would like to express my grateful appreciation to my colleagues D. T. Hodges, F. F. Chen, M. McColl, V. Arunasalam, D. Boyd, T. Hartwick, A. Y. Wong, D. Hutchinson, H. Fetterman, D. R. Cohn, A. Frank, I. H. Hutchinson, B. Stallard, and J. Gustincic for kindly supplying information used in this chapter. The work reported from UCLA was performed with Th. de Grauuw, J. Gustincic, A. Mase, W. A. Peebles, and A. Semet and was supported by NSF ENG 5-14452, DOE-DMFE EY-76-C-03-0010 PA 26 and US AFOSR F49620-76-C-0012.

REFERENCES

Arunasalam, V., and Brown, S. C. (1965). *Phys. Rev.* **140**, A47.
Arunasalam, V., Cano, R., Hosea, J. C., and Mazzucatto, E. (1977). *Phys. Rev. Lett.* **39**, 888.
Ault, E. R., and Ikezi, H. (1970). *Phys. Fluids* **13**, 2874.
Bekefi, G. (1966). "Radiation Processes in Plasmas." Wiley, New York.
Belland, P., and Véron, D. (1973). *Opt. Commun.* **9**, 146.
Blaney, T. G., Cross, N. R., and Jones, R. G. (1978). *Proc. Int. Conf. Submillimeter Waves Their Appl., 3rd, Guildford, England.*
Boyd, D. A., Stauffer, F. J., and Trivelpiece, A. W. (1976). *Phys. Rev. Lett.* **37**, 98.
Boyd, D. A., Celata, C. M., Davidson, R. C., Liu, C. S., Mok, V. C., Rutgers, W. R., Stauffer, F. J. and Trivelpiece, A. W. (1977a). *In Proc. Int. Conf., 6th, Berchtesgāda, 1976,* Vol. 1, p. 399. IAEA, Vienna.

Boyd, D. A., Celata, C. M., Rutgers, W. R., Stauffer, F. J., and Tait, G. D. (1977b). "Status of Ohmic Heating in PLT," Vol. 3.

Boyd, D. A., Celata, C. M., and Stauffer, F. J. (1977c). *Proc. Soc. Photo-Opt. Instrum. Eng.* **105.**

Boyd, D., Stauffer, F. J., and Tait, G. (1979). Private communication.

Brueckner, K. A., and Jorna, S. (1974). *Rev. Mod. Phys.* **46**, 325.

Carlson, E. R., Schneider, M. V., and McMaster, T. F. (1978). *IEEE Trans. Microwave Theory Tech.* **MTT-26**, 706.

Celata, C. M., and Boyd, D. A. (1977). *Nucl. Fusion* **17**, 735.

Chao, C., Bernick, B. L., Nakaji, E. M., Ying, R. S., Weller, K. P., and Lee, D. H. (1977). *IEEE MTT-S Int. Microwave Symp. Digest.*

Clifton, B. J., Lindley, W. T., Chick, R. W., and Cohen, R. A. (1977). *Proc. Biennial Cornell Eng. Conf., 3rd, Ithaca, New York.*

Cohn, D. R., Fuse, T., Button, K. J., Lax, B., and Drozdowicz, Z. (1975). *Appl. Phys. Lett.* **27**, 280.

Costley, A. E., and TFR Group (1977). *Phys. Rev. Lett.* **38**, 1477.

Costley, A. E., Hastie, R. J., Paul, J. W. M., and Chamberlain, J. (1974). *Phys. Rev. Lett.* **33**, 758.

Daiku, Y., Mizuno, K., and Ono, S. (1978). *Proc. Int. Conf. Submillimeter Waves Their Appl., 3rd, Guildford, England.*

Danielewicz, E. J. Jr., and Weiss, C. O. (1978). *J. Quantum Electron.* **QE-14**, 705.

Degnan, J. J. (1973). *Appl. Opt.* **12**, 1026.

Dodel, G., and Kunz, W. (1978). *Infrared Phys.* **18**, 773.

Drozdowicz, Z., Temkin, R. J., Button, K. J., and Cohn, D. R. (1976). *Appl. Phys. Lett.* **28**, 328.

Drozdowicz, Z. *et al.* (1977), *IEEE J. Quantum Electron.* **QE-13**, 413.

Drummond, W. E., and Rosenbluth, M. N. (1962). *Phys. Fluids* **5**, 1507.

Efthimion, P. C., Arunasalam, V., and Hosea, J. C. (1978). *Bull. Am. Phys. Soc.* **23**, 901.

Engelmann, F., and Curatolo, M. (1973). *Nucl. Fusion* **13**, 497.

Ernst, W. P. (1971). *Proc. IEEE Nucl. Sci. Symp., San Francisco, California.*

Evans, E. E., James, B. W., Peebles, W. A., and Sharp, L. E. (1976). *Infrared Phys.* **16**, 193.

Evans, D. E., Sharp, L. E., Peebles, W. A., and Taylor, G. (1977). *IEEE J. Quantum Electron.* **QE-13**, 54.

Fetterman, H. R., Clifton, B. J., Tannenwald, P. E., and Parker, C. D. (1974). *Appl. Phys. Lett.* **24**, 70.

Fetterman, H. R., Tannenwald, P. E., Clifton, B. J., Parker, C. D., Fitzgerald, W. D., and Erickson, N. R. (1978). *Appl. Phys. Lett.* **33**, 151.

Fetterman, H. R., Tannenwald, P. E., Parker, C. D., Woskoboinikow, P., Praddaude, H. C., and Mulligan, W. J. (1979). *Appl. Phys. Lett.* **34**, 123.

Galantowicz, T. A., Danielewicz, E. J. Jr., Foote, F. B., and Hodges, D. T. (1978). *Int. Conf. Lasers.*

Golant, M. B., Alekseenko, Z. T., Korotkova, Z. S., Lunkina, L. A., Negirev, A. A., Petrova, O. P., Rebrova, T. B. and Savel'ev, V. S. (1969), *Prib. i Tekh. Eksp.*, **3**, 231.

Grossman, J., Peebles, W. A., Semet, A., and Luhmann, N. C., Jr. (1979). *Rev. Sci. In.* (to be published).

Gustincic, J. J., deGrauuw, Th., Hodges, D. T., and Luhmann, N. C. Jr. (1977). *IEEE Trans. Microwave Theory Tech.* (submitted); also post-deadline paper at *1977 IEEE MTT-S Int. Microwave Symp., 1977.*

Gustincic, S. S. (1976). IEEE 76 CH 1152-8 MTT.

Gustincic, J. J. (1977a). *Proc. Soc. Photo-Opt. Instrum. Eng.* **105**, 40.

Gustincic, J. J. (1977b). *IEE MTT-S Int. Microwave Symp. Digest* IEEE 77 CH 1219-5 MTT.

Gustincic, J. J. (1978). Private communication.

Hagström, C., and Lidholm, S. (1977). Res. Lab. of Electronics and Onsala Space Observatory Res. Rep. No. 130.

Hartwick, T. S., Hodges, D. T., Burke, D. H., and Foote, F. B. (1976). *Appl. Opt.* **15**, 1919.

Heald, M. A., and Wharton, C. B. (1965). *In* "Plasma Diagnostics with Microwaves." Wiley, New York.

Hodges, D. T., Foote, F. B., and Reel, R. D. (1976). *Appl. Phys. Lett.* **29**, 662.

Hosea, J. C., and Jobes, F. C. (1975). Princeton Plasma Physics Laboratory Rep. MATT-1176.

Hosea, J., Arunasalam, V., and Cano, R. (1977). *Phys. Rev. Lett.* **39**, 408.

Hutchinson, I. H. (1978). *Bull. Am. Phys. Soc.* **23**, 861.

Hutchinson, I. H., and Komm, D. S. (1977). *Nucl. Fusion* **17**, 1077.

Hutchinson, D. P., Vandersluis, K. L., and Ma, C. H. (1978). *Proc. Int. Conf. Submillimeter Waves Their Appl., 3rd, Guildford, England.*

Iizuka, K., Ogura, H., Yen, J. L., Nguyen, V. H., and Weedmark, J. R. (1976). *Proc. IEEE* **64**, 1493.

Ishibashi, T., Ino, M., Makimura, T., and Ohmori, M. (1977). *Electron Lett.* **13**, 299.

Jackson, J. D. (1962). "Classical Electrodynamics." Wiley, New York.

Jahoda, F. C., and Sawyer, G. A. (1971). *Methods Exp. Phys.* **9B**, 1.

Kahl, G. D., and Wedemeyer, E. H. (1964). *Phys. Fluids* **7**, 596.

Kerr, A. R., Grange, J. A., and Lichtenberger, J. A. (1978). NASA Tech. Memorandum T9616.

Krall, N. A., and Trivelpiece, A. W. (1973). "Principles of Plasma Physics." McGraw-Hill, New York.

Kräutle, H., Sauter, E., and Schultz, G. V. (1977). *Infrared Phys.* **17**, 437.

Kräutle, H., Sauter, E., and Schultz, G. V. (1978). *Infrared Phys.* **18**, 705.

Kunze, H. J. (1968). *In* "Plasma Diagnostics" (W. Lochte-Holtgreven, eds.). North-Holland Publ., Amsterdam.

Lawson, J. D. (1957). *Proc. Phys. Soc. London* **B70**, 6.

Lichtenberg, A. J., Sesnic, S., and Trivelpiece, A. W. (1964). *Phys. Rev. Lett.* **13**, 387.

Lidholm, S. (1977). Res. Lab. of Electronics and Onsala Space Observatory Res. Rep. No. 129.

Lidholm, S. and de Grauuw, Th. (1979). Private communication.

Luhmann, N. C. Jr., Peebles, W. A., Semet, A., deGrauuw, Th., and Gustincic, J. (1978). *Infrared Phys.* **18**, 777.

McColl, M. (1977). *Proc. Soc. Photo-Opt. Instrum. Eng.* **105**, 24.

McColl, M., and Hodges, D. T. (1977). *Appl. Phys. Lett.* **30**, 5.

McColl, M., Hodges, D. T., Chase, A. B., and Garber, W. A. (1977). *Proc. Ann. Symp. Freq. Contr., 31st.*

Meddas, B. J. H., and Taylor, R. J. (1974). M.I.T. Rep. PRR 7411. Also issued by Association Euratrom-FOM Jutphas as Rynhurzen Rep. 74-85 (1974).

Mizuno, K., Daiku, V., and Ono, S. (1977). *IEEE Trans. Microwave Theory Tech.* **MTT-25**, 470.

Mizuno, K., Ohmori, M., Miyazawa, K., Morimoto, M., Kodaira, S., and Ono, S. (1978). *Infrared Phys.* **18**, 401.

Montgomery, D., and Tidman, D. (1964). "Plasma Kinetic Theory." McGraw-Hill, New York.

Murphy, R. A., Bozler, C. O., Parker, C. D., Fetterman, H. R., Tannenwald, P. E., Clifton, B. J., Donnelly, J. P., and Lindley, W. T. (1977). *IEEE Trans. Microwave Theory Tech.* **MTT-25**, 494.

Murphy, R. A., and Clifton, B. J. (1978). *Int. Electron Devices Meet. Digest, IEEE Cat. No. 78 CH 1234-3ED*, p. 124.

Myers, B. R., and Levine, M. A. (1978). *Rev. Sci. Instrum.* **49**, 610.

Ohmori, M., Ishibashi, T., and Ono, S. (1977). *IEEE Trans. Electron Devices* **ED-24**, 1323.

Okabayashi, M., and Arunasalam, V. (1977). *Nuclear Fusion* **17**, 497.

Ono, M., Porkolab, M., and Chang, R. P. H. (1977). *Phys. Rev. Lett.* **38**, 962.

Robinson, L. C. (1973). *Methods Exp. Phys.* **10**, 287.

Rose, D. J., and Clark, M. C. Jr. (1961). "Plasmas and Controlled Fusion." MIT Press, Cambridge, Massachusetts.

Rutgers, W. R., and Boyd, D. A. (1977). *Phys. Lett.* **62A**, 498.

Sauter, E., Rosen, H., and Schultz, G. (1978). *Proc. AGARD Conf. Millimeter Submillimeter Wave Propagation Circuits, Munich.*

Sauthoff, N. R., Von Goeler, S., and Stodiek, W. (1978). Princeton Plasma Physics Laboratory Rep. PPPL-1379.

Semet, A., and Luhmann, N. C. Jr. (1976). *Appl. Phys. Lett.* **28**, 659.

Semet, A. Luhmann, N. C. Jr., Peebles, W. A., Hodges, D. T., Foote, F., and Reel, R. D. (1978). *APS Topical Conf. High Temp. Plasma Diagnostics, 2nd.*

Simonen, T. C. *et al.* (1978). *Proc. Int. Conf. Plasma Phys. Controlled Nucl. Fusion Res., 7th, Innsbruck, Austria.*

Slusher, R. E., and Surko, C. M. (1978). *Phys. Rev. Lett.* **40**, 400.

Smith, P. N. (1965). *IEEE J. Quantum Electron.* **QE-1**, 343.

Staats, P. A., Hutchinson, D. P., Vander Sluis, K. L., and Thomas, D. A. (1978). *APS Topical Conf. High Temp. Plasma Diagnostics, 2nd,* Los Alamos Scientific Lab. Rep. LA-7160-C.

Stallard, B. W., Frank, A. M., and Hunt, A. L. (1978). Private communication.

Stauffer, F. J., and Boyd, D. A. (1978). *Infrared Phys.* **18**, 755.

Steiner, G. (1977). M.S. Thesis, Univ. of Toronto, Toronto, Canada.

Stix, T. H. (1962). "The Theory of Plasma Waves." McGraw-Hill, New York.

Surko, C. M., and Slusher, R. E. (1976). *Phys. Rev. Lett.* **37**, 1747. TFR Group (1975). *Proc. 7th European Conf. Controlled Fusion and Plasma Physics, Lausanne, Switzerland.*

TFR Group (1978). *Phys. Rev. Lett.* **41**, 113.

Trubnikov, B. A. (1958). *Sov. Phys.-Dokl.* **3**, 136.

Vander Sluis, K. L., Hutchinson, D. P., and Staats, P. A. (1978). *Proc. Int. Conf. Submillimeter Waves Their Appl., 3rd, Guildford, England.*

Véron, D. (1974). *Opt. Commun.* **10**, 95.

Véron, D., Certain, J., and Crenn, J. P. (1977). *J. Opt. Soc. Am.* **67**, 964.

Véron, D. Belland, P., and Beccaria, M. J. (1978). *Int. Conf. Submillimeter Waves Their Appl., 3rd, Guildford, England.*

Wolfe, S. M., Button, K. J., Waldman, J., and Cohn, D. R. (1976). *Appl. Opt.* **15**, 2645.

Woskoboinikow, P., Mulligan, W. J., Praddaude, H. C., and Cohn, D. R. (1978). *Appl. Phys. Lett.* **32**, 527.

Woskoboinikow, P., Praddaude, H. C., Mulligan, W. J., Cohn, D. R., and Lax, B. (1979). *J. Appl. Phys.* **50**, 1125.

Wrixon, G. T., and Kelly, W. M. (1978). *Infrared Phys.* **18**, 413.

Yamada, M., Arunasalam, V., Efthimion, P., Gaulke, B., Hosea, J., and Mazzucato, E. (1978). *Bull. Am. Phys. Soc.* **23**, 901.

CHAPTER 2

Submillimeter Interferometry of High-Density Plasmas

D. Véron

67

List of Symbols

a, b	wave amplitude	μ_e	extraordinary refractive index
c	speed of light	μ	permeability of vacuum
d	beam diameter	φ	phase shift
d_0	beam waist diameter	ψ	phase modulation
e	elementary charge	ω	radiation frequency
f	focal length	ω_{ce}	cyclotron frequency
g	grating or mesh constant	ω_p	plasma frequency
j	current density	B_T	toroidal induction
m_e	mass of the electron	$B_{//}$	component of the poloidal induction
n	electron density		parallel to the probing beam
n_0	center electron density	F	fringe number
n_c	cutoff density	F_0	maximum fringe number
r	plasma radius	I	beam intensity
r_0	plasma maximum radius	I_0	maximum beam intensity
α	angle of refraction	J	current intensity
β	blaze angle	J_0	total current intensity
γ	normalized plasma radius r/r_0	S_D	signal for density measurement
ε_0	dielectric constant of vacuum	S_F	signal for Faraday rotation measure-
λ	radiation wavelength		ment
μ	refractive index	S_R	reference signal
μ_0	ordinary refractive index	Ω	Faraday rotation angle

High-density plasmas are of interest in thermonuclear fusion research, for which several approaches are presently considered. Nevertheless, this chapter will deal only with density measurements in tokamak-type machines (Furth, 1975), because their plasma characteristics are in a range for which submillimeter interferometers are particularly well suited. In tokamaks, the plasma is magnetically confined in a toroidal vacuum chamber in which hydrogen gas is introduced. Plasma heating is primarily achieved by the Joule effect produced by inducing a high-intensity current into the plasma. Other means, such as fast neutrals injection or high-frequency radiation, are used for additional heating. In Table I, the main parameters of several tokamaks are summarized. The plasma radius is r_0, while r_M is the major radius of the torus, and B_T and J_0 are, respectively, the toroidal field and the current intensities. The average electron density $(\bar{n})$, the maximum electron and ion temperatures, T_e and T_i, are also given.

The first part of this chapter gives principles of density measurements with electromagnetic radiation and also indicates how to select the proper wavelength. Then phase modulation is introduced, and modulation techniques are discussed. This shall be followed by paragraphs about special requirements for large plasma machine interferometers, Gaussian beam propagation, and components. Next, specific systems are described, and sources of

TABLE I

PARAMETERS AND TYPICAL PLASMA PROPERTIES FOR SEVERAL TOKAMAK MACHINES

Tokamak (Country)	r_0 (m)	r_M (m)	B_T (tesla)	J_0 (10^3 A)	T_e (eV)	T_i (eV)	$\bar{n}$ (10^{20} m^{-3})
Alcator (U.S.A.)	0.095	0.54	6.5	200	1200	800	5.3
TFR (France)	0.20	0.98	6	400–600	2500	1000	0.9
FT (Italy)	0.22	0.80	10	1000			
Doublet III (U.S.A.)	0.45 × 1.35	1.45	2.6	4000			
JET (Europe)	1.3 × 2.0	2.8	2.8	3000			
TFTR (U.S.A.)	0.85	2.5	5.2	2500			

error discussed. Finally, the last paragraph deals with future development of laser sources and components. Poloidal field measurement by using Faraday rotation is also discussed, since it can be combined with an interferometric system.

Rationalized MKSA units are used throughout the chapter.

I.   Introduction to Plasma Interferometry

A.   PRINCIPLE OF DENSITY MEASUREMENT BY USING ELECTROMAGNETIC WAVES

The index of refraction of a medium is easily measurable by using electromagnetic waves for which the medium is transparent. High-density plasmas are transparent as long as the plasma frequency ω_p is smaller than the frequency ω of the probing wave. In the case of a fully ionized, collisionless, magnetized plasma and for a beam normal to the magnetic field direction, two values of the index of refraction can be derived (Heald and Wharton, 1965). The first one, μ_0, corresponds to a polarized wave with its electric field E parallel to the magnetic field B, or ordinary wave:

$$\mu_0 = [1 - (\omega_p^2/\omega^2)]^{1/2}. \tag{1}$$

The second one, μ_e, refers to a wave with its electric field normal to the magnetic field, or extraordinary wave:

$$\mu_e = \{1 - (\omega_p^2/\omega^2)[(\omega^2 - \omega_p^2)/(\omega^2 - \omega_p^2 - \omega_{ce}^2)]\}^{1/2}, \tag{2}$$

where ω_{ce} is the electron cyclotron frequency eB/m_e, e being the elementary charge, and m_e the mass of the electron.

When the beam propagates along the magnetic field direction, then two circularly polarized waves (left hand and right hand) can be considered. The corresponding values of the refractive index are

$$\mu_\pm = \{1 - (\omega_p^2/\omega^2)[\omega/(\omega \pm \omega_{ce})]\}^{1/2}.$$

In a tokamak, there usually is a component $B_\parallel$ of B along the beam direction, and

$$\mu_\pm \simeq \{1 - (\omega_p^2/\omega^2)[\omega/(\omega \pm \omega_{ce\parallel})]\}^{1/2}, \tag{3}$$

where $\omega_{ce\parallel} = eB_\parallel/m_e$. This formula is approximate, but the Faraday effect induced by the poloidal magnetic field in a tokamak plasma can be derived from it if the condition given by Eq. (108), according to De Marco and Segre (1972), is fulfilled (Section VIII.D).

Let us consider the simple case for which the refractive index is given by Eq. (1). The condition for plasma transparency is

$$\omega > \omega_p \quad \text{or} \quad \omega > (ne^2/\varepsilon_0 m_e)^{1/2}, \tag{4}$$

where n is the electron density and ε_0 the permittivity of vacuum. Equation (4) can be written as $n < n_c$ where

$$n_c = \omega^2(\varepsilon_0 m_e/e^2), \tag{5}$$

n_c being the cutoff density, that is to say the value of n above which the beam is reflected back by the plasma.

By introducing the wavelength λ of the probing beam,

$$n_c = (4\pi^2 c^2/\lambda^2)(\varepsilon_0 m_e/e^2), \tag{6}$$

where c represents the speed of light. By using numerical values,

$$n_c = (1.11/\lambda^2) \times 10^{15} \quad \text{m}^{-3}. \tag{7}$$

Some particular numbers are given in Table II.

By combining Eqs. (1) and (5), and with $\omega_p^2 = ne^2/(\varepsilon_0 m_e)$, it follows that

$$\mu_0 = [1 - (n/n_c)]^{1/2}. \tag{8}$$

It will be shown in Section I.B that, in most practical cases, $n \ll n_c$ and the refractive index is well approximated by

$$\mu_0 \simeq 1 - (n/2n_c) = 1 - [\omega_p^2/(2\omega^2)]. \tag{9}$$

If the plasma is placed in one arm of an interferometer (Fig. 1), the phase difference between the two arms is changed by φ, e.g.,

$$\varphi = (2\pi/\lambda) \int_{Z_1}^{Z_2} [\mu_v - \mu_0(Z)] \, dZ. \tag{10}$$

TABLE II

CUTOFF DENSITY FOR SOME WAVELENGTHS

$\lambda(10^{-3}$ m):	2	1	0.337	0.195	0.119
$n_c(10^{20}$ m^{-3}):	2.8	11	98	290	780

Here, μ_v is the refractive index of vacuum and $Z_2 - Z_1$ the path length inside the plasma. Since $\mu_v = 1$, by using Eqs. (9) and (6),

$$\varphi = \pi/\lambda n_c \int_{Z_1}^{Z_2} n(Z)\, dZ$$

$$= [\lambda e^2/(4\pi c^2 \varepsilon_0 m_e)] \int_{Z_1}^{Z_2} n(Z)\, dZ \tag{11}$$

or

$$\varphi = 2.82 \times 10^{-15}\lambda \int_{Z_1}^{Z_2} n(Z)\, dZ. \tag{12}$$

The corresponding fringe number is

$$F = \varphi/(2\pi) = 4.49 \times 10^{-16}\lambda \int_{Z_1}^{Z_2} n(Z)\, dZ. \tag{13}$$

The integrated line density $\int_{Z_1}^{Z_2} n(Z)\, dZ$ is readily derivable from these expressions. In order to obtain a thorough knowledge of the density distribution, it is necessary to probe the plasma along different chords. If the plasma

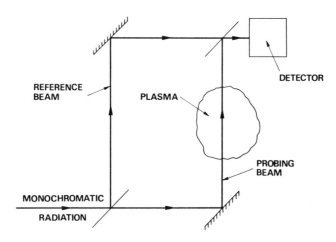

FIG. 1.   Principle of an interferometer.

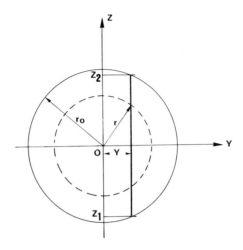

FIG. 2.   Coordinate system for the calculation of the phase shift along a chord.

density is constant along concentric circles, the expression for φ becomes, for a chord passing at the distance Y from the plasma center (Fig. 2)

$$\varphi(Y) = [2\pi/(\lambda n_c)] \int_Y^{r_0} [n(r)/(r^2 - Y^2)^{1/2}]r \, dr. \tag{14}$$

By inverting this integral, the spatial distribution $n(r)$ of the electron density can be deduced:

$$n(r) = -(\lambda n_c/\pi^2) \int_r^{r_0} [d\varphi(Y)/dY] [dY/(Y^2 - r^2)^{1/2}]. \tag{15}$$

In the case of more complicated density profiles, it is a common practice to look for the best fit of measured values of φ obtained along each chord with numbers calculated from given density profiles.

B.   Why Is Submillimeter Radiation Suitable for Interferometry?

Probing the plasma along different chords of its cross section introduces a more stringent requirement than that of simply choosing a wavelength for which $n_c > n$. The density gradient along the plasma diameter produces a refractive effect and the beam does not propagate along a straight line. Since the plasma is enclosed in a vacuum tank with windows limited in size, it is clear that refractive effects must be kept small (Fig. 3). The angle of refraction α varies with the position of the probing beam. For an axisymmetric parabolic

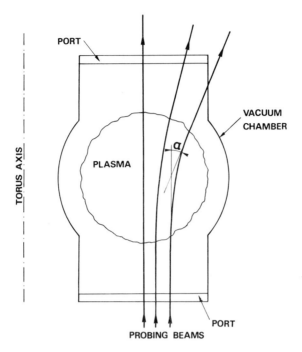

FIG. 3. Refractive effect in a plasma column.

density profile, with a center density n_0, the maximum value of α is given to a good approximation by (Shmoys, 1961)

$$\alpha_m = \sin^{-1}(n_0/n_c) \simeq n_0/n_c = [e^2/(4\pi^2 c^2 \varepsilon_0 m_e)]n_0 \lambda^2 \tag{16}$$

or

$$\alpha_m = 8.97 \times 10^{-16} n_0 \lambda^2. \tag{17}$$

The relative variation of α with the normalized plasma radius $\gamma = r/r_0$ is given Fig. 4. Its maximum value is for a probing distance from the center equal to about 0.7 times the plasma radius.

It will be shown (Section IV) that the diameter d of the probing beam at the window is related to the distance Z_0 of the median plane of the plasma to the window by the equation

$$d = 2(\lambda Z_0/\pi)^{1/2}. \tag{18}$$

In practical cases, the diameter of the window cannot be made much larger than $2d$. The result is that a reasonable upper limit for α_m is set by

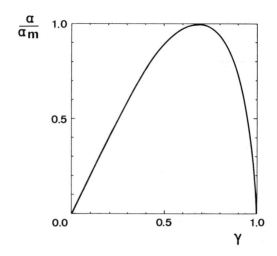

FIG. 4.   Variation of the relative angle of refraction as a function of the normalized radius $\gamma = r/r_0$.

writing that the displacement of the beam at the window should not exceed d. So, according to Fig. 5 and using the small angle approximation,

$$Z_0 \alpha_M \leq 2(\lambda Z_0/\pi)^{1/2}. \tag{19}$$

Introducing Eq. (17) and solving for λ,

$$\lambda \leq 1.16 \times 10^{10}(Z_0 n_0^2)^{-1/3} \quad \text{m}. \tag{20}$$

As an example, the following numbers:

$$Z_0 = 1 \quad \text{m}, \qquad n_0 = 10^{20} \quad \text{m}^{-3}$$

lead to

$$\lambda \leq 5 \times 10^{-4} \quad \text{m}.$$

The corresponding cutoff density is [Eq. (7)]

$$n_c = 40 \times 10^{20} = 40 n_0.$$

This clearly shows that, as stated before, refraction is a much more important constraint than cutoff, and, for present tokamak plasma machines, which operate in the 10^{20}- to 10^{21}-m^{-3} density range, radiation with small enough wavelengths has to be used for interferometric measurements.

Since visible (argon) or near infrared (carbon dioxide) lasers deliver high powers and sensitive detectors are available in this region of the spectrum, one can think about using such sources for plasma interferometry. However, some difficulties arise with short wavelengths, as is shown below. The plasma

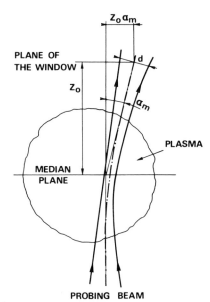

FIG. 5.   Beam displacement at the vacuum chamber window.

shift φ corresponds to a path length change λF [Eq. (13)]. In practice, vibrations of optical elements are almost unavoidable. As a consequence, the total path length variation is $l = \lambda F + \Delta l$, where Δl is caused by small displacements of optics. In principle, Δl could be reduced by servocontrolling the position of the optical elements. An alternative to this method is to use simultaneously two different wavelengths (Section VI.A). Yet, it will be shown that a careful design of the support and proper choice of the probing wavelength could efficiently lower the ratio $\Delta l/\lambda F$ to a small enough value.

If the required relative precision of the line density measurement is $\eta \times 10^{-2}$, η being expressed as percent of the fringe number F_0 obtained along a diameter of the plasma column, then the relation $\Delta l < \eta \times 10^{-2}\lambda F_0$ should be satisfied. For a parabolic density profile of radius r_0 and center density n_0,

$$F_0 = [1/(2\lambda n_c)] \int_{-r_0}^{+r_0} n_0[1 - (r^2/r_0^2)] \, dr \tag{21}$$

and

$$F_0 = \tfrac{2}{3}[r_0 n_0/(\lambda n_c)]. \tag{22}$$

Then, the preceding condition becomes

$$\tfrac{2}{3}r_0(n_0/n_c) > 10^2(\Delta l/\eta). \tag{23}$$

Introducing the value of n_c given by Eq. (7) and solving for λ,

$$\lambda > 4.08 \times 10^8 [\Delta l/(\eta r_0 n_0)]^{1/2}. \tag{24}$$

To proceed it is necessary to make some assumptions. For instance, it is reasonable to assume that Δl could not be reduced to less than 10^{-6} m during the observation time, which is of the order of 1 s. If, in addition, a precision of 1% is to be expected, $\eta = 1$, and

$$\lambda > 4.08 \times 10^5 (r_0 n_0)^{-1/2} \quad \text{m}. \tag{25}$$

Now, taking $r_0 = 0.25$ m and $n_0 = 10^{20}$ m^{-3}, then $\lambda > 8.16 \times 10^{-5}$ m.

The results of the previous study can be summarized by the following double inequality:

$$4.08 \times 10^5 (r_0 n_0)^{-1/2} < \lambda < 1.16 \times 10^{10} (Z_0 n_0^2)^{-1/3}, \tag{26}$$

which is represented graphically in Fig. 6. It is clear from this figure that the adequate wavelength range is not very large, especially for high-density plasmas.

As a concluding remark, it has to be emphasized that the above calculated limits for λ depend on the assumptions made about the permissible deviation

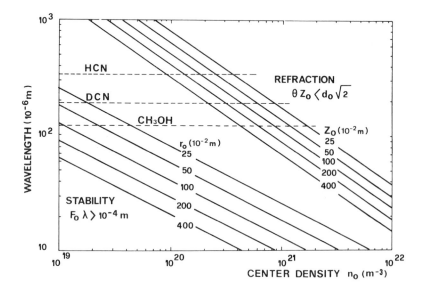

FIG. 6. Curves for the choice of the wavelength of the probing beam. Above the upper set of lines, refraction effects are too large. Below the lower set of lines, stability requirements are too stringent. The HCN, DCN, and CH₃OH laser lines are shown.

of the probing beam on one hand and the mechanical stability of the interferometer on the other hand. It must be made clear that these assumptions
can be modified according to each specific case. In particular, the number
chosen for η is somewhat subjective, and the minimum value for Δl can be
strongly affected by the size of the supporting structure or its effect reduced
by the use of more or less sophisticated compensating devices (see Section
VI.A).

II. Phase Modulation

In a simple interferometer, the probing beam interferes with a reference
beam whose phase is fixed (Fig. 1). The amplitudes of the waves can be
represented by

$$x = a \cos(\omega t - \varphi), \tag{27}$$

$$x_R = b \cos \omega t, \tag{28}$$

respectively, where φ is the phase difference due to the plasma.

By combining these waves, the resulting power is proportional to $(x + x_R)^2$,
that is to say,

$$a^2 \cos^2(\omega t - \varphi) + b^2 \cos^2 \omega t + ab[\cos(2\omega t - \varphi) + \cos \varphi]. \tag{29}$$

Only the slowly varying component $ab \cos \varphi$ is of interest. In practical
devices, the other components are automatically eliminated because the
detector cannot respond to such high frequencies. In principle, the quantity
to be measured, φ, can be deduced from our knowledge of $ab \cos \varphi$. However,
this scheme suffers from at least two important drawbacks: (a) it is sensitive
to amplitude variation of the waves, and (b) it is not possible to determine the
sign of φ. Phase modulation is a technique which can be used to overcome
these difficulties and to increase the sensitivity.

A. PRINCIPLE OF PHASE MODULATION

If the phase of the reference wave [Eq. (28)] is made equal to $\omega t + \psi$,
where ψ is a function of time, the detected signal, in the absence of plasma, is
modulated like $\cos \psi$. In the presence of plasma, the useful part of the signal is

$$S = ab \cos(\varphi + \psi). \tag{30}$$

By mixing part of the power of the two beams of the interferometer on a
second detector before the probing beam enters the plasma (Fig. 7), then a
reference signal S_R is obtained:

$$S_R = a'b' \cos \psi. \tag{31}$$

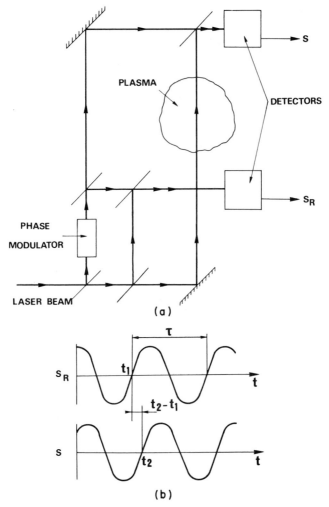

FIG. 7.   (a) Principle of a phase-modulated interferometer. (b) The signals S and S_R are shown for the particular case $\psi = \Delta\omega\, t$. The phase shift φ is proportional to $t_2 - t_1$.

The zero crossings (with positive derivative) of S_R and S occur for $\psi_1 = 2k_1\pi$ and $\varphi + \psi_2 = 2k_2\pi$, respectively, where k_1, k_2 take values $1, 2, 3, \ldots$. It follows that

$$\varphi = \psi_1 - \psi_2 + 2\pi(k_2 - k_1).$$

By assuming that ψ can be approximated by a linear function of t, qt, during the observation time, $\psi_1 = qt_1$ and $\psi_2 = qt_2$, then

$$\varphi = q(t_1 - t_2) + 2\pi(k_2 - k_1). \tag{32}$$

Introducing the pseudoperiod τ of the function cos ψ, defined by $\psi(t + \tau) = \psi(t) + 2\pi$,

$$\varphi = 2\pi[(t_1 - t_2)/\tau] + 2\pi(k_2 - k_1). \qquad (33)$$

Usually, $k_2 = k_1$, and $\varphi = 2\pi(t_1 - t_2)/\tau$.

The result is that the measurement of φ is not anymore dependent on the amplitude of the signal and is simply deduced from time measurements, which are easily made with high accuracy. Moreover, the sign of $d\varphi/dt$ is unambiguously given by the sign of $(d/dt)[(t_1 - t_2)/\tau]$. Nevertheless, it should be noted that φ is determined only once for each time interval τ. This is not a real inconvenience if τ is chosen small enough compared to the time scale variation of φ.

B. MODULATION TECHNIQUES

1. Moving Mirror

One of the simplest ways of changing the phase of the reference beam is to reflect it from a moving mirror assembly (Fig. 8). In this case, if v is the component of the velocity vector of the mirror parallel to the reflected beam, then the optical path variation is $2vt$ and

$$\psi = 2\pi(2vt/\lambda) = 2\pi(t/\tau), \qquad (34)$$

in which λ is the wavelength of the radiation. As an example, if the desired time resolution is 10^{-4} s (this means that $\tau = 10^{-4}$ s) and if $\lambda = 3.37 \times 10^{-4}$ m (HCN laser), then

$$v = \lambda/2\tau = 1.7 \quad \text{m s}^{-1}.$$

Even much higher speeds are feasible in the laboratory (Peterson and Jahoda, 1971), but, for plasmas with time duration of 1 s or more, the distance that has to be covered by the mirror might be somewhat too long to be really practical. To overcome this difficulty, a vibrating mirror can be used (Olsen, 1971).

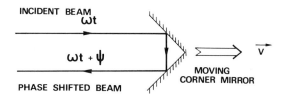

FIG. 8.   Moving corner mirror for phase shifting of a laser beam.

2. Rotating Cylindrical Grating

Continuous variation of ψ with time can be obtained by frequency shifting the reference beam by a small amount $\Delta\omega$. Then

$$\psi = \Delta\omega t = (2\pi/\tau)t. \tag{35}$$

Note that this expression of ψ is the same as that given in Eq. (34), and shows that a change in path length is equivalent to a frequency shift. An attractive way of shifting the frequency of the beam is to use the Doppler effect (Blanc and Véron, 1972; Véron, 1974) generated by a rotating grating. Let us consider a cylinder with grooves machined parallel to its axis (Fig. 9) with a blaze angle β. The incident beam is focused so as to reduce its diameter to a few millimeters at the surface of the grating. If the radius ρ of the cylinder is large enough, the curvature of the grating can be neglected, and, to a good approximation, the whole beam can be set at the blaze angle, although the grating may have to be slightly tilted to enable the diffracted beam to be separated from the incident beam in the case of a Mach–Zehnder interferometer (Section VI.C.2). The diffracted beam experiences a Doppler shift equal to

$$\Delta\omega = (2\omega/c)(2\pi\rho N \sin \beta), \tag{36}$$

where N is the angular velocity of the cylinder in revolutions per second. Since $\omega/c = 2\pi/\lambda$ and $\Delta\omega = 2\pi/\tau$,

$$N = (1/\tau)[1/(2\pi\rho)][\lambda/(2 \sin \beta)]. \tag{37}$$

By introducing the total number of grooves $G = 2\pi\rho/g$, g being the grating constant $\lambda/(2 \sin \beta)$, the preceding expression reduces to

$$N = 1/(\tau G). \tag{38}$$

The grating used in the device reported by Véron et al. (1977) has the following characteristics:

$$\rho = 6 \times 10^{-2} \quad \text{m}, \quad \beta = 54°, \quad G = 1800 \quad (\text{for} \quad \lambda = 3.37 \times 10^{-4} \quad \text{m}).$$

For a typical time resolution, $\tau = 10^{-4}$ s,

$$N = 5.56 \quad \text{rps}.$$

The rotating speed of the grating could be easily increased by more than an order of magnitude allowing time resolution as low as a few microseconds to be reached.

This technique has the advantage of simplicity, reliability, and high efficiency. A complete interferometric setup using such a technique is described in Section VI.C.

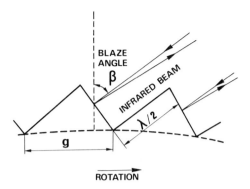

FIG. 9.   Cylindrical rotating blazed grating for Doppler shifting the beam frequency (after Véron, 1974).

3.   Two-Laser System

Here each beam of the interferometer is fed with its own laser. The two lasers are tuned such that their frequencies differ by $\Delta\omega$ (Wolfe *et al.*, 1976). As in the previous case, the function ψ is expressed by Eq. (35). Beat frequencies of several megahertz can be obtained, as shown by Wolfe *et al.* (1976) with 1-m-long cavity, CO_2 pumped, methyl alcohol lasers (see Fig. 23). The frequency shift $\Delta\omega$ as a function of the cavity length variation ΔL is

$$\Delta\omega = \omega \, \Delta L/L, \tag{39}$$

where L is the total cavity length.

For the alcohol laser with $L = 1$ m,

$$\Delta\omega/2\pi = 2.5 \times 10^{12} \, \Delta L.$$

This shows that a length variation of only 10^{-6} m leads to a frequency change of 2.5 MHz. It follows that a high degree of stability of the laser mount is necessary, although $\Delta\omega$ does not need to be constant with time (see Section VI.B).

III.   Special Requirements for Large Plasma Machine Interferometers

The electron density is one of the most fundamental parameters of the plasma to be measured. Present plasma machines are highly sophisticated, and their technology is very complicated. As a consequence, they are costly to run, and maximum information and data have to be recorded for each shot. It follows that the diagnostic systems and, in particular, the interferometer should operate reliably. Moreover, the starting process of the interferometer or, more precisely, of the laser source and detection system should be as simple as possible to allow the apparatus to be turned on whenever the plasma machine is ready to operate. This implies the use of a powerful enough laser source to avoid the need for sensitive detectors, which are associated with the troublesome manipulation of liquid helium.

As an example, the following subsection shall be devoted to the description of the HCN laser source of the TFR interferometer (see also Section VI.C.2), which seems to fulfil most closely the preceding requirements at the present time.

A. WAVEGUIDE HCN LASER

The resonant cavity consists of a 3.2-m-long, 5.4×10^{-2}-m-diameter pyrex waveguide tube (Belland *et al.*, 1975, 1976b) and two plane reflectors against its ends (Fig. 10). One of the reflectors is an aluminized glass mirror, and the other is a metal mesh, which ensures proper coupling of the output power. The window is a mylar sheet whose thickness $(10^{-4}$ m) is chosen to minimize the reflection loss. Its transmission coefficient is about 0.9. The gas is a mixture of nitrogen (6%), methane (17%), and helium (77%) at a total pressure near 1.5 Torr and a flow rate of about 1.7×10^{-6} m³ s⁻¹ at STP. The discharge is fed by a current-stabilized power supply. The current intensity is 1.3 A. The cathode consists of a tantalum cylinder heated by the discharge current itself. The combination of gas mixture, hot cathode, and stabilized power supply gives a uniformly striated discharge, which is adequate for the production of stable output. The glass tube is surrounded by an oil jacket, whose temperature is maintained at about 130°C. This

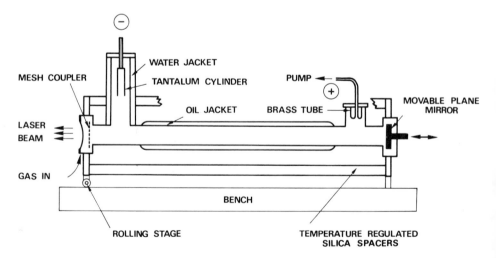

FIG. 10. Thermally stabilized waveguide discharge excited laser (after Véron *et al.*, 1978).

feature prevents any polymer deposit on the tube wall, while it does not affect the output power (Belland and Véron, 1973). The cathode is located in a water-cooled side arm, which collects most of the polymers released from the discharge. This part of the laser can easily be dismantled for cleaning without disturbing the optical alignment. This operation can be performed in a few minutes and needs to be done for every 50 h of operation.

The whole laser assembly is supported by an aluminum bench. The mirror (or anode) end is tightly fixed to the bench, while the mesh (or cathode) end is free to move along the bench with the help of a rolling stage. Variations of the distance between the reflectors are kept well below 10^{-6} m by means of three spacers, made of silica tubes, whose temperature is controlled by a continuous flow of water, so as to maintain the laser cavity at optimum tuning. The mirror and mesh may have to be replaced after a few hundred hours of running time. The mechanical design is such that any new alignment of the optics of the laser is not necessary after they have been replaced. Several parallel copper wires are placed in front of the mirror to fix the direction of the beam polarization.

The operation of the laser is simplified by the use of a single compressed cylinder containing premixed gases, and by having all the adjustments preset. Good stability of the output power needs a warming up period of about one hour. Nevertheless, during this time, remote tuning of the cavity from the control room allows the laser to be used a few minutes after being turned on if necessary.

Owing to optimization of most parameters and to the use of the discharge tube as a waveguide, which increases the efficiency of the lasing medium, the laser just described has a total output power of 0.15 W. It shall be shown in Section VI.C.2 that this laser can feed an eight-channel interferometer operating with room-temperature pyroelectric detectors.

B. OPTICAL COMPONENTS

Another important feature of the interferometer is its ability to be aligned with visible light. The large number of optical components (about 100 for the eight-channel TFR interferometer, see Section VI.C.2) that have to be fitted within a restricted space makes the alignment in the infrared almost impossible. It is then highly desirable to have components with good optical properties in the visible. This excludes some plastic materials, such as polyethylene, which is not transparent to visible light, but crystal quartz, TPX, and, obviously, metal mirrors are widely used. Metal meshes and parallel wire grids are also quite convenient, although they do not reflect very much visible light, but their transparency is satisfactory. Special mention should be made about thin plastic foils, such as mylar, which look attractive for their good optical properties and their negligible cost. They could, in principle, be used to make beam splitters, but they are very sensitive to microphonic effects. Since the environment of tokamak machines is somewhat noisy, such materials must be avoided in the construction of the interferometer. Details about optical elements shall be given in Section V.

C. Supporting Structure

Optical elements of the interferometer may be fixed to the vacuum chamber or to the surrounding structure of the plasma machine. Nevertheless, the mechanical stress during a shot is such that a disturbance of the order of a fraction of a wavelength of the radiation may generate spurious phase shifts and introduce an error in the plasma density measurement. Although the use of a compensating system (feedback control of mirror position or the use of two different wavelengths; see Section VI.A) may overcome this difficulty, it is preferable to reduce the possibility of a disturbance to the optical elements by mounting them on a frame independent from the machine.

There is no unique solution for designing such a vibration-free support, but it should be emphasized that extreme care has to be taken to avoid effects of time-varying magnetic fields produced by the coils of the machine. As a consequence, proper use of insulating materials is necessary so that induced eddy currents cannot appear in the frame.

IV. Optical Propagation of Laser Beams

A. Introduction to Gaussian Beam Theory

Submillimeter radiation has the interesting property that it can be propagated by using suitable components, in the same way as visible light. Although waveguiding is also feasible and, in some cases, desirable, this section shall deal only with optical propagation because present submillimeter interferometric setups do not use waveguides.

It has been experimentally shown that infrared lasers that are uniformly coupled over their whole cross section (Belland *et al.*, 1975) produce a Gaussian beam, at least at a distance of a few meters from the end of the waveguide discharge tube (Fig. 11). It follows that Gaussian beam theory of propagation can be applied to design the optical part of the interferometer. Most of the following information is deduced from the study made by Kogelnik and Li (1966). For the convenience of experimentalists, the following formulas refer to the intensity distribution, which is directly measurable, rather than to the field distribution, as in the previously mentioned reference.

The radial distribution of the intensity I is a Gaussian curve:

$$I = I_0 \exp(-4u^2/d^2) \tag{40}$$

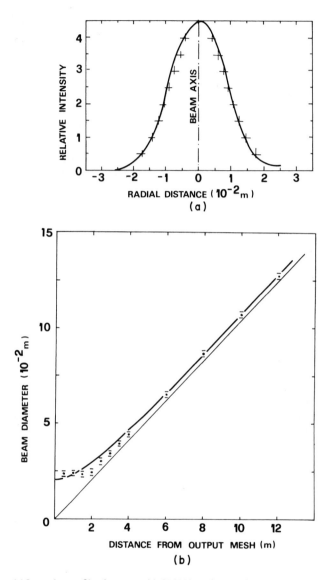

FIG. 11.   (a) Intensity profile of a waveguide HCN laser beam. The solid curve is experimental. Crosses correspond to the best fit calculated Gaussian profile. (b) Diameter of a waveguide HCN laser beam as a function of distance from the coupling mesh. The dashed line is the best fit calculated curve for an ideal Gaussian beam (after Belland *et al.*, 1975).

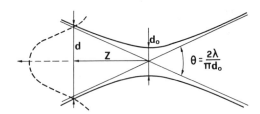

FIG. 12.   Main parameters of a Gaussian beam, from Eq. (41).

in which I_0 is the peak intensity, u the radial distance from the axis, and d the diameter at the $1/e$ point of the intensity profile.

As it propagates through space, the beam expands, and its diameter at a distance Z from the beam waist is

$$d = [d_0^2 + (4\lambda^2/\pi^2)(Z^2/d_0^2)]^{1/2}. \tag{41}$$

In this equation, d_0 is the diameter at the beam waist (Fig. 12).

The beam contour is a hyperbola with asymptotes inclined at an angle:

$$\theta = 2\lambda/(\pi d_0). \tag{42}$$

Finally, the radius of curvature R of the wave front is

$$R = Z\{1 + [\pi^2 d_0^4/(4\lambda^2 Z^2)]\}. \tag{43}$$

When a focusing element (mirror or lens with focal length f) is used, the beam waists d_1 and d_2 satisfy the following equation:

$$\frac{1}{d_2^2} = \frac{1}{d_1^2}\left(1 - \frac{Z_1}{f}\right)^2 + \frac{1}{f^2}\left(\frac{\pi d_1}{2\lambda}\right)^2, \tag{44}$$

while the distances Z_1 and Z_2 of the corresponding waists to the focusing element are such that

$$Z_2 - f = (Z_1 - f)\{f^2/[(Z_1 - f)^2 + (\pi d_1^2/(2\lambda))^2]\}. \tag{45}$$

B.   USEFUL FORMULAS FOR THE DESIGN OF SUBMILLIMETER
      INTERFEROMETERS

The windows through which the probing beam has to pass cannot generally be very large because of mechanical constraints. Similarly, the size of most optical elements is also limited. In addition, when dealing with elements as costly as crystal quartz beam splitters, it is desirable to keep their dimensions to reasonable values. To have a better knowledge of what "reasonable values" means, it is useful to consider how much power is excluded from a Gaussian beam of diameter d upon traversing a diaphragm of diameter D.

The total power P transported by the beam is given by a proper integration of Eq. (40):

$$P = 2\pi I_0 \int_0^\infty u \exp(-4u^2/d^2) \, du = \tfrac{1}{4}\pi I_0 d^2. \tag{46}$$

The power transmitted through the diaphragm is

$$P_D = 2\pi I_0 \int_0^{D/2} u \exp(-4u^2/d^2) \, du$$

$$= \tfrac{1}{4}\pi I_0 d^2 [1 - \exp(-D^2/d^2)]. \tag{47}$$

By combining Eqs. (46) and (47)

$$P_D/P = 1 - \exp(-D^2/d^2). \tag{48}$$

Figure 13 shows that this ratio comes close to 1 for $D > 2.2d$.

It should be noted, however, that this simple calculation does not take into account diffraction effects at the edge of the diaphragm which may increase the losses. It follows that, as a general rule, a safe limit to the diameter D of windows and other optical elements is given by

$$D > 2.2d. \tag{49}$$

To have the best spatial resolution in probing the plasma, it is necessary to place the beam waist in the median plane of the plasma and to make its diameter as small as possible. On the other hand, a small waist diameter corresponds to a large divergence of the beam, as Eq. (42) indicates. It then becomes quite obvious that there is an optimum beam waist diameter. If it is

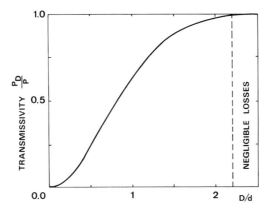

FIG. 13.   Calculated transmissivity of a diaphragm with diameter D for a Gaussian beam with diameter d. Losses are considered as negligible for $D/d > 2.2$.

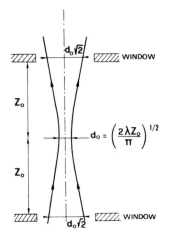

FIG. 14.   Determination of the minimum value of the beam diameter at the vacuum chamber window.

assumed that the window is the limiting aperture for the beam, which is usually the case, then it is important to reduce the diameter of the beam at this position by as much as possible. By differentiating Eq. (41) with respect to d_0, it is clear that the minimum value of d at the window is

$$d_{min} = 2(\lambda Z_0/\pi)^{1/2}, \tag{50}$$

where Z_0 is the distance from the median plane of the plasma to the window (Fig. 14). The corresponding value of d_0 is

$$d_0 = (2\lambda Z_0/\pi)^{1/2} \tag{51}$$

If $\lambda = 3.37 \times 10^{-4}$ m (HCN laser), then

$$d_{min} = 2.1 \times 10^{-2}(Z_0)^{1/2} \quad \text{m}, \qquad d_0 = 1.5 \times 10^{-2}(Z_0)^{1/2} \quad \text{m},$$

and, according to Eq. (49), the power loss at the window is negligible if

$$D > 4.6 \times 10^{-2}(Z_0)^{1/2} \quad \text{m}$$

The beam coming out of the laser diverges and a focusing element has to be placed in the beam path in order to obtain a beam waist inside the plasma. The beam characteristics before focusing are entirely defined by the laser, while the beam characteristics after focusing depend on d_0 calculated from Eq. (51). If only one focusing element is to be used, the diameter d of the beam should satisfy the following relations (Fig. 15):

$$d = [d_1^2 + (4\lambda^2/\pi^2)(Z_1^2/d_1^2)]^{1/2}, \tag{52}$$

$$= [d_0^2 + (4\lambda^2/\pi^2)(Z_0'^2/d_0^2)]^{1/2}. \tag{53}$$

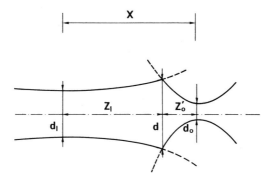

FIG. 15.   Matching of two Gaussian beams with a focusing element, using Eqs. (52)–(55).

The sum

$$Z_1 + Z'_0 = X \tag{54}$$

represents the distance between the beam waists. Equations (52)–(54) and (49) define the position and size of the focusing element, whose focal length can be calculated by combining Eqs. (44) and (45):

$$f = (Z'_0 d_1^2 - Z_1 d_0^2)/(d_1^2 - d_0^2). \tag{55}$$

It might also be interesting to consider the relation

$$d_1/d_0 = [(f - Z_1)/(f - Z'_0)]^{1/2}. \tag{56}$$

Equations (52), (53) and (54) are general formulas used to solve beam-matching problems. They have been applied here to the case in which only one lens (or spherical mirror) is employed. Successive applications of the same equations can solve more complicated cases for which several lenses (or spherical mirrors) are placed in series (see Section VI.C.2.a).

V.   Components

A.   MIRRORS

Standard aluminized glass or all metal mirrors are adequate for reflecting and focusing submillimeter radiation. Their loss coefficient is of the order of 10^{-2}.

B.   WINDOWS AND BEAM SPLITTERS

Various materials can be used to make windows and beam splitters. However, only a few are suitable for designing interferometers for large plasma machines, as already mentioned in Section III.B.

TABLE III

OPTICAL CONSTANTS OF CRYSTAL QUARTZ IN THE FAR INFRARED[a]

Frequency $(10^{-2} \text{ m})^{-1}$	Refractive indices		Absorption coefficients $(10^{-2} \text{ m})^{-1}$	
	Ord.	Ext.	Ord.	Ext.
20.2	2.1073	2.1541		
25.2	2.1076	2.1561		
30.2	2.1076	2.1560	0.10	0.10
35.3	2.1083	2.1564		
40.3	2.1093	2.1573		
45.4	2.1105	2.1580	0.15	0.12
50.4	2.1114	2.1590		
55.4	2.1124	2.1602		
60.5	2.1134	2.1615	0.32	0.21
65.5	2.1147	2.1629		
70.6	2.1159	2.1644		
75.6	2.1175	2.1662	0.47	0.37
80.6	2.1190	2.1679		
85.7	2.1209	2.1699		
90.7	2.1228	2.1718	0.61	0.56
95.8	2.1248	2.1739		
100.8	2.1269	2.1762		

[a] After Russel and Bell, 1967.

Crystalline quartz and sapphire are among the best materials for windows, since plastics (as TPX) are not acceptable for high-vacuum devices. The characteristics of crystal quartz in the far infrared region are given in Table III, after Russel and Bell (1967a). The numbers clearly indicate that the absorption is low, mainly for long wavelength radiation. If the windows are made of plates with parallel surfaces, multiple reflections occur inside the material. The reflectivity of the plate can be minimized either by using the resonant effect between waves reflected by both surfaces or placing the windows at the Brewster angle. Yet, the Brewster angle has a high value (about 65°), leading to large-sized windows. Therefore, it is more convenient and also cheaper to use the resonant effect, although it implies a precise calibration of the thickness. The latter is computed by using the Fresnel formulas (Born and Wolf, 1970). Either the ordinary or extraordinary ray can be used. In the second case, absorption losses are smaller (at least for wavelengths shorter than 3×10^{-4} m), but with the plate being cut parallel to the optical axis of the crystal, the electric field vector of the wave has to be parallel to this axis.

TABLE IV

OPTICAL CONSTANTS OF SAPPHIRE IN THE FAR INFRARED[a]

Frequency $(10^{-2}$ m$)^{-1}$	Refractive indices		Absorption coefficients $(10^{-2}$ m$)^{-1}$	
	Ord.	Ext.	Ord.	Ext.
20.2	3.0688	3.4111		
25.2	3.0698	3.4129		
30.2	3.0704	3.4134	0.4	0.5
35.3	3.0720	3.4163		
40.3	3.0740	3.4187		
45.4	3.0752	3.4232	1.7	2.2
50.4	3.0770	3.4260		
55.4	3.0795	3.4294		
60.5	3.0822	3.4334	3.6	4.0
65.5	3.0843	3.4391		
70.6	3.0870	3.4444		
75.6	3.0906	3.4510	4.9	7.6
80.6	3.0941	3.4569		
85.7	3.0982	3.4625		
90.7	3.1019	3.4689	7.2	12.7
95.8	3.1060	3.4766		
100.8	3:1103	3.4836		

[a] After Russel and Bell, 1967.

Sapphire is not as good as crystalline quartz as far as the absorption coefficient is concerned (Russel and Bell, 1967b). Its higher refractive index (Table IV) necessitates a somewhat better precision in determining the thickness, but its great advantage is that windows can be welded to the port and the whole assembly baked so as to match high-vacuum requirements.

Crystalline quartz is also a good material for making beam splitters. The orientation of the optical axis must be carefully controlled in order to avoid any possible change in the linear polarization of the beam. Three cases of propagation, either purely ordinary or purely extraordinary, are possible (Fig. 16):

(a)   the optical axis is parallel to the incident plane, and the electric field vector must be normal to the incident plane. Propagation is then purely ordinary, or

(b)   the optical axis is normal to the incident plane. Then the electric field can either be parallel to the incident plane (ordinary mode) or normal to the incident plane (extraordinary mode).

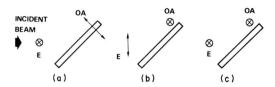

FIG. 16.   Different cases of beam propagation inside a quartz beam splitter, either purely ordinary or purely extraordinary. Shown are the optical axis (0A) and electric field vector (E). See the text for discussion of different parts.

For radiation that is only slightly absorbed (as for the HCN laser for instance) the reflectivity of the beam splitters can be calculated with a fairly good accuracy from the Fresnel formulas by neglecting the absorption. The transmittivity is affected more by the absorption as shown by calculations carried out by Frank (1976). For a plate that is a few millimeters thick, reflectivity and transmittivity vary rapidly with the angle of incidence, and

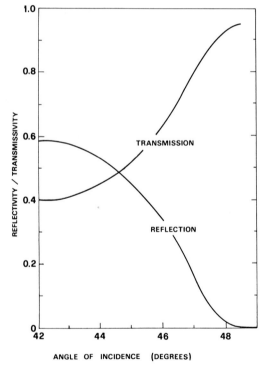

FIG. 17.   Reflectivity and transmissivity of a quartz plate for a polarized HCN laser beam with its electric field vector normal to the plane of incidence, with 3×10^{-3} m thickness, $\lambda = 3.37 \times 10^{-4}$ m, index $= 2.156$, and abs. coeff. $= 10$ m^{-1} (after Frank, 1976).

thus can be adjusted to the desired value by slightly tilting the plate (Frank 1976), as shown in Fig. 17. Quartz beam splitters are employed most of the time at angles of incidence near 45°. Maximum reflectivity of 0.6 can then be reached when the electric field vector is normal to the incident plane. On the other hand, reflectivity cannot be larger than 0.2 for a beam with its electric field vector parallel to the plane of incidence. Since reflection coefficients up to 0.5 are usually required, quartz beam splitters are not well suited for that particular case. This difficulty can be overcome by using meshes or wire grids. The first are insensitive to polarization, although some variation of their properties with the direction of polarization may result from their utilization at large angles of incidence. The second are strongly polarization sensitive. Ulrich *et al.* (1970) have computed the power reflectance of metal meshes (Fig. 18) for normal incidence. The absorption losses of such meshes are quite low, but care must be taken to avoid too large diffraction losses

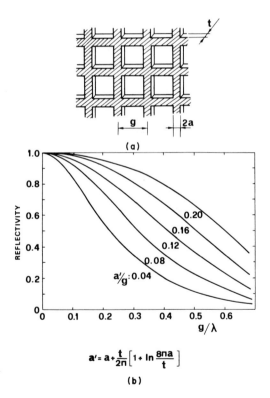

$$a' = a + \frac{t}{2n}\left[1 + \ln\frac{8na}{t}\right]$$

(b)

FIG. 18. Metal meshes: (a) definition of mesh parameters; (b) reflectivity of a metal mesh as a function of beam wavelength and mesh parameters, for normal incidence (after Ulrich *et al.*, 1970).

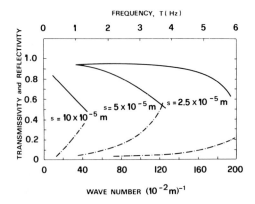

FIG. 19.   Wire grid characteristics for 45° incidence as a function of wire spacing s and wave number. The electric field vector of the beam is parallel to the wire. The curves shown are for transmissivity (·–·) and reflectivity (—) (after Costley *et al.*, 1977).

which occur when wire spacing becomes larger than the wavelength (Sakaï *et al.*, 1969). Copper, nickel, and gold meshes are commercially available.

Free-standing wire grids are even more attractive because their optical properties can be adjusted by simply varying the wire spacing. Details of the manufacturing process are given by Costley *et al.* (1977), together with their characteristics at 45° incidence (Fig. 19). The direction of the wires can be chosen so as to match the direction of the electric field vector.

Meshes and wire grids are not microphonic, but, as already mentioned (Section III.B), their reflection coefficient in the visible region of the spectrum is very poor, and this must be taken into account in the alignment process of a complicated assembly of components as, for example, in a multichannel system.

C. LENSES

In principle, lenses can be made of any material transparent to sub-millimeter radiation and, if possible, to visible light. High efficiency requires the use of materials not only with low absorption coefficient, but also with small refractive index in order to minimize reflection losses on the interfaces. For normal incidence, reflection losses amount to

$$[(\mu - 1)/(\mu + 1)]^2. \tag{57}$$

As an example, for $\mu = 2.1$ (crystalline quartz), the losses exceed 0.12 per interface.

Plastics are more suitable, and the absorption coefficients of three of them are given Fig. 20, after Chantry *et al.* (1971). Nevertheless, only one, Poly-4-

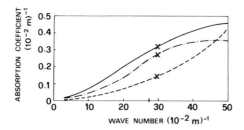

FIG. 20. Absorption coefficient of three plastic materials as a function of wave number: TPX (—), polypropylene (·–·), and polyethylene (– – –). The crosses indicate HCN laser measurements (after Chantry *et al.*, 1971).

methylpentene-1, or TPX, presents two interesting features: (a) it is transparent to visible light, and (b) it has the same refractive index (1.45) for both regions of the spectrum. Reflection losses are quite low, 0.03 per interface. As a consequence, TPX is widely used for making lenses for submillimeter radiation.

Antireflection coatings, developed for infrared filters (Armstrong and Low, 1974), could be applied to lenses made of high refractive index materials, such as crystal quartz. Figure 21 gives an example of transmittance enhancement of quartz by proper coating with a thin polyethylene layer. Although this technique looks promising, it has not been extensively used up to now to improve the performances of optical components other than filters.

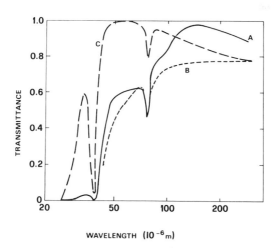

FIG. 21. Transmittance of a 8.6×10^{-4} m thick quartz sample as a function of wavelength. Curve A, coated with 25×10^{-6} m polyethylene; B, uncoated; C, coated with 9×10^{-6} m polyethylene (after Armstrong and Low, 1974).

VI. Description of Specific Systems

As suggested previously, there are several ways of designing interferom-eters for electron density measurements in large plasma machines, although they all proceed from the same basic ideas. The purpose of the present section is not to give an exhaustive list of what has been done on the subject, but to describe some of the devices that have been operated or are being designed, as typical examples of the application of infrared radiation to plasma interferometry. At this point, the following remark should be made about the choice of the proper radiation wavelength.

Mechanical vibrations of the supporting structure of the interferometer lead to spurious phase shifts (Section I). Nevertheless, the construction of a vibration-free frame can be avoided by using simultaneously two wave-lengths, both beams employing the same optical elements. It follows that the lower limit of the wavelength of the probing radiation discussed in Section I can be drastically relaxed, and powerful, commercially available, near infrared lasers can be used as sources.

The following subsection shall be devoted to this type of interferometer, while far infrared, single-wavelength interferometers shall be described in the subsequent subsections.

A. Two-Wavelength Interferometer

If the vibrations correspond to an optical path length change δ, a phase shift $2\pi\delta/\lambda$ is added to the phase shift φ due to the plasma. The resulting phase shifts ϕ_1 and ϕ_2 for the two wavelengths λ_1 and λ_2 are, respectively,

$$\phi_1 = \varphi_1 + (2\pi\delta/\lambda_1), \qquad \phi_2 = \varphi_2 + (2\pi\delta/\lambda_2). \tag{58}$$

Adding these equations, after multiplying them by λ_1 and $-\lambda_2$, respec-tively, we obtain

$$\phi_1\lambda_1 - \phi_2\lambda_2 = \varphi_1\lambda_1 - \varphi_2\lambda_2. \tag{59}$$

By using Eq. (11), in which the factor $e^2/(4\pi c^2\varepsilon_0 m_e)$ has been replaced by K,

$$\phi_1\lambda_1 - \phi_2\lambda_2 = K(\lambda_1^2 - \lambda_2^2) \int_{z_1}^{z_2} n(z)\, dz, \tag{60}$$

and the line density is then given by

$$\int_{z_1}^{z_2} n(z)\, dz = (\phi_1\lambda_1 - \phi_2\lambda_2)/[K(\lambda_1^2 - \lambda_2^2)], \tag{61}$$

thus eliminating the effect of vibrations. The short-wavelength limit discussed in Section I can be relaxed by using this technique, at the expense of additional complication. In particular, the optical components must be designed in such a way as to satisfy to the requirements for both wavelengths, and two detectors must be used for each channel. Nevertheless, the case in which $\lambda_2^2 \ll \lambda_1^2$ is of interest because Eq. (61) simplifies to

$$\int_{z_1}^{z_2} n(z)\,dz = (\phi_1\lambda_1 - \phi_2\lambda_2)/(K\lambda_1^2). \tag{62}$$

This means that the choice of a wavelength λ_2 such as $\lambda_2^2 \ll \lambda_1^2$ allows the plasma phase shift to be neglected for λ_2, and ϕ_2 is a direct measurement of δ [Eq. (58)].

This principle has been applied by Gibson et al. (1966) for the measurement of line density in Zeta. The two wavelengths 6.3×10^{-7} m and 3.39×10^{-6} m of the He–Ne laser were used. The ratio $(\lambda_2/\lambda_1)^2 = 0.03$ is small enough to permit application of Eq. (62).

As suggested by Gibson and Reid (1964), the signal corresponding to visible radiation could be used to keep the optical path of the interferometer constant by means of a feedback loop and a piezoelectric system driving one of the mirrors. More recently Baker and Shu-Tso Lee (1978) have designed a two-wavelength system for Doublet III, using the visible light of a He–Ne laser and the 1.06×10^{-5}-m radiation of a CO_2 laser. The schematic diagram of the apparatus is shown Fig. 22. The vertical dimension of the plasma is

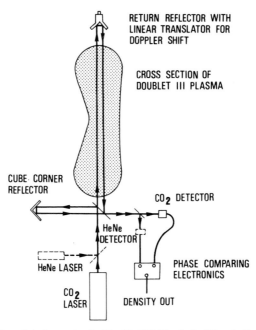

FIG. 22.   Dual laser interferometer for Doublet III. The dashed lines indicate the optical path out of the plane of the paper (after Baker and Shu-Tso Lee, 1978).

3 m. This gives an idea of the total size of the interferometer. The upper corner mirror is translated along a vertical axis, thus shifting the frequency of the radiation by the Doppler effect (see Section II). The He–Ne laser beam is sensitive only to mechanical vibrations, while the CO_2 laser beam is phase shifted, both by vibration and by the plasma. Since the wavelength ratio is well approximated by $\frac{67}{4}$, the frequencies of the beat signals corresponding to each beam are in the ratio $\frac{4}{67}$. By dividing the frequency of the signal monitored by the infrared detector by 4 and the frequency of the signal monitored by the visible detector by 67, the resulting signals S_1' and S_2' have the same frequency. It follows that the phase shift due to the plasma is directly measured by comparison of the time lag between the zero crossings of S_1' and S_2', with S_2' playing the role of the reference signal S_R [Eq. (31)], as already explained in Section II.A. Thus the effect of vibrations is automatically eliminated. The authors claim that a sensitivity better than 10^{-2} fringe is reached under test conditions. In terms of line density, this corresponds to 2×10^{18} electrons per square meter along the total beam path (that is to say for a roundtrip, since this interferometer is of the Michelson type).

B.   METHYL ALCOHOL LASER INTERFEROMETER FOR ALCATOR

In the Alcator device, plasma densities up to 10^{21} m^{-3} are achieved. The distance Z_0 of the median plane of the plasma to the window is of the order of 0.25 m and the plasma radius 0.1 m. Figure 6 shows that the 1.19×10^{-4}-m wavelength of the methyl alcohol laser is well suited for this machine.

Phase modulation is obtained by using two waveguide methyl alcohol lasers (see Section II.B.3) as shown Figs. 23 and 24. A single cw, CO_2 laser is used to optically pump them. The CO_2 radiation is coupled into the far infrared laser cavity by a hole in the end mirror. The output coupler is a capacitive grid, made by depositing aluminum coating through a metal mesh on a crystal quartz disk. As discussed previously (Section II.B.3), the stability of the beat signal frequency is directly related to the stability of the

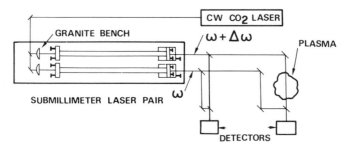

FIG. 23.   Two-frequency, beat-modulated, methyl alcohol laser interferometer for Alcator (after Wolfe *et al.*, 1976).

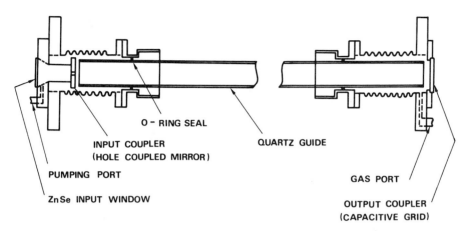

O - RING SEAL

INPUT COUPLER
(HOLE COUPLED MIRROR)

QUARTZ GUIDE

PUMPING PORT

ZnSe INPUT WINDOW

GAS PORT

OUTPUT COUPLER
(CAPACITIVE GRID)

FIG. 24.   Methyl alcohol laser (after Wolfe, private communication).

cavity length. The effect of thermal expansion is reduced by mounting both lasers on the same piece of granite. Nevertheless, the most important factor affecting the stability of the whole system is the frequency drift of the CO_2 laser beam, which leads to some loss of submillimeter power.

The detectors are photoconductive, gallium-doped, germanium crystals, cooled to 4.2 K. Their NEP is of the order of 10^{-10} W $Hz^{-1/2}$ at 10^6 Hz. The choice of these high-sensitivity detectors is important for the design of the whole interferometer. Since the output power of each laser is 2×10^{-3} W, the efficiency of the optical components can be relatively low without affecting the quality of the measurement.

Windows are made of Z-cut crystal quartz disk, 5×10^{-3} m thick, with a 5×10^{-2}-m-diameter aperture. The power loss for each window is of the order of 40%. The focusing elements are lenses made of TPX, whose optical properties are given in Section V.C. The optical components are supported by two shelves (Fig. 25) resting on foam sheets to damp the vibrations. Plane mirrors and corner reflectors are placed in such a way that the beam waist is in the median plane of the plasma and the length adjusted. Beam splitters are nickel meshes with a reflection coefficient near 0.5.

The two modulation signals are filtered and amplified. The zero crossing of the reference signal S_R sets the output high of a single flip–flop, which is set back to zero by the next zero crossing of the measuring signal S. The output of the flip–flop has a pulse width equal to the time lag $t_1 - t_2$ and a pseudo-period τ equal to that of the beat signal (see Section II.A). By integrating the pulse train generated by the flip–flop circuit with a low-pass filter, a signal proportional to $(t_1 - t_2)/\tau$ is obtained, from which the phase φ is readily deduced [Eq. (33)]. It is clear that this averaging technique gives a result

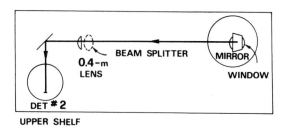

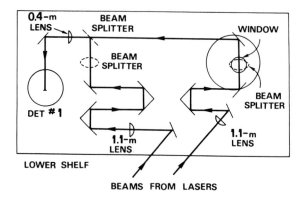

FIG. 25.   Alcator interferometer arrangement (after Wolfe, private communication).

that does not require a constant modulation frequency. When φ reaches 2π, the output automatically reverts to zero and is displayed as stripes, each of them corresponding to one fringe. In principle, a sensitivity of 10^{-2} fringe could be reached. In fact, owing to some residual vibrations of the support, it is limited to about 10^{-1} fringe, which corresponds to a line density of 2×10^{18} m^{-2}.

C.   HCN Laser Interferometers for TFR

The density in TFR (Equipe TFR, 1977) lies in the 10^{20}-m^{-3} range. The plasma radius and the parameter Z_0 (see Section I) are 0.2 m and 0.8 m, respectively, and the HCN laser (3.37×10^{-4}-m wavelength) is a suitable choice as a source.

1.   One-Channel Interferometer

The application of submillimeter radiation for tokamak plasma density measurements has been first demonstrated by Véron (1974) with the operation of a single-channel Michelson interferometer fed by a 0.02-W HCN laser.

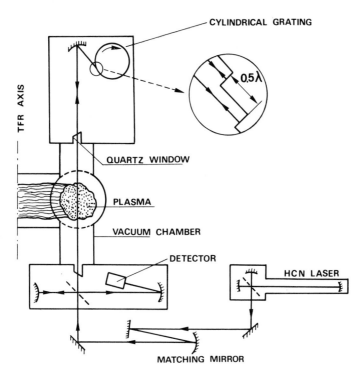

FIG. 26.   Single-channel HCN interferometer for TFR—not to scale (after Véron, 1974).

Figure 26 schematically represents the experimental arrangement. The laser output beam is focused by a concave mirror so that the beam waist is in the median plane of the plasma. Before entering the vacuum chamber, the beam passes through a mylar beam splitter. Its thickness (5×10^{-5} m) is such that its reflection coefficient is close to 50%. The reference beam of the interferometer is then reflected back toward the beam splitter by a convex mirror, whose curvature ensures proper beam matching. The probing beam is focused onto the rotating grating located on the upper part of the vacuum vessel, comes back through the plasma, and finally recombines with the reference beam. The resulting beat signal (Section II.B.2) is monitored by a pyroelectric detector. The reference signal is obtained by taking advantage of the fact that the beat frequency τ^{-1} [Eq. (38)] is equal to the rate at which grooves of the grating pass any fixed point near its circumference. The reference signal could then be obtained by imaging the surface of the grating on a photodiode. However, since visible light is sensitive to imperfections of the grating, an averaging technique is used; an image of about ten adjacent grooves is formed by a lens in the plane of a mask (Fig. 27), which is a fixed

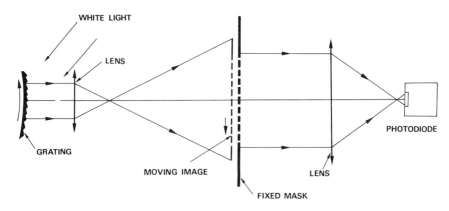

FIG. 27.   Optical arrangement to generate a reference signal for the single-channel TFR interferometer.

image of the grooved surface. It follows that when bright stripes of the moving image form on the transparent part of the mask, light is passing through the mask and vice versa. The result is the production of modulated light at frequency τ^{-1}, which is then transformed into a reference signal by the photodiode. Data acquisition is similar to that described later (Section VI.C.2). All parts of the interferometer are attached to the vacuum chamber. As a result, spurious phase shifts (up to 10% of the total plasma shift), caused by a slight displacement of the interferometer, are noticeable when the machine operates at toroidal field values larger than 4 tesla. The mylar beam splitter also leads to some microphonic effects.

2.   Multichannel Interferometer

In a Michelson interferometer, some of the radiation is reflected back into the laser, thus producing interferences with the emitted beam. This is not a real inconvenience for a one-channel interferometer. However, in a multichannel setup fed by a single laser, this may lead to cross talk between channels, and a Mach–Zehnder arrangement is better suited. Such a device has been designed and operated on TFR 400 (Véron et al., 1977) and later on TFR 600. The general layout of the optical system is shown in Fig. 28. The beam emitted from the 0.15-W laser (see Section III.A) is focused with the help of a telescope. Quartz beam splitters are used to feed each channel with beams of equal intensity and polarized so that their electric field vector is parallel to the toroidal field of the machine. Eight measuring channels and one reference channel are thus operated simultaneously. Plane mirrors M are properly placed to make the path length Z_2 between the mirror M_2 of the telescope and the median plane of the plasma equal for each channel. So,

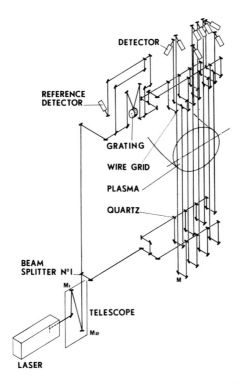

FIG. 28. Multichannel HCN interferometer for TFR 600. Schematic of optical elements and beams arrangement—not to scale. The optical path between beam splitter 1 and one detector is nearly 5 m.

with the proper choice of the telescopic mirrors and their spacings, the beam waists can be brought into the median plan of the plasma for all channels. All mirrors are made of aluminized glass.

a. *Telescopic System.* The size of the beam waist is calculated by applying Eq. (51) in which $Z_0 = 0.8$ m. The result is $d_0 = 1.3 \times 10^{-2}$ m. The diameter of the beam at mirror M_2 of the telescope (Fig. 29) is given by Eq. (41), which, for $d_0 = 1.3 \times 10^{-2}$ m and $Z = Z_2 = 4.935$ m, leads to $d = 8.2 \times 10^{-2}$ m.

On the other hand, the waist of the beam leaving the laser has a diameter $d = 2.2 \times 10^{-2}$ m. If only the mirror M_2 had to be used to ensure proper beam matching, a distance of about 8 m between the laser mesh coupler and the mirror M_2 would be necessary. This distance is in fact conveniently reduced by the use of a convex mirror M_1, as shown in Fig. 29.

The calculations of the focal lengths f_1 and f_2 of the spherical mirrors M_1 and M_2 are made by applying the formulas given in Section IV. The results

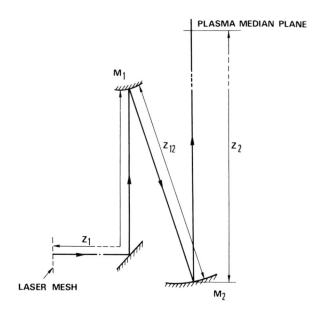

FIG. 29.   Telescope for beam matching.

are $f_1 = -0.45$ m, $f_2 = 1$ m, $Z_{12} = 0.835$ m, and $Z_1 = 1.35$ m, Z_{12} and Z_1 being, respectively, the distance between M_1 and M_2 and the distance between the laser mesh and mirror M_1. The curve drawn in Fig. 30 shows that a small variation ΔZ_{12} of the distance Z_{12} leads to a large change ΔZ_2 of Z_2. It follows that the position of the beam waist can be adjusted by slightly moving the mirror M_1, mounted on a translation stage.

b.   *Windows.*   The beam diameter, at the windows, is equal to 1.9×10^{-2} m, as deduced from Eq. (50). Most of the windows have a clear diameter of 4.3×10^{-2} m, which satisfies inequality (49). Nevertheless, since the ports on TFR are wedge shaped, the clear diameter of one of the windows is somewhat smaller, but the transmitted power is still large enough and the sensitivity of the corresponding channel is not affected.

The windows are made of crystal quartz. They are slightly tilted (3°) with respect to normal incidence, in order to eliminate spurious reflections of the helium–neon laser light used for alignment of the system. Since absorption losses are smaller for the extraordinary ray (Table III), windows are cut parallel to the optical axis, and the latter is parallel to the electric field vector of the beam. The thickness of the windows (3.903×10^{-3} m) is calculated so as to reduce the reflection coefficient to zero. Figure 31 shows how this coefficient changes for a slight deviation of the thickness or of the angle of incidence from their optimum values. The measured transmission coefficient

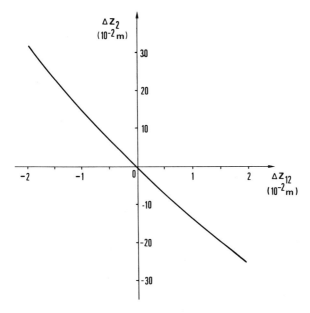

FIG. 30.   Adjusting curve of the telescope shown in Fig. 29.

is 0.96 for the HCN laser radiation. Vacuum seals are made of indium. The flanges are air cooled to prevent fusion of the seals when baking the chamber. Reliability of this type of seal has been proved to be satisfactory over periods of years.

c.   *Beam Splitters.*   Most beam splitters are also made of natural crystal quartz and are cut in such a way that the beam propagates according to the extraordinary mode. Thicknesses in the neighborhood of $2 \times 10^{-3}, 3 \times 10^{-3}$, and 4×10^{-3} m are used, depending on the size of the beam splitters (the larger being obviously the thicker), to permit good manufacturing and to ensure sufficient mechanical strength. The clear diameter is determined by using inequality (49). The larger beam splitter (1 in Fig. 28) is made of three parts, assembled on a well-machined aluminum support, because of difficulties encountered in finding big enough crystals. It should be mentioned that the quality of the beam is not seriously affected by this divided beam splitter. The calculated reflection coefficient as a function of thickness for beam splitters placed at 45° to the direction of propagation (for the HCN laser radiation propagating according to the extraordinary mode) is shown in Fig. 32.

At the recombining points, the electric field vector is parallel to the plane of incidence. Under these conditions, the optimum value (0.5) of the reflection

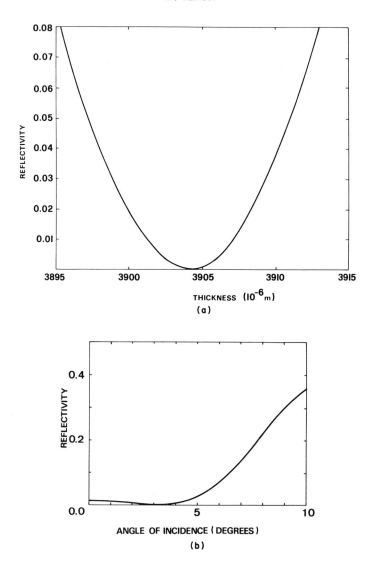

FIG. 31.   Reflectivity of a quartz window for the extraordinary ray (the optical axis is parallel to the plane of the plate): (a) as a function of thickness, with $\mu_e = 2.156$, E perpendicular to the incident plane, and an incidence of $3°$; (b) as a function of incidence. (This last curve is after Frank, 1976.)

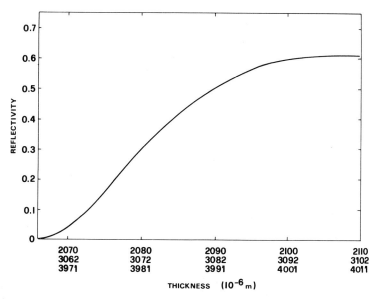

FIG. 32.   Reflectivity of quartz beam splitters as a function of thickness for different ranges corresponding to the three horizontal scales. The electric field vector of the beam and the optical axis are normal to the plane of incidence (45°), with $\mu_e = 2.156$.

coefficient required for maximum efficiency of the interferometer cannot be reached by a quartz plate (see Section V.B). It then follows that wire grids are used in this particular case. Since they are placed close to the detectors (Fig. 28), their alignment, which cannot be made with the help of visible light, is easily feasible with the infrared radiation.

    d. *Rotating Grating Assembly.*   Phase modulation is achieved by the rotating grating technique, for which details are given in Section II.B.2. The incident beam is focused onto the grating by a 31.2×10^{-2}-m focal length concave mirror M_1 (Fig. 33) which reduces the diameter of the beam to 5×10^{-3} m at the wheel. Since the blaze angle is 54°, the beam covers a length of 8.5×10^{-3} m of the grooved surface, that is to say more than 40 grooves, a large enough number to avoid any intensity fluctuations of the diffracted beam which could be expected if only a few grooves were involved. On the other hand, this number is small enough so the angle of incidence does not deviate too much from the blaze angle over the whole cross section of the beam. The diffracted beam is then focused by a mirror M_2, identical to M_1, in such a way as to give the same divergence and diameter at the recombining beam splitters as those of the probing beams.

    A direct current motor is used to rotate the grating. A typical modulation frequency is 10^4 Hz.

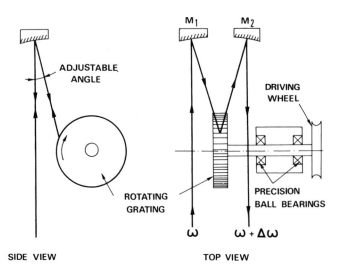

FIG. 33.   Rotating grating assembly.

e.   *Detectors.*   Whereas the average number of optical elements placed in series between the laser output and each detector is as high as 20, the overall efficiency of the interferometer is 40%. This corresponds to a loss coefficient of 0.04 per element, assuming that the total losses are equally distributed among all elements. This high efficiency, added to the large power delivered by the laser source (Section III.A), allows the use of room-temperature detectors. They are triglycine sulfate (TGS) crystals (Baker *et al.*, 1972) which exhibit pyroelectric properties. Therefore they are sensitive to temperature changes due to absorption of infrared radiation. Typical noise equivalent power (NEP) curves are given in Fig. 34. Commercially available TGS detectors are covered with a thin metal layer on both sides to collect charges arising from the pyroelectric effect. There is experimental evidence that part of the incoming infrared radiation is reflected back by this metal layer, leading to important power loss. In order to increase the efficiency of the detector, a thin, calibrated, quartz plate is placed in front of the crystal, parallel to its surface (Fig. 35). Multiple reflections then occur between the plate and the crystal, part of the beam power being absorbed at each reflection. By adjusting both the thickness of the quartz plate and its distance from the crystal surface, it is experimentally shown that the total reflection of the crystal and quartz plate assembly can be reduced to nearly zero (Duverger *et al.*, 1975). Since the absorption in the quartz plate is negligible, it follows that most of the power is absorbed by the crystal. The result is a net increase in the efficiency of the detector by nearly a factor of 2.

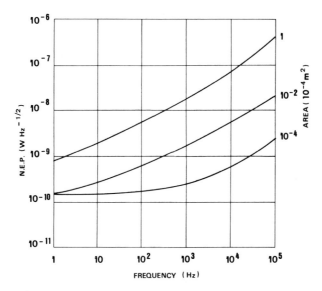

FIG. 34.   Noise equivalent power (NEP) of different-sized triglycine sulfate detectors as a function of frequency (after Baker *et al.*, 1972).

The crystal has a diameter of 2×10^{-3} m. In order to concentrate most of the power on it, the beam is focused by a 1.5×10^{-1}-m focal length concave mirror. The crystal output is connected to a preamplifier, which is followed by an amplifier with a gain coefficient of 10^2. No filter is used. The signal-to-noise ratio at the modulation frequency (10^4 Hz) is about 100. Each detector and the associated electronics assembly is enclosed in a 8×10^{-3}-m-thick copper shield.

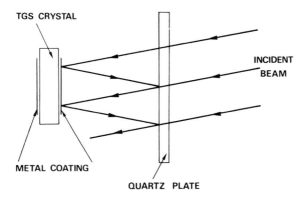

FIG. 35.   Multiple reflection quartz window for efficiency enhancement of a triglycine sulfate (TGS) detector.

f. *Supporting Structure.*   The supporting structure consists of three main parts (Fig. 36):

(1)   the lower part L which supports the beam splitters and mirrors that distribute the power to the eight probing beams;

(2)   the upper part U containing beam splitters, the mirrors distributing the power of the reference beam to each channel, the rotating grating and associated optics, and the detectors;

(3)   the vertical column V which rigidly connects the two previous sections.

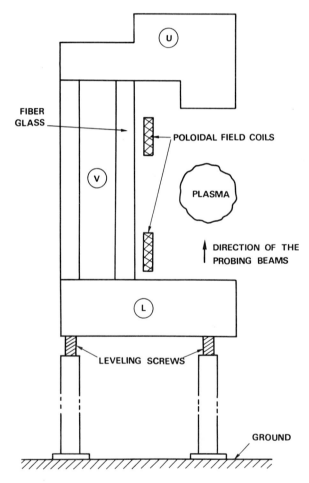

FIG. 36.   Schematic of the supporting structure of the multichannel TFR 600 interferometer.

Most of the structure is made of aluminum. Fiberglass spacers are used whenever it is necessary to avoid closed loops of conductive material and, also, for part of the vertical column section that passes very close to the poloidal field coils, thus reducing induced eddy currents and consequently the resulting electromagnetic forces.

Section L is supported by a tripod rigidly bolted to the ground, by means of three leveling screws. Wheels and rails allow section L to be moved for assembling and dismantling the interferometer. There are no mechanical connections between the machine and the interferometer frame.

The system is aligned prior to mounting on the TFR. It is then dismantled and reassembled on the machine. This operation is carried out without seriously disturbing the alignment, with the aid of well-adjusted centering pins that allow exact repositioning of each section with respect to the others.

g. *Data Acquisition, Results.* The sine wave generated by each detector is amplified and transformed into a pulse train. Each pulse of the reference signal initiates a counting process of pulses delivered by a clock, whose frequency exceeds that of the beat signal by three orders of magnitude. The counting process is then stopped by the following pulse delivered by the measuring signal. The resulting number is proportional to the time interval $t_1 - t_2$ [Eq. (33)] and, therefore, to the phase shift φ. Its value is stored in a ferrite memory. The same operation is repeated for each period of the beat signal.

The number of memories is 2000, divided in two sets of 1000 for each channel. The total temporal storage capability is then $2000T$ s, T being the storage period. The minimum value of T is obviously equal to τ, but it can be increased up to 8τ for longer plasma pulses. The resulting loss of accuracy of the measurement is partly compensated by the possibility of choosing T independently for each set of 1000 memories, thus allowing a finer analysis either at the beginning or at the end of the plasma pulse.

Examples of typical displays of fractional phase shifts are given in Fig. 37 for four of the eight channels of the interferometer previously built for TFR 400. On the same figure, the total phase shifts given by the on-line computer are also shown. A line density profile, obtained on TFR 400 is given in Fig. 38. The error bars are equal to the standard deviation resulting from the shot-to-shot variations of the plasma.

Maximum sensitivity of the interferometer is equal to 5×10^{16} m^{-2} for each channel and is limited by the noise level of the detectors. When the machine is fired, residual vibrations, added to this noise, lead to an actual sensitivity of 10^{-2} fringe corresponding to a line density of 6.6×10^{16} m^{-2}. It follows that the apparatus can be operated over a wide density range. Figure 39 shows that small fluctuations (relaxations) are easily detectable.

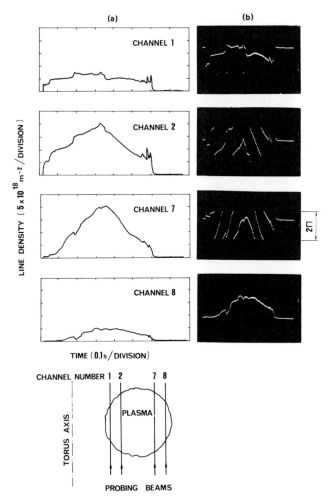

FIG. 37.   Line density recorded on four of the eight channels of the TFR 400 HCN interferometer: (a) line density as given by the on-line computer, as a function of time, for the same
channels; (b) oscilloscope display of fringes stored in ferrite memories (after Véron *et al.*, 1977).

D.   HCN INTERFEROMETER FOR FT

A single-channel interferometer is in operation on the FT machine
(Pieroni, 1977). Since it is very similar to that built for TFR (Section VI.C.1),
it will not be described here. For probing the plasma across the whole of its
cross section, a moving channel interferometer has been chosen instead of
the true multichannel device. The rotating grating is divided in sections. The
grooves on each section have a different profile. Therefore, the diffracted

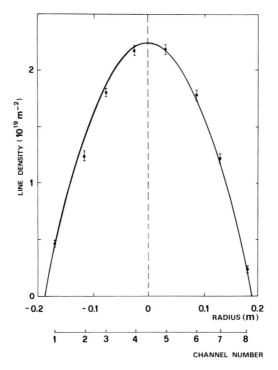

FIG. 38.   Line density profile of the TFR 400 plasma. The error bars represent the standard deviation due to shot-to-shot variations calculated over 20 shots. The dashed line represents the center of the plasma column and the solid curve is the best fit analytically (after Véron *et al.*, 1977).

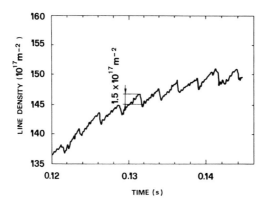

FIG. 39.   Saw-tooth density fluctuations detected on the center channel of the multichannel TFR 600 interferometer. Note that fluctuations as small as 1.5×10^{17} m^{-2} are clearly visible.

beam has a different direction for each section, and experiences a different frequency shift. Subsequently, a parabolic mirror reflects the beam along different chords of the plasma cross section. At the exit window, another parabolic mirror sends each beam onto a single detector. Since the beat frequency is different for each chord, it is easy to separate the signal pertaining to a given chord.

VII. Sources of Error

Errors in density measurements result from the combination of several phenomena, namely refraction, extraordinary mode propagation, Faraday rotation, lack of mechanical stability, and source and detector noise.

A. REFRACTION

Refraction of the probing beam as it passes through the plasma causes two kinds of error. First, the probing beam is no longer collinear with the reference beam and their wave surfaces are not exactly superimposed, thus introducing some additional phase shift. Second, the beam trajectory inside the plasma deviates from a straight chord and the length of the path in the plasma is modified.

1. *Error Due to the Angular Deviation of the Probing Beam*

This error will be estimated with the help of Fig. 40. Since the angle of refraction α is small (Section I.B), the axis of the outgoing beam is well approximated by a straight line issued from the projection A of the incident beam on the median plane of the plasma. The detector is located at B, at a distance $AB = Z$ from the median plane. Let M and M' be the centers of curvature of the incident and outgoing beam wave fronts S and S', respectively. Note that M and S are also the center of curvature and the wave front of the reference beam, since it is supposed to be matched with the undeviated probing beam. The radius of curvature of S and S' is then

$$R = MB = M'C = M'H.$$

The path difference measured on the axis of the incident (or reference) beam is

$$
\begin{aligned}
\mathbf{BC} = \mathbf{BA} + \mathbf{AC} &= \mathbf{BA} + \frac{\mathbf{AK}}{\cos \alpha} \simeq \mathbf{BA} + \mathbf{AK}(1 + \tfrac{1}{2}\alpha^2) \\
&= \mathbf{BA} + (\mathbf{AM'} + \mathbf{M'K})(1 + \tfrac{1}{2}\alpha^2) \\
&\simeq \mathbf{BA} + (\mathbf{AH} + \mathbf{HM'} + \mathbf{M'C} \cos \alpha')(1 + \tfrac{1}{2}\alpha^2) \\
\mathbf{BC} &= -Z + [Z - R + R(1 - \tfrac{1}{2}\alpha'^2)](1 + \tfrac{1}{2}\alpha^2) \\
&= -Z + (Z - \tfrac{1}{2}R\alpha'^2)(1 + \tfrac{1}{2}\alpha^2). \qquad (63)
\end{aligned}
$$

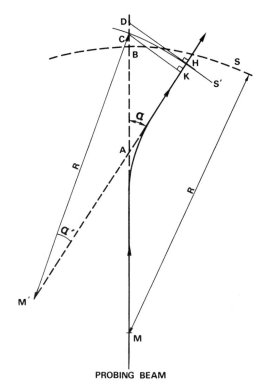

PROBING BEAM

FIG. 40.   Scheme for the evaluation of the error ε_1 due to angular deviation of the probing beam.

Using the small angle approximation,

$$\sin \alpha'/\sin \alpha = \alpha'/\alpha = Z/R,$$

and, retaining only up to second-order terms, Eq. (63) becomes

$$BC = \tfrac{1}{2}Z\alpha^2[1 - (Z/R)]. \tag{64}$$

The corresponding fringe number is

$$\Delta F = \frac{BC}{\lambda} = \frac{Z}{\lambda}\frac{\alpha^2}{2}\left(1 - \frac{Z}{R}\right). \tag{65}$$

By using the value of R given by Eq. (43) and also introducing the optimum beam waist d_0 deduced from Eq. (51),

$$\Delta F = \frac{Z}{\lambda}\frac{\alpha^2}{2}\frac{Z_0^2}{Z^2 + Z_0^2}. \tag{66}$$

Finally, α is replaced by its maximum value α_m from Eq. (16), and

$$\Delta F = \frac{1}{\lambda} \frac{n_0^2}{2n_c^2} \frac{Z_0^2}{Z[1 + (Z_0^2/Z^2)]}. \tag{67}$$

It is interesting to compare ΔF to the number of fringes F_0 obtained on a plasma diameter [Eq. (22)]:

$$\varepsilon_1 = \frac{\Delta F}{F_0} = \frac{3}{4} \frac{n_0}{r_0 n_c} \frac{Z_0^2}{Z[1 + (Z_0^2/Z^2)]}. \tag{68}$$

Since $n_c = (1.11/\lambda^2) \times 10^{15}$ [see Eq. (7)],

$$\varepsilon_1 = 6.76 \times 10^{-16} \frac{n_0 \lambda^2}{r_0} \frac{Z_0^2}{Z[1 + (Z_0^2/Z^2)]}. \tag{69}$$

This relation gives only an approximate value of ε_1 since it results from considering the path difference on the axis of the probing beam, and not on its entire diameter. But at least it indicates how the error varies with the machine parameters and with the radiation wavelength. It can be seen that the location of the detector is important. For instance, since the detector must be placed outside the vacuum vessel, $Z > Z_0$ and ε_1 decreases as Z increases from a maximum value:

$$\varepsilon_{1\max} = 6.76 \times 10^{-16} \frac{n_0 \lambda^2}{r_0} \frac{Z_0}{2} \tag{70}$$

obtained for $Z = Z_0$. If the detector is located far enough from the plasma, then the ratio $Z_0^2/\{Z[1 + (Z_0^2/Z^2)]\}$ reduces to a small value and the error is negligible (Fig. 41).

For the following typical numbers, which apply to TFR: $n_0 = 10^{20}$ m^{-3}, $r_0 = 2 \times 10^{-1}$ m, $\lambda = 3.37 \times 10^{-4}$ m, and $Z_0 = 8 \times 10^{-1}$ m,

$$\varepsilon_{1\max} = 1.5\%.$$

Since the detectors are located in fact at a distance $Z = 1.8$ m of the median plane

$$\varepsilon_1 = 1.1\%.$$

2. Error Due to Path Length Change inside the Plasma Column

On Fig. 42, IJ represents the trajectory of the undeviated beam inside the plasma cross section when no plasma is present, while IJ' is the actual trajectory in the plasma. Assuming that the density profile is a parabola, the density along IJ also varies as a parabolic function, with its maximum at A equal to $n_0(1 - \gamma^2)$, where $\gamma = OA/r_0$.

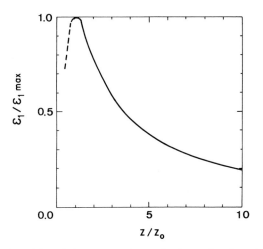

FIG. 41.   Variation of the relative error ε_1 (normalized) as a function of the distance Z/Z_0 between the detector and the plasma median plane.

The total fringe number calculated along the chord IJ is then [Eq. (21)]

$$F = [n_0/(2\lambda n_c)](1 - \gamma^2) \int_{Z_1}^{Z_2} [1 - (Z^2/Z_1^2)] \, dZ \qquad (71)$$

or, by noting that $-Z_1 = Z_2 = AJ = r_0(1 - \gamma^2)^{1/2}$,

$$F = \tfrac{2}{3}[n_0 r_0/(\lambda n_c)](1 - \gamma^2)^{3/2}. \qquad (72)$$

In fact, the measured fringe number corresponds to the actual trajectory IJ' in the plasma. Since α is small, this number can be approximated by the number calculated along the chord $I''J''$ passing through the middle of AA' which differs from F [Eq. (72)] by

$$\Delta F = -2[n_0 r_0/(\lambda n_c)](1 - \gamma^2)^{1/2}\gamma \, \Delta\gamma. \qquad (73)$$

Here,

$$\Delta\gamma = \tfrac{1}{2}(AA'/r_0) \simeq \tfrac{1}{2}(AJ\alpha/r_0) = \tfrac{1}{2}(1 - \gamma^2)^{1/2}\alpha$$

and

$$\Delta F = -[n_0 r_0/(\lambda n_c)](1 - \gamma^2)\gamma\alpha. \qquad (74)$$

The relative variation with respect to F_0 [Eq. (22)] is

$$\varepsilon_2 = \Delta F/F_0 = -\tfrac{3}{2}(1 - \gamma^2)\gamma\alpha. \qquad (75)$$

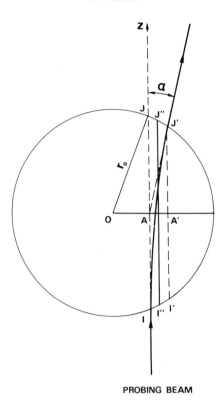

PROBING BEAM

FIG. 42.   Diagram for the evaluation of the error ε_2 due to path length change inside the plasma column.

The term $(1 - \gamma^2)\gamma$ is maximum for $\gamma = 0.58$, which is close to the value for maximum α. It follows that

$$\varepsilon_{2\max} = -0.58\alpha_m$$

or, by using Eq. (17),

$$\varepsilon_{2\max} = -5.2 \times 10^{-16} n_0 \lambda^2. \tag{76}$$

It is interesting to note that ε_1 and ε_2 have opposite signs so that they partly compensate for each other. For instance, for the TFR, $n_0 = 10^{20}$ m^{-3}, $\lambda = 3.37 \times 10^{-4}$ m, $\varepsilon_2 = -0.6\%$, and the resulting error is

$$\varepsilon = \varepsilon_1 + \varepsilon_2 = 0.5\%.$$

It follows that the total error due to the refraction phenomenon is small enough to be neglected. Refraction, however, decreases the detectable power, and, consequently, it may reduce the sensitivity.

B. EXTRAORDINARY MODE PROPAGATION

Usually the probing beam is such that its electric field is parallel to the toroidal magnetic field B_T (ordinary wave) or normal to it (extraordinary wave). Assuming that the ordinary mode is chosen, it is interesting to have an idea of the error resulting from a slight misalignment of the electric field with respect to B_T.

Since it has been shown that $n \ll n_c$ (Section I.B), μ_0 and μ_e are well approximated by

$$\mu_0 = 1 - (\omega_p^2/2\omega^2)$$

and

$$\mu_e = 1 - \frac{\omega_p^2}{2\omega^2} \frac{1}{1 - [\omega_{ce}^2/(\omega^2 - \omega_p^2)]}.$$

For the case of the TFR where $\omega = 5.59 \times 10^{12}$ (HCN laser), $\omega_p = 3.99 \times 10^{11}$ ($\bar{n} = 5 \times 10^{19}$ m^{-3}), and $\omega_{ce} = 1.06 \times 10^{12}$ ($B_T = 6$ T),

$$\mu_e \simeq 1 - (\omega_p^2/2\omega^2)\{1 + [\omega_{ce}^2/(\omega^2 - \omega_p^2)]\}$$
$$= 1 - (1 + 3.6 \times 10^{-2})(\omega_p^2/2\omega^2). \tag{77}$$

According to Eq. (10), the phase shift created by the plasma is proportional to $\omega_p^2/2\omega^2$ for μ_0 and to $1.036\omega_p^2/(2\omega^2)$ for μ_e. Since those two values differ by only 3.6%, it is clear that for a wave nearly ordinary the phase shift given by Eq. (10) is valid, the relative error being approximately 4×10^{-4} per degree of misalignment.

C. FARADAY ROTATION

Faraday rotation results from the presence of a magnetic field component $B_{//}$ parallel to the beam direction and produced by the current induced in the plasma (see Fig. 45). The linearly polarized probing wave entering into the plasma can be treated as the superposition of two circularly polarized waves, rotating in opposite directions and traveling at different speeds according to the two values of the refractive index [Eq. (3)]. Their projection on the x axis, parallel to the direction of the electric field of the reference beam, can be represented by

$$x_+ = a \cos(\omega t - \varphi_+) \quad \text{for the left-handed wave,}$$
$$x_- = a \cos(\omega t - \varphi_-) \quad \text{for the right-handed wave.}$$

By combining these waves with the reference one,

$$x = b \cos(\omega + \Delta\omega)t,$$

the signal monitored by a square law detector is

$$S = 2ab[\cos(\Delta\omega\, t + \varphi_+) + \cos(\Delta\omega\, t + \varphi_-)]. \tag{78}$$

Then the Faraday rotation angle is given by

$$\Omega = \tfrac{1}{2}(\varphi_+ - \varphi_-). \tag{79}$$

It will be shown (Section VIII.D) that the refractive index $\mu_\pm$ [Eq. (3)] can be written to a good enough accuracy as follows [see the remark about the validity of Eq. (3), Section I.A)]:

$$\mu_\pm = 1 - (\omega_p^2/2\omega^2)[1 \mp (\omega_{ce//}/\omega)] \tag{80}$$

with $\omega_{ce//} = eB_{//}/m_e$.

So, by comparing Eqs. (9) and (80),

$$\mu_0 = \tfrac{1}{2}(\mu_+ + \mu_-).$$

Consequently, the phase shift φ can be expressed by

$$\varphi = \tfrac{1}{2}(\varphi_+ + \varphi_-). \tag{81}$$

Combining Eqs. (79) and (81),

$$\varphi_+ = \varphi + \Omega, \qquad \varphi_- = \varphi - \Omega.$$

Then the detected signal S becomes

$$S = 4ab \cos \Omega \cos(\Delta\omega\, t + \varphi), \tag{82}$$

showing that Faraday rotation does not affect the phase shift measurement, but only reduces the amplitude of the beat signal in the ratio $\cos \Omega$.

D.   MECHANICAL STABILITY AND SOURCE AND DETECTOR NOISE

The mechanical stability has been extensively discussed in the preceding sections of this chapter (I.B, III.C, VI.A–C) and therefore will not be studied in this subsection.

The source and detector noise may introduce errors in the measurement only if they have components in the neighborhood of the modulation frequency. It should be noted, however, that the modulation frequency of the measuring signal changes slightly when the line density varies. It follows that extreme care has to be taken in using narrow-band filters, since they can introduce an additional phase shift due to this frequency change.

VIII.   Future Developments

Speaking about power, there is a competition between sources and detectors. Indeed, there is a natural trend toward an increase of the power of lasers and a decrease of the NEP of detectors. However, as stated in Section

III, another important factor for the choice of sources and detectors for tokamak interferometers is reliability.

So, a good system results from the best compromise among laser power, detector NEP, and reliability of the whole setup. It follows that the development of lasers and detectors has to be considered simultaneously (Subsection VIII.A). In Subsection VIII.B, waveguiding of laser beams is introduced in view of possible applications to future large tokamaks. The advantages of line-density gradient measurement are outlined in Subsection VIII.C, and, finally, the feasibility of the simultaneous measurement of line density and poloidal field-induced Faraday rotation is discussed in Subsection VIII.D.

A. LASER SOURCES AND DETECTORS

1. *Lasers*

Up to now, cw submillimeter sources satisfying the requirements specified in Section III have not been very numerous. Optically pumped gas lasers are presently being widely developed (Hodges *et al.*, 1976), but they still need to be technically improved to attain good enough reliability. Although these lasers seem simpler to build than discharge-excited ones, they have to be pumped by a CO_2 laser, and the entire system has to be considered as the source. Then difficulties in stabilizing the whole assembly may arise from the use of a powerful CO_2 laser whose wavelength can be selected by an intracavity grating (Section VI.B) requiring fine adjustment. However, the optically pumped methyl alcohol laser already delivers powers up to 0.15 W at the 1.19×10^{-4}-m wavelength (Hodges *et al.*, 1976), and attainment of long-term stability and reliability is most likely within the reach of present-day technology.

As mentioned in Section III.A, the development of the discharge-excited HCN laser is now at a satisfactory stage. More recently, a DCN laser has been successfully operated (Véron *et al.*, 1978). Its design is similar to that of the HCN laser. It delivers up to 0.25 W at a wavelength of 1.95×10^{-4} m, its total output power being increased to 0.40 W when the cavity is tuned for the simultaneous emission of two wavelengths at 1.95×10^{-4} and 1.90×10^{-4} m. This laser is of great interest because there is no other powerful source in this frequency range. Moreover, its output characteristics fit the requirements for interferometric measurements on large tokamaks, such as JET and TFTR (Table I), which are at the present being developed.

In addition, small-sized HCN and DCN lasers have proved to be operational (Belland *et al.*, 1976a; Bruneau *et al.*, 1978). They deliver several milliwatts (up to 8 for the DCN laser) with a discharge length of only about half a meter. Such sources could be used either for single-channel interferometers or for multichannel ones with low-noise detectors, such as helium-cooled crystals or Schottky barrier diodes.

So, adequate sources already exist in the wavelength range from about 10^{-4} to 3×10^{-4} m, whereas some technical improvements are needed for the optically pumped alcohol laser.

2. Detectors

Pyroelectric crystals are very good room-temperature detectors (Fig. 34). They are sensitive to temperature changes produced by absorption of the incident energy, and, therefore, they operate over a wide range of radiation frequencies. Their main drawback is that their sensitivity decreases as the modulation frequency increases.

Schottky barrier diodes can be used in the submillimeter range, but they do not seem very reliable or at least not reliable enough to be employed in a tokamak interferometer at the present time. Their great advantage over pyroelectric crystals is that their sensitivity has a constant value up to modulation frequencies in the megahertz range or more.

Liquid-helium-cooled detectors (InSb, GaAs) combine low NEP with fast response. They could be used to monitor interferometric signals, but they should be made more practical to operate either by improving the holding time of their cryostat or by inserting them in a closed loop cw helium liquefier.

Both Schottky barrier and helium-cooled detectors may have NEPs several orders of magnitude better than pyroelectric detectors, so they may lead to the use of low-power, small-sized lasers.

B. WAVEGUIDING

The access to large tokamak machines, such as JET and TFTR, will be restricted for reasons of safety. Moreover, the expected high flux of radiations from their plasma may prevent proper working of detectors and associated electronics. Consequently, lasers and detectors may have to be located far away from the machine, at distances up to several tens of meters, and even if only small refraction angles are to be experienced by the probing beams, large displacements may result from their long paths. Increase of the beam diameter due to divergence must also be considered. It follows that optical propagation may be difficult to use, and waveguiding could be an alternative technique.

Calculations (Crenn, 1979) show that losses undergone by a beam propagating in hollow cylindrical waveguides are smaller for dielectric materials than for metals, at least for relatively large bore tubes, as shown in Fig. 43. Experiments performed with an HCN laser beam ($\lambda = 3.37 \times 10^{-4}$ m) show that the beam power is best coupled into the guide if a beam waist is formed at the entrance of the guide and if its diameter d_0 is such that

$$0.35 < d_0/D < 0.5,$$

D being the diameter of the guide.

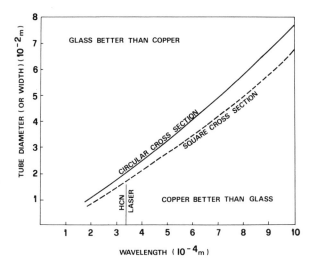

FIG. 43.   Curves for the fundamental mode in glass and copper waveguides corresponding to equal losses (calculated). The diagram clearly shows that oversized glass waveguides are better than copper waveguides (after Crenn, 1979).

The measured loss coefficient for a glass (or lucite) tube, whose diameter is 4×10^{-2} m, is 1.6×10^{-2} m^{-1}. The calculated loss coefficient is 0.7×10^{-2} m^{-1}. The discrepancy between these numbers can be at least partly explained by the existence of higher-order modes or by some mechanical imperfections of the guides. However, sufficiently large ($D > 4 \times 10^{-2}$ m) dielectric tubes have a low enough loss coefficient to be considered for beam propagation.

The degree of polarization of the beam as it comes out of the guide is also an important factor. Experiments show that only circular glass tubes and square metal tubes are satisfactory: the degree of polarization is of the order of 98% for those guides, while it goes down to less than 90% for a circular brass tube, all measurements being made with 2.5-m-long guides (Fig. 44). The numbers previously given are relative to guides with the beam entering parallel to their axis. In an actual interferometer, owing to refraction, beams coming out of the plasma enter at a slight angle with respect to guide axis, which causes an increase in the losses, and the degree of polarization decreases. Results concerning different guides are given in Fig. 44. It should be noted that guides used for those experiments have a characteristic dimension (diameter for the circular ones, diagonal length of their cross section for the square ones) of about 5×10^{-2} m. Focusing of the outgoing beam is also of concern. Indeed, the size of the detectors is usually much smaller than that of the guide, and then the beam diameter has to be considerably reduced by lenses or concave mirrors. When the beam does not enter the guide parallel

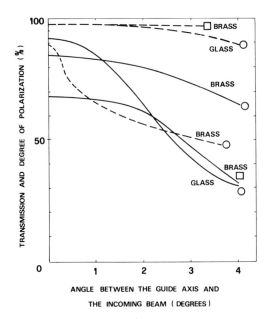

FIG. 44.   Transmission (solid lines) of different waveguides, 2.5 m long, for the HCN radiation. Also indicated is the degree of polarization of the output beam (dotted lines). The degree of polarization is defined as the ratio of the output power with the same polarization as the incoming beam to the total output power. Shown are square guides (□) and circular guides (○) (after Crenn, 1979).

to its axis, the importance of higher-order modes increases, and, also, the mean axis of the outgoing beam slightly deviates from tube axis. It follows that the power reaching the detector decreases much faster than the total output power. Further studies have to be made in order to increase the efficiency of focusing techniques. However, circular glass waveguides, which are cheap and easy to use, appear to be promising for beam propagation in the next generation of tokamak machines. They can also be conveniently evacuated or filled with dry gas to eliminate atmospheric absorption.

C.   DIRECT MEASUREMENT OF LINE DENSITY GRADIENT

Referring to the multichannel device described in Section VI.C.2, the signal S_k delivered by the detector of channel number k can be used as a reference signal for the adjacent channel S_{k-1}. Then measurement of the time lag between the zero crossings of S_k and S_{k-1} directly leads to the determination of the difference in line density between channel k and k^{-1}. The complete profile of line density can thus be deduced from differential measurements between adjacent channels. Absolute measurements need to be done only on one channel for calibration of the profile. This system exhibits an interesting feature: for large tokamaks, building a vibration-free frame may be difficult, but the optical elements corresponding to two adjacent channels are close enough to experience similar mechanical movement, leading to the same spurious phase shift for both channels. Since only the phase difference between these channels is to be measured, the spurious phase shift is automatically eliminated. The absolute measurement could be made by using a two-wavelength system, as described in Section VI.A, on one particular channel only.

D. FARADAY ROTATION

After the first successful measurement of poloidal field-induced Faraday rotation in the TFR plasma (Kunz et al., 1978), the interest in this phenomenon as a practical tool to determine the current profile in the plasma is growing, although theoretical considerations on the subject have been published as early as 1972 (De Marco and Segre, 1972).

In this subsection the amplitude of the Faraday rotation shall be evaluated, and then the possibility of combining its measurement with an interferometric system shall be discussed.

In the plasma column, the induction B depends on the current distribution. If it is assumed to be parabolic, the current intensity inside a circle of radius r (Fig. 45) is

$$J = 2\pi \int_0^r j_0[1 - (r^2/r_0^2)]r \, dr = \pi j_0 r^2\{1 - [r^2/(2r_0^2)]\}, \tag{83}$$

where j_0 is the current density at the center and r_0 the radius of the current distribution, assumed to be equal to the plasma radius.

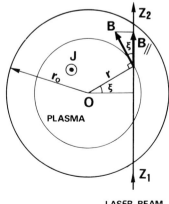

FIG. 45. Diagram for the calculation of the integral of the $B_{//}$ poloidal field component along the beam path $Z_1 Z_2$.

The current J can also be expressed as a function of the total intensity J_0 and of the ratio $\gamma = r/r_0$:

$$J = 2J_0\gamma^2(1 - \tfrac{1}{2}\gamma^2). \tag{84}$$

The induction B at a distance r from the plasma center is then

$$B = [\mu J_0\gamma/(\pi r_0)](1 - \tfrac{1}{2}\gamma^2), \tag{85}$$

where μ is the permeability. Then

$$B_{//} = B \cos \xi = [\mu J_0\gamma/(\pi r_0)](1 - \tfrac{1}{2}\gamma^2) \cos \xi \tag{86}$$

The maximum value of $B_{//}$ is for $\gamma \simeq 0.8$ and $\cos \xi = 1$, and for the TFR parameters, $J_0 = 6 \times 10^5$ A and $r_0 = 0.2$ m,

$$B_{//max} = 0.65 \quad \text{T.}$$

It follows that the corresponding value of $\omega_{ce//} = eB_{//}/m_e$ is 1.15×10^{11}.

Since $\omega = 5.59 \times 10^{12}$, $\omega_{ce//} \ll \omega$ and the index of refraction $\mu_{\pm}$ for the left-hand and right-hand circularly polarized waves given by Eq. (3) can be approximated by [See the remark about the validity of Eq. (3), Section I.A]:

$$\mu_{\pm} = 1 - (\omega_p^2/2\omega^2)[1 \mp (\omega_{ce//}/\omega)]. \tag{87}$$

The plane of polarization of the probing beam then rotates by the angle

$$\Omega = (\pi/\lambda) \int_{Z_1}^{Z_2} (\mu_+ - \mu_-) \, dZ \tag{88}$$

$$= (\pi/\lambda) \int_{Z_1}^{Z_2} (\omega_p^2\omega_{ce//}/\omega^3) \, dZ. \tag{89}$$

Since

$$\omega_{ce//} = eB_{//}/m_e \quad \text{and} \quad \omega_p = [ne^2/\varepsilon_0 m_e]^{1/2},$$

$$\Omega = [\lambda^2 e^3/(8\pi^2 c^3\varepsilon_0 m_e^2)] \int_{Z_1}^{Z_2} nB_{//} \, dZ \tag{90}$$

or

$$\Omega = 2.62 \times 10^{-13}\lambda^2 \int_{Z_1}^{Z_2} nB_{//} \, dZ. \tag{91}$$

This equation shows how the Faraday rotation angle Ω is related to the product $nB_{//}$ in each point of the probed chord. Inversely, the magnetic field $B_{//}$, and consequently the current profile, can be deduced, in principle, from the knowledge of both Ω and n. Equation (91) is strictly valid only if inequality

(108) is fulfilled. Nevertheless, it is accurate enough to give an estimate of the expected value of Ω for tokamak plasmas.

To evaluate Ω numerically n is supposed to be constant over the whole beam path and equal to its mean value $\bar{n}$ along this path. Straightforward calculations show that $\int_{Z_1}^{Z_2} B_{//} \, dZ$ is a maximum for $\gamma = 0.7$. The value of $\bar{n}$ along the corresponding chord is $\simeq n_0/3$. It then follows that the maximum angle of rotation Ω is

$$\Omega_{max} = 2.3 \times 10^{-20} J_0 n_0 \lambda^2. \tag{92}$$

By introducing the typical values for the TFR, $J_0 = 6 \times 10^5$ A, $n_0 = 10^{20}$ m^{-3}, and $\lambda = 3.37 \times 10^{-4}$ m, then

$$\Omega_{max} = 0.16 \quad \text{rad.} \tag{93}$$

This angle is small, and its accurate measurement requires somewhat elaborated techniques. Two suggested arrangements are described next. In the following, phase shifts induced by the reflection, by the transmission of a wave by a grid beam splitter, or by an analyzer are neglected for the sake of simplicity. Calculations show that the results are not seriously affected by these phase shifts (Crenn, private communication).

1. *Arrangement 1*

This arrangement is described by referring to Fig. 46. The components of the waves normal to the plane of the figure are referred to as components x, while the components parallel to the plane of the figure are referred to as components y.

The wave entering the plasma is assumed to be purely ordinary, that is to say it has only one component parallel to the toroidal magnetic field B_T and is represented by

$$x_1 = a \cos \omega t, \qquad y_1 = 0. \tag{94}$$

This linearly polarized wave can be treated as the superposition of two circularly polarized waves, left- and right-handed, respectively:

$$x_{1+} = \tfrac{1}{2}a \cos \omega t, \qquad y_{1+} = \tfrac{1}{2}a \sin \omega t \tag{95}$$

and

$$x_{1-} = \tfrac{1}{2}a \cos \omega t, \qquad y_{1-} = -\tfrac{1}{2}a \sin \omega t. \tag{96}$$

It is clear that

$$x_{1+} + x_{1-} = x_1, \qquad y_{1+} + y_{1-} = y_1,$$

showing that the set of Eqs. (95) and (96) is equivalent to Eqs. (94). As they came out of the plasma, the components of the two circularly polarized waves

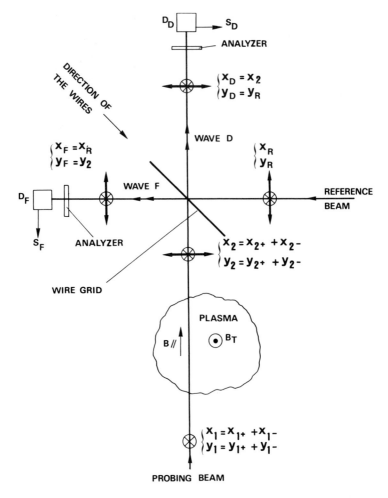

FIG. 46.   Arrangement 1 for simultaneous measurement of line density and Faraday rotation. (See text for discussion.)

are phase shifted by the amount φ_+ and φ_- for the x, or ordinary components, and φ'_+ and φ'_- for the y, or extraordinary components. They are then represented by

$$x_{2+} = \tfrac{1}{2}a \cos(\omega t - \varphi_+), \qquad y_{2+} = \tfrac{1}{2}a \sin(\omega t - \varphi'_+) \qquad (97)$$

and

$$x_{2-} = \tfrac{1}{2}a \cos(\omega t - \varphi_-), \qquad y_{2-} = -\tfrac{1}{2}a \sin(\omega t - \varphi'_-). \qquad (98)$$

The reference wave, which is frequency shifted by the amount $\Delta\omega$, is linearly polarized such that its components x_R and y_R are equal:

$$x_R = b\cos(\omega + \Delta\omega)t, \qquad y_R = b\cos(\omega + \Delta\omega)t. \tag{99}$$

The recombining beam splitter is a wire grid which is assumed to be completely transmitting for the x components and completely reflecting for the y components. It follows that waves F and D (Fig. 46) are, respectively, represented by

$$x_F = x_R, \qquad y_F = y_2 = y_{2+} + y_{2-}$$

and

$$x_D = x_2 = x_{2+} + x_{2-}, \qquad y_D = y_R$$

That is to say,

$$x_F = b\cos(\omega + \Delta\omega)t, \qquad y_F = \tfrac{1}{2}a[\sin(\omega t - \varphi'_+) - \sin(\omega t - \varphi'_-)] \tag{100}$$

and

$$x_D = \tfrac{1}{2}a[\cos(\omega t - \varphi_+) + \cos(\omega t - \varphi_-)], \qquad y_D = b\cos(\omega + \Delta\omega)t. \tag{101}$$

The components x_F and y_F (or x_D and y_D) are at right angle with each other, so they have to pass through an analyzer inclined at $45°$ with respect to their direction to produce interferences. The resulting waves reaching the detectors D_F and D_D are then, respectively,

$$\tfrac{1}{4}a\sqrt{2}[\sin(\omega t - \varphi'_+) - \sin(\omega t - \varphi'_-)] + \tfrac{1}{2}b\sqrt{2}\cos(\omega + \Delta\omega)t$$

and

$$\tfrac{1}{4}a\sqrt{2}[\cos(\omega t - \varphi_+) + \cos(\omega t - \varphi_-)] + \tfrac{1}{2}b\sqrt{2}\cos(\omega + \Delta\omega)t.$$

Finally, the resulting signals are

$$S_F = \tfrac{1}{2}ab[\sin(\Delta\omega\, t + \varphi'_+) - \sin(\Delta\omega\, t + \varphi'_-)], \tag{102}$$

$$S_D = \tfrac{1}{2}ab[\cos(\Delta\omega\, t + \varphi_+) + \cos(\Delta\omega\, t + \varphi_-)]. \tag{103}$$

By introducing

$$\varphi'_+ = \varphi' + \Omega', \qquad \varphi'_- = \varphi' - \Omega'$$

and

$$\varphi_+ = \varphi + \Omega, \qquad \varphi_- = \varphi - \Omega,$$

the expressions for S_F and S_D become

$$S_F = ab \sin \Omega' \cos(\Delta\omega\, t + \varphi'), \tag{104}$$

$$S_D = ab \cos \Omega \cos(\Delta\omega\, t + \varphi). \tag{105}$$

In the simple case for which $\Delta\mu = \mu_e - \mu_0$ is small enough so that $\varphi \simeq \varphi'$ and, consequently, $\Omega = \Omega'$,

$$S_F = ab \sin \Omega \cos(\Delta\omega\, t + \varphi), \tag{106}$$

$$S_D = ab \cos \Omega \cos(\Delta\omega\, t + \varphi). \tag{107}$$

It is clear that the amplitude of S_F is proportional to the Faraday rotation angle Ω if Ω is small, and that the phase shift φ due to the density can be deduced from S_D by comparing it to a reference signal S_R proportional to $\cos \Delta\omega\, t$ as described in Section VI.C. Since S_F and S_D have the same phase, S_D can be used as a reference signal in a lock-in amplifier to measure the amplitude of S_F with better accuracy. The Faraday rotation system can be calibrated with the aid of a half-wave plate placed on the probing beam path, in the absence of plasma, and rotated by a known angle. An actual setup built according to a similar scheme has been operated using a rotating half-wave plate to simulate the plasma. With an HCN laser beam of a few milliwatts and a beat frequency of 10^4 Hz, a sensitivity of 2×10^{-3} rad has been achieved (Crenn, private communication) with an integration time constant of 2×10^{-3} s. This shows that this apparatus is well suited to operate on a tokamak machine, since angles of about two orders of magnitude larger are expected [Eq. (93)]. Nevertheless, if $\mu_e - \mu_0$ is not small enough, then corrections to the measurements have to be made, mainly for the Faraday rotation. According to De Marco and Segre (1972), the value of Ω can be attributed only to $\int_{Z_1}^{Z_2} B_{//}\, dZ$, as expressed by Eq. (91), if the condition

$$\int_{Z_1}^{Z_2} (\omega/c)\, \Delta\mu\, dZ \ll 1 \tag{108}$$

is fulfilled. Since

$$\Delta\mu = \mu_0 - \mu_e = [\omega_p^2/(2\omega^2)][\omega_{ce}^2/(\omega^2 - \omega_p^2)]$$

and $\omega^2 \gg \omega_p^2$, condition (108) becomes

$$\int_{Z_1}^{Z_2} [\omega_p^2 \omega_{ce}^2/(2c\omega^3)]\, dZ \ll 1 \tag{109}$$

or, numerically, after introducing $\omega_{ce} = eB_T/m_e$ in which B_T is the toroidal induction,

$$2.45 \times 10^{-11} \lambda^3 B_T^2 \bar{n}\, \Delta Z < 0.1, \tag{110}$$

where $\int_{Z_1}^{Z_2} n \, dZ$ has been replaced by $\bar{n} \, \Delta Z$ and the upper limit of the integral [Eq. (108)] was set at 0.1.

Inequality (110) can be finally written as

$$B_T^2 \bar{n} < \lambda^{-3} \Delta Z^{-1} \times 4.1 \times 10^9.$$

For the TFR, in which $\Delta Z = 0.4$ m and $\lambda = 3.37 \times 10^{-4}$ m, the condition is

$$B_T^2 \bar{n} < 2.7 \times 10^{20}.$$

If $\bar{n} = 5 \times 10^{19}$,

$$B_T < 2.3 \quad \text{T.}$$

A slight reduction of λ leads to a substantial increase in the upper limit for B_T. For instance, using the DCN laser radiation (1.95×10^{-4} m), the upper limit for B_T is nearly 5.2 T.

It should be noted, however, that if the condition of De Marco and Segre is not fulfilled, the measurement is still possible, but it is then more complicated to deduce the actual value of the Faraday rotation.

2. Arrangement 2

The principle behind arrangement 2 was first proposed by Dodel and Kunz (1978). A continuous rotation of the electric field vector of the probing beam is obtained by using the rotating grating technique (Section II.B.2) as follows: The laser beam is split into two beams of equal intensity. One of them is frequency shifted with the help of a rotating grating (Fig. 47). Then each beam is traversing a quarter-wave plate, which can be made of crystal quartz oriented in such a way that two, respectively, left-handed and right-handed circularly polarized waves come out. Those waves can be represented by the following set of equations:

$$x_+ = a \cos \omega t, \qquad\qquad y_+ = a \sin \omega t, \qquad\qquad (111)$$

$$x_- = a \cos(\omega + \Delta\omega)t, \qquad y_- = -a \sin(\omega + \Delta\omega)t, \qquad (112)$$

assuming that any possible phase difference is compensated.

By recombining the two waves with a polarization insensitive beam splitter, the components of the resulting wave are

$$\begin{aligned} x &= 2a \cos(\tfrac{1}{2} \Delta\omega \, t) \cos(\omega + \tfrac{1}{2} \Delta\omega)t, \\ y &= -2a \sin(\tfrac{1}{2} \Delta\omega \, t) \cos(\omega + \tfrac{1}{2} \Delta\omega)t, \end{aligned} \qquad (113)$$

showing that it is a linearly polarized wave whose representative vector rotates at a frequency $\tfrac{1}{2} \Delta\omega$. The detector D_R located behind an analyzer monitors the signal

$$S_R = a^2 \cos \Delta\omega \, t, \qquad\qquad (114)$$

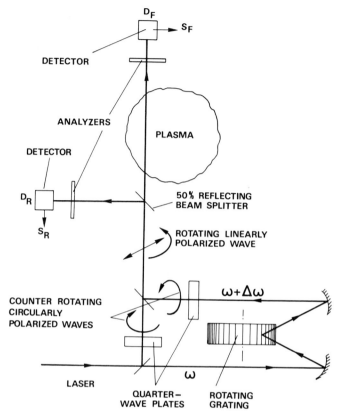

FIG. 47. Setup for generating a rotating linearly polarized wave and for Faraday rotation measurement.

assuming that half the beam intensity is reflected by the beam splitter. After traversing the plasma the two-counter rotating circular waves are

$$x_+ = \tfrac{1}{2}a\sqrt{2}\cos(\omega t - \varphi_+), \qquad y_+ = \tfrac{1}{2}a\sqrt{2}\sin(\omega t - \varphi'_+), \quad (115)$$

$$x_- = \tfrac{1}{2}a\sqrt{2}\cos[(\omega + \Delta\omega)t - \varphi_-],$$
$$y_- = -\tfrac{1}{2}a\sqrt{2}\sin[(\omega + \Delta\omega)t - \varphi'_-]. \quad (116)$$

Detector D_F monitors either a signal

$$S_F = a^2\cos(\Delta\omega\,t + \varphi_+ - \varphi_-) = 2a^2\cos(\Delta\omega\,t + 2\Omega) \quad (117)$$

or

$$S'_F = -a^2\cos(\Delta\omega\,t + \varphi'_+ - \varphi'_-) = -2a^2\cos(\Delta\omega\,t + 2\Omega'), \quad (118)$$

depending on the orientation of the analyzer (either along $0x$ or along $0y$).

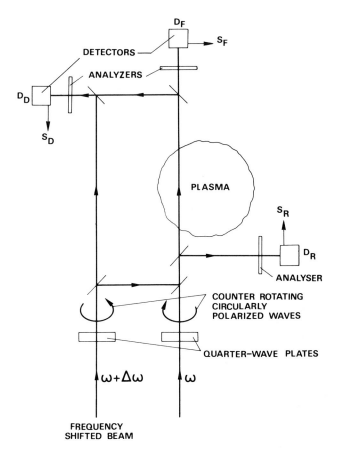

FIG. 48.   Arrangement 2 for the simultaneous measurement of line density and Faraday rotation.

By using the technique outlined in Section VI.C.2, angle 2Ω can be deduced from the time lag between the zero crossings of signals S_R and S_F (or S_F'). The same condition as stated before [Eq. (108)] about the significance of Ω also applies here. The setup is almost insensitive to vibrations since the two-counter rotating waves follow the same path, except for the small portion of the optical circuit where they are generated, and which can be made very rigid. It has already been shown in the laboratory (Dodel and Kunz, 1978) that a similar arrangement, with a 0.15-W HCN laser source and two pyroelectric detectors, was able to reach a sensitivity of 2×10^{-2} rad (3×10^{-3} fringe), using only a crude visual display method. A sensitivity of at least one order of magnitude better should be attainable with suitable electronics (Section VI.C.2) and lower NEP detectors.

The complete setup for measuring both Faraday rotation and line density is schematically represented in Fig. 48. The signal S_D monitored by detector D_D is proportional to

$$\cos(\Delta\omega\, t + \varphi) + \cos(\Delta\omega\, t + \varphi'),$$

that is to say, with the condition $\varphi' \simeq \varphi$,

$$2\cos(\Delta\omega\, t + \varphi).$$

The signal monitored by the detector D_F is similar to that given by Eq. (117).

This system requires beam splitters insensitive to polarization. It should be possible to make them with a combination of two wire grids at right angles with different spacings. Compensators made of two quartz wedges could also be used together with ordinary beam splitters so as to produce the desired circularly polarized waves.

IX. Conclusion

It has been demonstrated in this chapter that submillimeter interferometry is at the present time one of the best methods for measuring the plasma density in tokamak machines. However, it must be pointed out that tokamaks have to be designed with access along sufficiently numerous lines of sight to allow the measurement of the complete density profile of the plasma column. Most of the problems associated with the use of submillimeter radiation have now been solved, or they are in the process of being solved. In particular, sources powerful enough to feed a multichannel system are presently available. Some of them already have the reliability required for long runs on tokamaks.

It has also been shown that the optical arrangement of submillimeter interferometers can be made efficient enough to allow the use of rugged detectors. Sensitivity to vibration can be reduced to a sufficiently low value by carefully designing the supporting structure or by using two different wavelengths. Finally, the association of a poloidal field measurement system with an interferometer has been studied and seems promising.

Acknowledgments

The author is grateful to all the contributors who gave him the permission to use valuable material already reported and more recent information not yet published. He is greatly indebted to Doros Andreou who kindly accepted the difficult task of critically reading the manuscript. Many thanks are also due to the Infrared Team of Fontenay-aux-Roses for their help and suggestions.

REFERENCES

Armstrong, K. R., and Low, F. J. (1974). *Appl. Opt.* **13**, 425–430.
Baker, D. R., and Shu-Tso Lee. (1978). *Rev. Sci. Instrum.* **49**, 919–922.
Baker, G., Charlton, D. E., and Lock, P. J. (1972). *Radio Eng. Electron.* **42**, 1–5.
Belland, P., and Véron, D. (1973). *Opt. Commun.* **9**, 146–148.
Belland, P., Véron, D., and Whitbourn, L. B. (1975). *J. Phys. D* **8**, 2113–2121.
Belland, P., Pigot, C., and Véron, D. (1976a). *Phys. Lett.* **56A**, 21–22.
Belland, P., Véron, D., and Whitbourn, L. B. (1976b). *Appl. Opt.* **15**, 3047–3053.
Blanc, P., and Véron, D. (1972). *Proc. Int. Conf. Fusion Technol. Eur.* 4938e, pp. 63–67, Luxembourg.
Born, M., and Wolf, E. (1970). "Principles of Optics." Pergamon, Oxford,
Bruneau, J. L., Belland, P., and Véron, D. (1978). *Opt. Commun.* **24**, 259–264.
Chantry, G. W., Fleming, J. W., Smith, P. M., Cudby, M., and Willis, H. A. (1971). *Chem. Phys. Lett.* **10**, 473–477.
Costley, A. E., Hursey, K. H., Neill, G. F., and Ward, J. M. (1977). *J. Opt. Soc. Am.* **67**, 979–981.
Crenn, J. P. (1979). *IEE Trans. Microwave Theory Tech.* **27**, 573–577.
De Marco, F., and Segre, S. E. (1972). *Plasma Phys.* **14**, 245–252.
Dodel, G., and Kunz, W. (1978). *Infrared Phys.* **18**, 773–776.
Duverger, B., Certain, J., and Véron, D. (1975). Unpublished.
Equipe TFR (1977). EUR-CEA-FC. 916, Fontenay-aux-Roses, France.
Frank, A. M. (1976). *Digest Int. Conf. Winter School Submillimeter Waves, 2nd* pp. 214–215. IEEE. Cat. No. 76 CH 1152–8, MTT.
Furth, H. P. (1975). *Nucl. Fusion* **15**, 487–534.
Gibson, A., and Reid, G. W. (1964). *Appl. Phys. Lett.* **5**, 195–197.
Gibson, A., Coxell, H., Powell, B. A., and Reid, G. W. (1966). *Plasma Phys.* **9**, 1–12.
Heald, M. A., and Wharton, C. B. (1965). "Plasma Diagnostics with Microwaves," pp. 12–50. Wiley, New York.
Hodges, D. T., Foote, F. B., and Reel, R. D. (1976). *Appl. Phys. Lett.* **29**, 662–664.
Kogelnik, M., and Li, T. (1966). *Appl. Opt.* **5**, 1550–1566.
Kunz, W., and TFR Group (1978). *Nucl. Fusion* **18**, 1729–1732.
Olsen, J. N. (1971). *Rev. Sci. Instrum.* **42**, 104–106.
Peterson, R. W., and Jahoda, F. C. (1971). *Appl. Phys. Lett.* **18**, 440–442.
Pieroni, L. (1977). C.N.E.N. Edizioni Scientifiche C. P. 65 00044 Frascati, Rome, Italy. 77.25/CC.
Russell, E. E., and Bell, E. E. (1967a). *J. Opt. Soc. Am.* **57**, 341–348.
Russell, E. E., and Bell, E. E. (1967b). *J. Opt. Soc. Am.* **57**, 543–544.
Sakai, K., Fukui, T., Tunawaki, Y., and Yoshinaga, H. (1969). *Jpn. J. Appl. Phys.* **8**, 1046–1055.
Shmoys, J. (1961). *J. Appl. Phys.* **32**, 689–695.
Ulrich, R., Bridges, T. J., and Pollak, M. A. (1970). *Appl. Opt.* **9**, 2511–2516.
Véron, D. (1974). *Opt. Commun.* **10**, 95–98.
Véron, D., Certain, J., and Crenn, J. P. (1977). *J. Opt. Soc. Am.* **67**, 964–967.
Véron, D., Belland, P., and Beccaria, M. J. (1978). *Infrared Phys.* **18**, 465–468.
Wolfe, S. M., Button, K. J., Waldman, J., and Cohn, D. R. (1976). *Appl. Opt.* **15**, 2645–2648.

CHAPTER 3

Dispersive Fourier Transform Spectrometry

J. R. Birch   and   T. J. Parker

I. Introduction

Dispersive Fourier transform spectrometry is a technique that has been developed recently for the direct determination of the optical constants of materials at millimeter and submillimeter wavelengths. It is a broad-band technique that uses a two-beam interferometer, usually of the Michelson type, with the specimen to be studied placed in one beam of the interferometer rather than near the detector as for conventional Fourier transform spectrometry. This asymmetry is the important feature of the technique, leading to its sensitivity to the phase of the interaction between the electromagnetic field and the specimen. In the conventional technique the phase shift caused by the specimen is present in both beams of the interferometer and consequently does not, to first order, affect the intensity delay pattern produced at the exit aperture of the interferometer. Thus only the power attenuation of the specimen is measured. When, however, the specimen is inserted in one beam, its refractive index causes the delay pattern to become asymmetric and shifted in path difference, and both the attenuation and phase shift of the specimen may be recovered from the recorded intensity delay pattern, allowing the optical constants to be calculated. The asymmetry also gives rise to an alternative name for the technique, "asymmetric Fourier transform spectrometry."

The relevance of such measurements to modern physics is a consequence of the role of the optical constants of a material as its fundamental electromagnetic parameters, totally describing the manner in which the electromagnetic field is coupled to the material. As Bell (1967a) wrote, "These constants have value not only for describing the wave progress but also for their intimate relation to the fundamental constitution of the material. The frequency dependence of the optical constants gives a large amount of information about the physical nature of the material." One might expect, therefore, that the experimental investigation of these constants would be of prime importance to many scientists, not only those interested in dielectric phenomena, but also those concerned with the applications and techniques of submillimeter waves. However, this has not been so. The object of most spectroscopic investigations of materials has been to determine the spectral variation of the power transmitted through, reflected from, or emitted by a specimen. The power absorption coefficient may be calculated from the measured power transmission using approximate expressions, but only infrequently is the effect of the specimen on the phase of the probing radiation sought, and then usually by indirect means such as the analysis of channel spectra or the use of the Kramers–Krönig relations.

In this chapter we shall review the development and present status of dispersive Fourier transform spectrometry, concentrating mainly on those aspects of technique that have led to improved measurement precision. This shall be illustrated with reference to the results of dispersive measurements on gases, liquids, and solids, many of which were previously unobtainable, that are being used to improve and develop our understanding in such diverse fields as the molecular dynamics of gases and liquids, lattice dynamics, and ferroelectric phenomena.

In the following we shall refer to dispersive Fourier transform spectrometry by the acronym DFTS and to Fourier transform spectrometry as FTS.

A. HISTORICAL BACKGROUND

The development of DFTS can be said to have its origins in the period between 1880 and 1890 when Michelson (1881) devised the interferometer that bears his name, and Rubens (1889a,b) began the systematic investigation of the infrared spectral region. He and his coworkers developed quantitative spectroscopic techniques for wavelengths longer than 25 μm and applied them to the study of the optical properties of a variety of materials. Although subsequent workers further refined and developed these techniques, opening up new fields of study such as molecular spectroscopy and quantitative chemical analysis, the development of truly precise techniques for the longer wavelength region beyond, say, 50 μm did not occur until the late 1950s, with the practical realization of the technique of Fourier transform spectrometry (Gebbie and Vanasse, 1956; Strong, 1957) following the enunciation of the throughput advantage for instruments possessing cylindrical symmetry by Jacquinot and Dufor (1948; Jacquinot, 1954) and of the multiplex advantage by Fellgett (1951, 1958), combined with the development of high-speed digital computers. This led to a rapid expansion in the study and use of the infrared spectrum, recorded by the bibliographies of Palik (1960, 1963) and Bloor (1970), in which a vast amount of information on the power transmission and reflection properties of materials from all phases of matter was gathered together.

Many of the spectra so obtained were analyzed by a variety of approximate techniques of differing degrees of complexity and applicability to give the optical constants of the material studied with correspondingly different errors. This was an unsatisfactory state of affairs as there was no single technique or group of related techniques that could be applied to any material of any degree of opacity, and, partly as a consequence of this, there were no systematic comparisons of the different techniques, their experimental problems, accuracies, or sources of error. This situation was changed in 1963 with the publication by Chamberlain *et al.* (1963) of the first DFTS measurement of the real refractive index of crystal quartz parallel to its optic axis between 20 and 55 cm^{-1}. This result is reproduced in Fig. 1. Their measurements achieved a reproducibility of about ± 0.01, or one part in 10^2, although

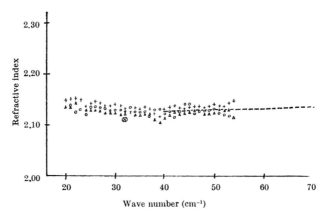

FIG. 1. Three independent determinations of the refractive index of crystalline quartz parallel to its optic axis between 20 and 55 cm^{-1}. The dashed line is taken from earlier work of Geick (1961), while the point marked × at 32 cm^{-1} is from Korff and Breit (1932). [From Chamberlain et al. (1963). Reproduced by permission of Macmillan Journals Ltd.]

it is apparent from the figure that much of this scatter is due to systematic differences between independent determinations, an error that is now understood and largely suppressed by improved experimental technique. The authors did not attempt to derive the absorption index from their measurements, being more interested at that time in the new information provided by the new technique.[1] Recent measurements on the same material (Passchier et al., 1977a), in which the real refractive index was determined with a precision of several parts in 10^4, illustrate the improvement in measurement techniques that have occurred in the first 15 years of DFTS.

In the period immediately following this initial work the subject developed through the work of Chamberlain and coworkers at the National Physical Laboratory, mainly on gases and liquids (e.g., Chamberlain et al., 1965a, 1967a, 1968) and of Bell and coworkers at Ohio State University on gases and solids (e.g., Bell, 1965, 1966; Russell and Bell, 1966; Sanderson, 1967). More recently, groups at the Max Planck Institute, Stuttgart, Westfield College, University of London, the University of Cologne, and the University of British Columbia, Vancouver, have made significant advances in the study of solids, while the group at the Rijksuniversiteit, Leiden, have promoted major experimental and theoretical developments in the field of liquid DFTS. Various aspects of DFTS have been considered in the review article by Bell (1970), which was presented at the Aspen International Conference on Fourier Spectroscopy, and in the books by Bell (1972) and Chantry (1971).

---

[1] J. E. Gibbs, private communication.

B. TRANSPARENCY AND OPACITY: THE CHOICE OF AN EXPERIMENT

The object of the type of spectrometric measurement described in this chapter is the extraction of the optical constants of a specimen from its measured complex transmission or reflection spectrum. The ease with which this may be done, and the accuracy achieved in the final result, are both critically dependent on the type of experiment performed which is, in itself, dictated by the degree of transparency or opacity of the specimen. The classification of a specimen as transparent or opaque, or somewhere in between, is, however, at best only a semiobjective judgment. One could call a specimen "transparent" if it transmits a sufficient fraction of the radiant energy incident on it for an accurate transmission measurement to be made. Depending on various experimental conditions this could correspond to a power transmission coefficient of 10^{-2} for the conventional double-pass measurement or 10^{-4} for a single-pass measurement. In principle, of course, a specimen may always be made sufficiently thin for it to become transparent, but it can be difficult to create extremely thin, homogeneous films whose properties correspond to those of the bulk medium.

In order to accommodate this range of specimens there are, essentially, three dispersive techniques that are illustrated schematically in Fig. 2. In the

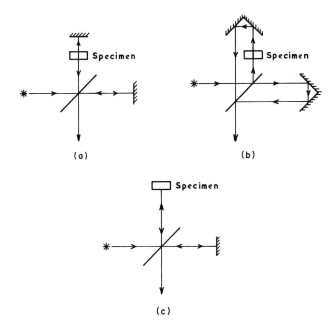

FIG. 2. The three basic interferometer configurations used in dispersive Fourier transform spectrometry: (a) double-pass transmission; (b) single-pass transmission; (c) reflection.

most simple of these the specimen is placed in the fixed mirror arm of the interferometer and the radiation in that arm passes through it twice. This is known as a *double-pass transmission measurement*. If, however, the specimen is opaque, a *reflection measurement* is required, and the specimen is made to replace the fixed mirror of the interferometer (or an equivalent operation is performed; see Sections V and VI). In the transmission experiment it is usual to measure phase shifts of tens of radians, or more, compared to tens of milliradians in the reflection experiment. Thus, transmission measurements, if they are possible, are usually preferred as they offer much greater phase accuracies than do reflection measurements. There are always specimens that, while being highly absorbing, are not highly reflecting and so are not particularly well suited to either of these two techniques. The third dispersive technique exists for them; it is the *single-pass transmission measurement*. The optics of the interferometer are arranged so that the radiation in the fixed mirror arm only passes through the specimen once. The insertion loss that is then measured is the square root of that measured in the double-pass system, and for low transmission levels, the former is very much greater than the latter. In practice, of course, there are many variants on these arrangements, especially in the case of liquids, where it becomes necessary to contain the specimen, but those shown in Fig. 2 illustrate the basic configurations.

C. Nondispersive Techniques

The measurement of the optical constants of materials at millimeter and submillimeter wavelengths did not, of course, have to wait for the development of DFTS. A wide variety of methods had long been used for analyzing the power transmission and reflection spectra of a specimen to give one or both of its optical constants. This area has been considered in detail by Bell (1967a), and in this subsection we shall briefly discuss the more widely used of these methods.

A reasonably transparent material would normally be measured in a transmission experiment using a plane parallel specimen. If its refractive index is known from some other measurement and provided that a channel spectrum due to internally reflected rays is not resolved, then its power absorption coefficient may be found from the use of approximate expressions given, among others, by Moss (1959) and Bell (1967a) and widely used by many workers. These are essentially statements of the law of exponential absorption within the specimen in which differing degrees of allowance are made for the unresolved internally reflected rays. When the refractive index is not known, the reflection losses of the transmitted radiation can be allowed for, to first order, by ratioing transmission measurements on different thickness specimens. Care must, however, be given to the choice of specimen thickness as this method is particularly susceptible to systematic errors.

The refractive index of a transparent solid has traditionally been found from measurements on a channel spectrum seen in transmission. If the order numbers of the local interference maxima or minima are known, the value of the real refractive index may be calculated by equating the optical thickness of the lamella, at the wavelength of the particular feature, to an even number of quarter-wavelengths for a maximum or to an odd number for a minimum. There are several criticisms that can be made of this simple, but widely used formulation of the method:

(a)   First, it ignores the phase shift that occurs on reflection from the surfaces of the specimen. This is usually a good approximation for specimens of a few millimeters thickness, but the method can be applied to very thin films of highly absorbing material in which case it would be necessary to allow for these phase shifts. Randall and Rawcliffe (1967) have presented equations that are appropriate to this situation.

(b)   The method as described depends on knowing the order number of the interference maxima and minima. For wave numbers below 50 to 100 cm^{-1} the order number of a spectrum can usually be determined unambiguously by inspection of the spectrum. This is generally only so at higher wave numbers if very thin specimens are studied, and in the situation of unknown order number, the wave number separation of adjacent features has been used to give a refractive index value, although Moss (1959) has criticized this as it suppresses any linear dependence of the refractive index on wavelength. Zwerdling (1970) has described a more general approach to the analysis of channel spectra of unknown order and the problems of the loss of the linear term.

(c)   It can be difficult to locate the precise wave number of the maxima and minima of the channel spectrum, especially if the overall level of the transmission spectrum is changing rapidly with wave number. Loewenstein and Smith (1971) have described a method of analysis that avoids this problem by working with a difference spectrum obtained from spectra in which the channel spectrum is and is not resolved. Thus, the problem of locating broad maxima and minima reduces to one of locating zero crossings, which can be performed more accurately. In addition, these authors describe how the absorption index may be found from the amplitude of the interference fringes, although this requires that the channel spectrum is fully resolved and that the effects of apodization and the finite spectral window of the spectrometer be considered. These authors also prefer to work in reflection rather than transmission as they have shown that for refractive indices below 3.8, which covers most materials of interest, the contrast of the reflection channel spectrum exceeds that of the transmission spectrum.

(d)   Adjacent maxima and minima in a channel spectrum are separated in wave number by approximately the reciprocal of the optical thickness of the specimen. Thus, the channel fringe method is limited in the spectral resolution that it can provide, especially when only one specimen is available for study.

These methods are only suitable for reasonably transparent materials. As a material becomes more and more absorbing so that impractically thin specimens would be required for a transmission measurement, it becomes necessary to turn to reflection methods for the derivation of the optical constants. In principle, the optical constants can be derived from any pair of independent reflection measurements, although the solution of the appropriate equations may have to be iterative and require automatic computation. Thus, the methods of Simon (1951), Avery (1952), and Lindquist and Ewald (1963), which rely on measurements at two or more angles of incidence or two polarizations and the use of precomputed charts or graphical constructions to give the optical constants, have been used. It is, however, disadvantageous to have to make two or more distinct measurements and many workers have preferred to use the Kramers–Krönig dispersion integral to compute the phase of the reflectivity from their intensity reflection measurements. The optical constants may then be calculated directly from the complex reflectivity. This method was first discussed by Robinson (1952) and Robinson and Price (1953) and has been widely applied in all parts of the electromagnetic spectrum even though errors arising from the necessary truncation of the infinite integral can be significant and although there are situations in which the use of the relationship may be erroneous or fail (Berreman, 1967; Cardona, 1969; Chambers, 1975; Young, 1977).

Internal reflection spectroscopy, also known as "attenuated total reflection spectroscopy," has proven to be a very sensitive method for the determination of optical constants and deserves mention for that reason, even though it has not, so far, been applied at the long wavelengths with which this chapter is mainly concerned. The method overcomes the lack of sensitivity of the normal reflection spectrum to the absorption index by using an angle of incidence greater than the critical angle of an interface. There are many advantages to the method: weak features may be studied through multiple reflections; the measured spectrum can be made to resemble the transmission spectrum; and there are no interface effects arising from the finite size of a specimen as the interaction occurs via a nonpropagating wave. However, practical limitations arising from dispersion within a specimen and the range of materials available for use as the internal reflection element have limited the application of the method to the near infrared and visible spectral regions. A detailed introduction to the method may be found in the work of Harrick (1967).

D. NOTATION

Notation and units have for a long time been a matter of controversy in electromagnetism. Within the present work it has been the authors' intention to adhere to SI units with two exceptions, both for reasons of common spectroscopic usage. Thus, the unit of inverse wavelength, the wave number, is expressed in cm^{-1}, and the unit of power absorption coefficient in neper cm^{-1}, although many workers express it in units of cm^{-1} only, dropping the dimensionless neper. Chamberlain and Chantry (1973) argue for its retention as a way of distinguishing between power absorption coefficient and absorbance by indicating that the former has been derived from the natural logarithm of a ratio. The present authors also find that the use of neper cm^{-1} is a convenient way of avoiding confusion between wave number and power absorption coefficient.

The convention followed for notation is generally that which has been favored by workers in the field of dispersive Fourier transform spectrometry. Thus, the following symbols appear frequently:

i $\sqrt{-1}$.

$\hat{}$ denoting a complex number $\hat{a}$ which, in its general algebraic form, is written with its imaginary part positive, $a + ib$,

$\hat{n}$ the complex refractive index, $\hat{n} = n + ik$,

n the real refractive index,

k the absorption index,

α the power absorption coefficient,

$\hat{\varepsilon}$ the complex relative permittivity, $\hat{\varepsilon} = \varepsilon' + i\varepsilon''$,

ω angular frequency,

$\tilde{\nu}$ wave number.

Vector quantities are denoted by bold face.

II. Electromagnetic Quantities

The interaction of a plane, monochromatic electromagnetic wave with a finite isotropic medium can be fully characterized by three parameters: the complex refractive index of the medium, the frequency of the radiation, and a factor describing the shape or form of the medium, which, in its most tractable form, would be the thickness of a plane parallel specimen. Given these parameters, the total attenuation and phase shift experienced by a wave on reflection from, or transmission through, the specimen may be calculated. This section, therefore, will show how the concept of a complex refractive index arises from Maxwell's equations for a homogeneous absorbing medium, deriving the form of the relationship between it and the conductivity, permeability, and permittivity of the medium before describing the complex form of the Fresnel formulas for the reflection and transmission of an interface between two dissimilar media and their limiting values for cases of practical interest.

Throughout this chapter the electromagnetic field vectors are shown as complex quantitities in order to simplify the representation of the phase relationships among different waves. Similarly, the conductivity, permeability, and permittivity are shown as being complex in order that allowance can be made for phase relationships that might exist between current density and electric field, magnetic induction and magnetic field, and electric displacement and electric field.

A.   THE COMPLEX REFRACTIVE INDEX

When Maxwell's equations are solved for a medium of conductivity $\hat{\sigma}$, relative permeability $\hat{\mu}$, and relative permittivity $\hat{\varepsilon}_1$, the electric field vector $\hat{\mathbf{E}}$ is constrained to satisfy the wave equation

$$\mathbf{V}^2\hat{\mathbf{E}} - \hat{\sigma}\mu_0\hat{\mu}\,(\partial\hat{\mathbf{E}}/\partial t) - \mu_0\hat{\mu}\varepsilon_0\hat{\varepsilon}_1(\partial^2\hat{\mathbf{E}}/\partial t^2) = 0, \tag{1}$$

where μ_0 and ε_0 are the permeability and permittivity of free space, respectively, and t is time (Slater and Frank, 1947; Born and Wolf, 1965). The relative permittivity has been written with a subscript to distinguish it from another permittivity to be introduced later in this section. A similar expression holds for the magnetic field vector, but we shall consider the behavior of the electric field only, by limiting the discussion to nonmagnetic materials in which the magnetic field does not couple strongly to the material. If, following the notation of Bell (1967a), we seek the condition under which the plane wave

$$\hat{\mathbf{E}} = \hat{\mathbf{E}}_0 \exp i(\hat{\mathbf{K}} \cdot \mathbf{r} - \omega t) \tag{2}$$

of angular frequency ω is a solution of Eq. (1), then it is found that the complex wave vector $\hat{\mathbf{K}}$ must be related to the constitutive parameters of the material by

$$\hat{\mathbf{K}} \cdot \hat{\mathbf{K}} = \omega^2\hat{\mu}\mu_0[\varepsilon_0\hat{\varepsilon}_1 + (i\hat{\sigma}/\omega)]. \tag{3}$$

The phase velocity of the plane wave of Eq. (2) is the velocity of propagation of a wave front of constant phase and is, therefore, $\partial\mathbf{r}/\partial t$ in

$$\hat{\mathbf{K}} \cdot (\partial\mathbf{r}/\partial t) = \omega. \tag{4}$$

For a homogeneous material the real and imaginary parts of the wave vector $\hat{\mathbf{K}}$ are parallel and lie in the direction of propagation. Equations (3) and (4) may therefore be rearranged to give the phase velocity of the wave as

$$\partial\mathbf{r}/\partial t = \{\hat{\mu}\mu_0[\varepsilon_0\hat{\varepsilon}_1 + (i\hat{\sigma}/\omega)]\}^{-1/2}, \tag{5}$$

but, by definition, this must also be equal to the velocity of light in the material. Hence the refractive index is a complex quantity given by

$$\hat{n} = c\{\hat{\mu}\mu_0[\varepsilon_0\hat{\varepsilon}_1 + (i\hat{\sigma}/\omega)]\}^{1/2}, \tag{6}$$

where c represents the speed of light. Remembering that we are only considering nonmagnetic materials, for which $\hat{\mu} = 1$, and writing c as $(\mu_0\varepsilon_0)^{-1/2}$ enables Eq. (6) to be written in the simpler form

$$\hat{n} = [\hat{\varepsilon}_1 + (i\hat{\sigma}/\omega\varepsilon_0)]^{1/2}. \tag{7}$$

This then is the basic equation with which DFTS is concerned; by a variety of related measurements the real and imaginary parts of this complex refractive index may be determined and then used to investigate the nature of the material by comparisons with calculations drawn from models of its relative permittivity and conductivity.

The role played by the complex refractive index in the propagation of the wave may be seen by writing it in terms of its real and imaginary parts:

$$\hat{n} = n + ik, \tag{8}$$

and substituting for Eqs. (3), (7), and (8) into Eq. (2) to obtain the alternate form of the plane wave

$$\hat{\mathbf{E}} = \hat{\mathbf{E}}_0 \exp(-i\omega t) \exp(-k\omega z/c) \exp(in\omega z/c), \tag{9}$$

where propagation is now considered as being in a specific direction z. Thus, the disturbance is a homogeneous plane wave, attenuated in the positive z direction with its phase advancing as z increases. The imaginary part k of $\hat{n}$ is the dissipative term and is known as the "absorption index," while the real part n is known as the "real refractive index" or, more often, just as the "refractive index." Jointly, n and k are known as the "optical constants" of the material.

Equation (7) for the complex refractive index may be written in the form

$$(\hat{n})^2 = \hat{\varepsilon} \tag{10}$$

to introduce the unsubscripted complex relative permittivity

$$\hat{\varepsilon} = \hat{\varepsilon}_1 + [i\hat{\sigma}/(\omega\varepsilon_0)] \tag{11}$$

which includes the effects of all processes determining the complex nature (the amplitude and phase) of the polarizability. If the real and imaginary parts of $\hat{\varepsilon}$ are given by

$$\hat{\varepsilon} = \varepsilon' + i\varepsilon'', \tag{12}$$

the following relationships between the real and imaginary parts of $\hat{n}$ and $\hat{\varepsilon}$ are simply derived:

$$\varepsilon' = n^2 - k^2, \tag{13}$$

$$\varepsilon'' = 2nk, \tag{14}$$

$$n^2 = \tfrac{1}{2}\{[(\varepsilon')^2 + (\varepsilon'')^2]^{1/2} + \varepsilon'\}, \tag{15}$$

$$k^2 = \tfrac{1}{2}\{[(\varepsilon')^2 + (\varepsilon'')^2]^{1/2} - \varepsilon'\}. \tag{16}$$

B. TRANSMISSION AND REFLECTION AT AN INTERFACE

The previous section dealt with the propagation of a plane wave through an isotropic medium, developing the equations that describe the manner in which the amplitude and phase of the wave change as it travels in the medium. In practical terms one is always concerned with finite specimens and it is necessary to include the effects of the interfaces between the specimen and its surroundings on the propagation of the wave. These will depend critically on the shape of the specimen and the geometry of the situation, but the starting point for any attempt to analyze such a system must always be Fresnel's equations.

1. Fresnel's Equations

Fresnel's equations relate the amplitude attenuation and phase shift suffered by an electromagnetic wave on reflection from and transmission through a plane interface between two semiinfinite media to the optical constants, the angle of incidence, and the polarization of the incident wave. They are derived from a consideration of the boundary conditions which apply at the interface, and are often given as real quantities although, in their most general form, they are complex quantities:

$$\hat{r}_\perp = (\hat{n}_1 \cos \hat{\theta}_1 - \hat{n}_2 \cos \hat{\theta}_2)/(\hat{n}_1 \cos \hat{\theta}_1 + \hat{n}_2 \cos \hat{\theta}_2), \tag{17}$$

$$\hat{r}_\parallel = (\hat{n}_1 \cos \hat{\theta}_2 - \hat{n}_2 \cos \hat{\theta}_1)/(\hat{n}_1 \cos \hat{\theta}_2 + \hat{n}_2 \cos \hat{\theta}_1), \tag{18}$$

$$\hat{t}_\perp = 2\hat{n}_1 \cos \hat{\theta}_1/(\hat{n}_1 \cos \hat{\theta}_1 + \hat{n}_2 \cos \hat{\theta}_2), \tag{19}$$

$$\hat{t}_\parallel = 2\hat{n}_1 \cos \hat{\theta}_1/(\hat{n}_1 \cos \hat{\theta}_2 + \hat{n}_2 \cos \hat{\theta}_1). \tag{20}$$

In these equations $\hat{r}$ is the amplitude reflection coefficient and $\hat{t}$ the amplitude transmission coefficient for the component of the incident wave polarized perpendicular to ($\perp$) and the component parallel to ($\parallel$) the plane of incidence. The subscript 1 refers to the medium from which the wave is incident and the subscript 2 to the other medium. The angles of incidence and refraction are $\hat{\theta}_1$ and $\hat{\theta}_2$, and, although it may seem strange to write them as complex quantities, Snell's law

$$\hat{n}_1 \sin \hat{\theta}_1 = \hat{n}_2 \sin \hat{\theta}_2 \tag{21}$$

indicates that, in general, $\hat{\theta}_1$ and $\hat{\theta}_2$ must be complex. In the special case of normal incidence the Fresnel equations simplify to

$$\hat{r}_\perp = \hat{r}_\parallel = \hat{r}_{12} = (\hat{n}_1 - \hat{n}_2)/(\hat{n}_1 + \hat{n}_2) \tag{22}$$

and

$$\hat{t}_\perp = \hat{t}_\parallel = \hat{t}_{12} = 2\hat{n}_1/(\hat{n}_1 + \hat{n}_2) \tag{23}$$

with the subscript order signifying incidence from medium 1 onto medium 2. It follows that for incidence from medium 2 onto medium 1

$$\hat{r}_{21} = -\hat{r}_{12} \tag{24}$$

and

$$\hat{t}_{21} = \hat{n}_2 \hat{t}_{12}/\hat{n}_1. \tag{25}$$

2.   The Exponential Form of Fresnel's Equations

Equations (22) and (23) are not the most useful forms of Fresnel's equations as they do not explicitly separate the attenuation and phase terms. A more convenient form uses exponential notation in which the complex attenuation of an electric field vector on reflection from an interface is written in the form

$$\hat{r}_{12} = r_{12} \exp i\phi^r_{12} \tag{26}$$

which explicitly shows the attenuation r_{12} and phase ϕ^r_{12} terms. It is readily shown from Eqs. (8) and (22) that, for normal incidence,

$$r_{12}^2 = [(n_1^2 - n_2^2 + k_1^2 - k_2^2)^2 + 4(n_2 k_1 - n_1 k_2)^2]/[(n_1 + n_2)^2 + (k_1 + k_2)^2]^2 \tag{27}$$

and

$$\tan \phi^r_{12} = 2(n_2 k_1 - n_1 k_2)/[n_1^2 - n_2^2 + k_1^2 - k_2^2]. \tag{28}$$

As the reflected and incident waves are in the same medium, Eq. (27) represents the power reflection coefficient of the interface.

When the incident medium is a vacuum, $\hat{n}_1 = 1 + i0$, and Eqs. (27) and (28) reduce to their more familiar forms

$$r_{12}^2 = [(n_2 - 1)^2 + k_2^2]/[(n_2 + 1)^2 + k_2^2], \tag{29}$$

and

$$\tan \phi^r_{12} = -2k_2/(1 - n_2^2 - k_2^2).\tag{30}$$

We are now in a position to derive the phase changes that occur on reflection from transparent and opaque media for normal incidence from a vacuum by considering the limiting behavior of Eq. (30) as $k_2 \rightarrow 0$ and ∞. In the former case

$$\lim_{k_2 \rightarrow 0} \phi^r_{12} = \lim_{k_2 \rightarrow 0} \arctan[-2k_2/(1 - n_2^2)]\tag{31}$$

and, in the latter,

$$\lim_{k_2 \rightarrow \infty} \phi^r_{12} = \lim_{k_2 \rightarrow \infty} \arctan(-2/-k_2).\tag{32}$$

As $n_2 > 1.0$ both angles lie in the third quadrant and in their limits tend to π rad. The result for the opaque medium is of particular relevance to dispersive reflection measurements, in which the complex reflectivity of an unknown specimen is compared to that of a metallized reference surface, for which it is usually assumed that ϕ^r_{12} is π rad.

When the wave is incident onto a vacuum from a more dense medium, $\hat{n}_2 = 1 + i0$, and the phase change on reflection is given by

$$\tan \phi^r_{12} = 2k_1/(n_1^2 + k_1^2 - 1)\tag{33}$$

and in the limits of transparent and opaque media ϕ^r_{12} now tends to zero.

In a similar manner to the preceding description of the complex reflectivity, it is simply shown that the exponential form for the complex transmission coefficient

$$\hat{t}_{12} = t_{12} \exp i\phi^t_{12}\tag{34}$$

leads to the expressions

$$t_{12}^2 = 4 \frac{[n_1(n_1 + n_2) + k_1(k_1 + k_2)]^2 + [n_2 k_1 - n_1 k_2]^2}{[(n_1 + n_2)^2 + (k_1 + k_2)^2]^2}\tag{35}$$

and

$$\tan \phi^t_{12} = (n_2 k_1 - n_1 k_2)/[n_1(n_1 + n_2) + k_1(k_1 + k_2)].\tag{36}$$

It is important to note that while r_{12}^2 is the power reflectivity of the interface t_{12}^2 is not its power transmission coefficient. The correct power transmission coefficient of the interface is $1 - r_{12}^2$.

The phase change on transmission for incidence from a vacuum is

$$\phi^t_{12} = \arctan[-k_2/(1 + n_2)]\tag{37}$$

and for incidence onto a vacuum is

$$\phi^t_{12} = \arctan \{k_1/[n_1(n_1 + 1) + k_1^2]\}.\tag{38}$$

For values of $\hat{n}$ that would be met in practical transmission experiments both angles represented by Eqs. (37) and (38) are close to zero.

III.   General Theory of Dispersive Fourier Transform Spectrometry

A.   COMPLEX INSERTION LOSS

In order to measure the optical properties of a particular specimen, which may be a solid, liquid, or gas, and which may be either very transparent or highly absorbing, a suitable optical configuration must first be chosen. The three basic optical arrangements used in DFTS are illustrated schematically in Fig. 2, and measurements on free-standing solid specimens may be carried out using these configurations with no further modifications. In the case of gases, dispersive transmission measurements are usually made by admitting the gas into a cell in the fixed arm of the interferometer, and a similar, evacuated, cell is installed in the other arm to preserve symmetry in the instrument for the background measurement. In the case of liquids, both reflection and transmission measurements have been made dispersively. For some of these measurements the liquids were free, gravity-held layers and for others they were contained in cells, but in each case the details of the optical configuration must be taken into account before the optical constants can be determined from the interferograms.

In the analysis which follows, we shall begin by deriving the Fourier integral for a generalized specimen placed in the fixed arm of an interferometer. This can be most conveniently done by following the approach of Chamberlain (1972a) and defining the complex insertion loss

$$\hat{L}(\tilde{v}) = L(\tilde{v}) \exp i\phi_L(\tilde{v}), \tag{39}$$

of a specimen in a particular optical configuration as the complex factor by which the amplitude of a wave is changed when a reference material such as a vacuum or perfect mirror is replaced by the specimen. This should not be confused with the complex transmission or reflection coefficient of the specimen, which is the complex factor that relates the emergent to the incident electric field amplitude. The relationship between $\hat{L}(\tilde{v})$ and the optical constants of the specimen shall be derived later for a number of important optical configurations, including those previously mentioned.

B.   THE FOURIER INTEGRAL FOR A GENERALIZED MEASUREMENT BY DFTS

Two different, but equivalent, approaches have been described in the literature to the problem of relating the interferograms recorded in DFTS to the optical constants of the specimen. Bell (1966) has derived the general result in terms of the impulse response function of the specimen, which is the Fourier transform of the complex insertion loss, and Chamberlain et al. (1969b) and Chamberlain (1972a) have derived the general result directly in terms of the complex insertion loss by using simple electromagnetic theory.

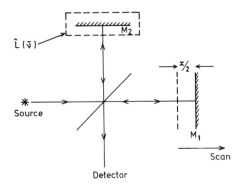

FIG. 3.   The generalized dispersive measurement. A specimen of complex insertion loss $\hat{L}(\tilde{\nu})$ is placed in the fixed mirror M_2 arm of a two-beam interferometer, either replacing that mirror or between it and the beam divider. M_1 is the moving mirror of the interferometer.

We shall use the latter approach as it follows in a convenient way from the development of the subject given in Section II. As it is more common in millimeter and submillimeter wave spectroscopy to characterize a wave by its reciprocal wavelength, known as wave number $\tilde{\nu}$, and related to the angular frequency by $\tilde{\nu} = \omega/(2\pi c)$, we shall develop the Fourier integral in terms of the conjugate quantities $\tilde{\nu}$ and optical path difference x.

In a typical Fourier spectrometer there will be many Fourier components present in the radiation beam passing through the instrument, and it follows from Eq. (9) that the amplitude of the electric vector of such a collimated beam of radiation propagating in free space may be expressed as a Fourier integral

$$\hat{\mathscr{E}} = \int_{-\infty}^{\infty} \hat{\mathbf{E}}_0(\tilde{\nu}) \exp[2\pi i \tilde{\nu}(z - ct)] \, d\tilde{\nu}. \tag{40}$$

Let us now consider the generalized dispersive measurement characterized by the complex insertion loss defined by Eq. (39) and illustrated schematically in Fig. 3. If a radiation beam defined by Eq. (40) is propagated into each arm of the interferometer, the output beam from the moving mirror arm can be written as

$$\hat{\mathscr{E}}_1 = \int_{-\infty}^{\infty} \hat{\mathbf{E}}_0(\tilde{\nu}) \exp[2\pi i \tilde{\nu}(z + x - ct)] \, d\tilde{\nu}, \tag{41}$$

where $x/2$ is the displacement of the moving mirror from the position of zero optical path difference. The output from the fixed arm, which contains the specimen, can similarly be written as

$$\hat{\mathscr{E}}_2 = \int_{-\infty}^{\infty} \hat{\mathbf{E}}_0(\tilde{\nu}) L(\tilde{\nu}) \exp i[2\pi \tilde{\nu}(z - ct) + \phi_L(\tilde{\nu}) + \phi_0(\tilde{\nu})] \, d\tilde{\nu}, \tag{42}$$

where $\phi_0(\tilde{v})$ is a small residual phase difference denoting lack of perfect symmetry in the symmetrical interferometer due to inadvertent imperfect alignment or some other asymmetry between the optical paths.

The resultant electric field amplitude in the recombined beam propagating toward the detector is

$$\hat{\mathscr{E}}_R = \hat{\mathscr{E}}_1 + \hat{\mathscr{E}}_2 = \int_{-\infty}^{\infty} \hat{\mathbf{g}}(\tilde{v}, x) \exp[2\pi i \tilde{v}(z - ct)] \, d\tilde{v}, \qquad (43)$$

where

$$\hat{\mathbf{g}}(\tilde{v}, x) = \hat{\mathbf{E}}_0(\tilde{v})[L(\tilde{v}) \exp i\phi(\tilde{v}) + \exp 2\pi i \tilde{v}x] \qquad (44)$$

and

$$\phi(\tilde{v}) = \phi_L(\tilde{v}) + \phi_0(\tilde{v}). \qquad (45)$$

The output intensity, $\hat{\mathscr{E}}_R \hat{\mathscr{E}}_R^*$, is proportional to

$$\int_{-\infty}^{\infty} \hat{\mathbf{g}}(\tilde{v}, x)\hat{\mathbf{g}}^*(\tilde{v}, x) \, d\tilde{v} = \int_{-\infty}^{\infty} \hat{\mathbf{E}}_0(\tilde{v})\hat{\mathbf{E}}_0^*(\tilde{v})[1 + L^2(\tilde{v})] \, d\tilde{v}$$

$$+ 2 \int_{-\infty}^{\infty} \hat{\mathbf{E}}_0(\tilde{v})\hat{\mathbf{E}}_0^*(\tilde{v})L(\tilde{v}) \cos[\phi(\tilde{v}) - 2\pi\tilde{v}x] \, d\tilde{v}, \quad (46)$$

where * indicates the complex conjugate. The first term in Eq. (46) is independent of x and is of no further interest, but the second term is the interference function, which is usually recorded as the interferogram, and may be written as the sum

$$I_s(x) = \int_{-\infty}^{\infty} \rho(\tilde{v}) \cos \phi(\tilde{v}) \cos 2\pi\tilde{v}x \, d\tilde{v} + \int_{-\infty}^{\infty} \rho(\tilde{v}) \sin \phi(\tilde{v}) \sin 2\pi\tilde{v}x \, d\tilde{v} \quad (47)$$

of an even part and an odd part, where

$$\rho(\tilde{v}) = 2\hat{\mathbf{E}}_0(\tilde{v})\hat{\mathbf{E}}_0^*(\tilde{v})L(\tilde{v}) \qquad (48)$$

is the recorded power spectrum and $\phi(\tilde{v})$ [Eq. (45)] describes the phase difference between the two partial beams. Taking the cosine and sine transforms of Eq. (47), we obtain

$$p(\tilde{v}) = \int_{-\infty}^{\infty} I_s(x) \cos 2\pi\tilde{v}x \, dx = \rho(\tilde{v}) \cos \phi(\tilde{v}) \qquad (49)$$

and

$$q(\tilde{v}) = \int_{-\infty}^{\infty} I_s(x) \sin 2\pi\tilde{v}x \, dx = \rho(\tilde{v}) \sin \phi(\tilde{v}). \qquad (50)$$

The computed spectrum

$$\hat{s}(\tilde{v}) = p(\tilde{v}) + iq(\tilde{v}) = \rho(\tilde{v}) \exp i\phi(\tilde{v}) \tag{51}$$

is thus complex, with a modulus

$$\rho(\tilde{v}) = [p^2(\tilde{v}) + q^2(\tilde{v})]^{1/2} \tag{52}$$

and a phase

$$\phi(\tilde{v}) = \arctan[q(\tilde{v})/p(\tilde{v})], \tag{53}$$

and it follows from Eqs. (39), (45) and (47)–(50) that the complex insertion loss is related to the complex Fourier transform of the interferogram by

$$2\hat{\mathbf{E}}_0(\tilde{v})\hat{\mathbf{E}}_0^*(\tilde{v})\hat{L}(\tilde{v}) \exp i\phi_0(\tilde{v}) = \int_{-\infty}^{\infty} I_s(x) \exp 2\pi i\tilde{v}x \, dx. \tag{54}$$

If the specimen is now removed, the interferometer returned to the symmetrical configuration, and a background interferogram $I_0(x)$ recorded, it can easily be shown that

$$I_0(x) = \int_{-\infty}^{\infty} \rho_0(\tilde{v}) \cos[\phi_0(\tilde{v}) - 2\pi\tilde{v}x] \, d\tilde{v}, \tag{55}$$

which is similar to the interference function in Eq. (46), with

$$\rho_0(\tilde{v}) = 2\hat{\mathbf{E}}_0(\tilde{v})\hat{\mathbf{E}}_0^*(\tilde{v}). \tag{56}$$

The cosine and sine transforms of Eq. (55) can now be computed, and we obtain

$$p_0(\tilde{v}) = \int_{-\infty}^{\infty} I_0(x) \cos 2\pi\tilde{v}x \, dx = \rho_0(\tilde{v}) \cos \phi_0(\tilde{v}) \tag{57}$$

and

$$q_0(\tilde{v}) = \int_{-\infty}^{\infty} I_0(x) \sin 2\pi\tilde{v}x \, dx = \rho_0(\tilde{v}) \sin \phi_0(\tilde{v}). \tag{58}$$

The computed background spectrum of the instrument is then given by

$$\hat{s}_0(\tilde{v}) = p_0(\tilde{v}) + iq_0(\tilde{v}) = \rho_0(\tilde{v}) \exp i\phi_0(\tilde{v}) \tag{59}$$

and has a modulus

$$\rho_0(\tilde{v}) = [p_0^2(\tilde{v}) + q_0^2(\tilde{v})]^{1/2} \tag{60}$$

and a phase

$$\phi_0(\tilde{v}) = \arctan[q_0(\tilde{v})/p_0(\tilde{v})]. \tag{61}$$

It then follows from Eqs. (56)–(59) that

$$2\hat{E}_0(\tilde{v})\hat{E}_0^*(\tilde{v}) \exp i\phi_0(\tilde{v}) = \int_{-\infty}^{\infty} I_0(x) \exp 2\pi i \tilde{v} x \, dx, \qquad (62)$$

so that the complex insertion loss can be obtained from Eqs. (54) and (62) as the ratio of the two complex Fourier transforms

$$\hat{L}(\tilde{v}) = \int_{-\infty}^{\infty} I_s(x) \exp 2\pi i \tilde{v} x \, dx \Big/ \int_{-\infty}^{\infty} I_0(x) \exp 2\pi i \tilde{v} x \, dx$$

$$= FT\{I_s(x)\}/FT\{I_0(x)\}. \qquad (63)$$

If it is not convenient to determine $\hat{L}(\tilde{v})$ directly from Eq. (63) by computing with complex numbers, $L(\tilde{v})$ and $\phi_L(\tilde{v})$ can be calculated separately from the sine and cosine Fourier transforms of $I_s(x)$ and $I_0(x)$ using Eqs. (49) and (50) and Eqs. (57) and (58), respectively. We then obtain

$$L(\tilde{v}) = [p^2(\tilde{v}) + q^2(\tilde{v})]^{1/2}/[p_0^2(\tilde{v}) + q_0^2(\tilde{v})]^{1/2}, \qquad (64)$$

$$\phi_L(\tilde{v}) = \text{arc tan}[q(\tilde{v})/p(\tilde{v})] - \text{arc tan}[q_0(\tilde{v})/p_0(\tilde{v})]. \qquad (65)$$

In the computation of the phase spectra ϕ and ϕ_0, their principal values lying between $\pm \pi/2$ radians would normally be obtained from Eqs. (53) and (61). It is usual practice to extend this range to $\pm \pi$ radians by reference to the signs of $q(\tilde{v})$ and $p(\tilde{v})$ in order to reduce phase branching.

C. THE TRANSMISSION AND REFLECTION OF A PLANE PARALLEL SPECIMEN

In order to consider the relationship between $\hat{L}(\tilde{v})$ and the optical constants of the specimen for particular optical configurations, we need to develop further the discussion of the interaction of an electromagnetic wave with an isotropic medium. We have already considered in Section II.A. how the amplitude and phase of such a wave change as it propagates through the medium, and in Section II.B, how the same quantities change when the wave interacts with the interface between two dissimilar media. In this section the discussion is extended to the situation that is the most easily realizable in practice: that of a plane parallel specimen immersed in another medium, and we shall show later that many important practical configurations can be considered as limiting cases of this. The specimen is assumed to be of infinite extent normal to the propagation direction in order to remove diffraction considerations.

A useful quantity to introduce at this stage is the complex propagation factor $\hat{a}$ of the medium given by

$$\hat{a} = \exp(-k\omega x/c) \exp(in\omega x/c), \qquad (66)$$

which represents the factor by which the complex amplitude of a wave is changed on traveling a distance x within the medium. It does not include

interface effects. Thus, Eq. (9) for the plane wave disturbance within a medium
becomes

$$\hat{\mathbf{E}} = \hat{a}\hat{\mathbf{E}}_0 \exp(-i\omega t). \tag{67}$$

In broad-band spectrometry at millimeter and submillimeter wavelengths,
the absorption index k of a medium is not widely used, as the attenuation
of a wave by a medium is usually characterized by its power absorption
coefficient $\alpha = 4\pi\tilde{\nu}k$. This can be used to give the alternative form for $\hat{a}$:

$$\hat{a} = \exp(-\tfrac{1}{2}\alpha x) \exp(i2\pi n\tilde{\nu}x). \tag{68}$$

Consider, now, a plane parallel, or lamellar, specimen of refractive index
$\hat{n}_2$ immersed in a medium of refractive index $\hat{n}_1$, as shown in Fig. 4. A wave
incident on it will suffer multiple internal reflections at the two interfaces,
and the resultant electric field amplitudes transmitted and reflected by the
lamella will be given by the infinite sums of all these partial waves transmitted
through or reflected from it. Hence, the complex amplitude transmission
coefficients $\hat{t}^L_{12}$ and $\hat{r}^L_{12}$, respectively, will be given by

$$\hat{t}^L_{12} = \hat{t}_{12}\hat{a}_2\hat{t}_{21} + \hat{t}_{12}\hat{a}^3_2\hat{r}^2_{21}\hat{t}_{21} + \hat{t}_{12}\hat{a}^5_2\hat{r}^4_{21}\hat{t}_{21} + \cdots \tag{69}$$

and

$$\hat{r}^L_{12} = \hat{r}_{12} + \hat{t}_{12}\hat{a}^2_2\hat{r}_{21}\hat{t}_{21} + \hat{t}_{12}\hat{a}^4_2\hat{r}^3_{21}\hat{t}_{21} + \cdots, \tag{70}$$

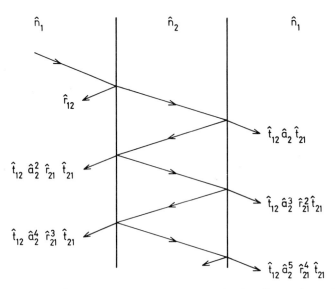

FIG. 4.   The rays reflected from and transmitted through a lamellar specimen of complex
refractive index $\hat{n}_2$ immersed in a medium of complex refractive index $\hat{n}_1$.

where $\hat{t}_{12}$, $\hat{t}_{21}$, $\hat{r}_{12}$, and $\hat{r}_{21}$ are the Fresnel transmission and reflection coefficients of the interfaces between the two media and are given by Eqs. (22) and (23), $\hat{a}_2$ the complex propagation factor for a distance equal to the thickness of the specimen, and the superscript L indicates that the specimen is lamellar. We shall now show how the configurations that are most conveniently used to determine the optical constants of solids by DFTS can be considered as limiting cases of Eqs. (69) and (70).

The first term in Eq. (69) is the single-pass transmission coefficient of the lamella, and that of Eq. (70) is its front surface reflection coefficient. For the purposes of DFTS, in which the phase sensitivity of the interferometer resolves the contributions of these waves as separate interference signatures in the interferogram, Eqs. (69) and (70) are the most useful forms of the complex transmission and reflection coefficients of a lamellar specimen. Often the interferogram will be recorded to path difference values such that only the first few of these signatures, at most, are recorded. Therefore, for DFTS it is not always particularly useful to deal with these infinite series as analytic expressions. These infinite sums do, however, have relevance in conventional FTS which lacks the phase sensitivity of DFTS, and when the sums are performed they lead to the other forms of Eqs. (69) and (70):

$$\hat{t}_{12}^{L} = \hat{a}_2(1 - r_{12}^2)/(1 - \hat{a}_2^2 \hat{r}_{12}^2) \tag{71}$$

and

$$\hat{r}_{12}^{L} = \hat{r}_{12}(1 - \hat{a}_2^2)/(1 - \hat{a}_2^2 \hat{r}_{12}^2), \tag{72}$$

which, by use of the exponential forms of $\hat{r}_{12}$ and $\hat{a}_2$, can be expanded into the forms derived by Bell (1967a) for the total power transmission and reflection coefficients of a lamellar specimen. Subsequent manipulation of these expressions leads to an understanding of the role played by the channel spectrum: how it may be used to give approximate values for the optical constants, and how the observed transmission and reflection are modified when the spectral resolution is insufficient to resolve the channel spectrum. These, however, are topics that are not particularly relevant to a discussion of DFTS. In deriving Eqs. (22), (23), and (68)–(70) we have, however, obtained the equations that enable us to determine the optical constants from the quantity measured in a DFTS experiment. The particular form taken by that measured quantity and the various numerical procedures necessary to extract the optical constants from it are dealt with in the following sections.

D.   Calculation of the Optical Constants from the Complex Insertion Loss

In this section the development of some of the equations that relate the measured complex insertion loss to the optical constants is outlined. As there are a large number of specimen configurations that can be used, especially for liquids, the section is not exhaustive in its coverage and is intended to be illustrative of the more generally met features of interferogram analysis. Although no assumptions are made as to the nature of the specimen material, the equations that are described are only really applicable to solid, self-supporting specimens. Analyses that are more specific to gases and liquids are found in the sections dealing with those materials and in the literature.

1. The Amplitude Reflectivity of an Opaque Solid

When the absorption coefficient of the specimen is so large that a trans-mission experiment is out of the question, the optical constants can be deter-mined by measuring the front surface complex amplitude reflection coefficient, and this is usually done with a thick specimen using the configuration illustrated in Fig. 2c. The principle of the method is that a background inter-ferogram $I_0(x)$ is first recorded with a symmetrical interferometer, and a specimen interferogram $I_s(x)$ is recorded after replacing the fixed mirror with the specimen in such a way that the two reflecting surfaces occupy the same plane. The amplitude reflectivity is described by Eq. (70), but, as the specimen is strongly absorbing, only the first term will contribute a measurable signal. The measurement is usually made with an evacuated interferometer so that $\hat{n}_1(v) = 1$, and, if we use the exponential notation of Section II.B.2, the amplitude reflectivity of the reference mirror can be written as

$$\hat{r}_m(\tilde{v}) = \exp i\pi \tag{73}$$

and that of the specimen as

$$\hat{r}(\tilde{v}) = r(\tilde{v}) \exp i\phi_r(\tilde{v}). \tag{74}$$

As the phase delay of the specimen, $\pi \leq \phi_r(\tilde{v}) \leq 2\pi$, the grand maxima of the two interferograms $I_s(x)$ and $I_0(x)$ are observed at very nearly the same position of the moving mirror, and both interferograms can be conveniently recorded from the same starting point which is then used as a reference point for calculating the phase spectrum. No further information is then required, and it follows from Eqs. (73) and (74) that the complex insertion loss is given by

$$\hat{L}(\tilde{v}) = r(\tilde{v}) \exp i[\phi_r(\tilde{v}) - \pi], \tag{75}$$

which can be related to the complex refractive index $\hat{n}_2(\tilde{v}) = \hat{n}(\tilde{v})$ of the specimen by inverting Eq. (22) to give

$$\hat{n}(\tilde{v}) = [1 - \hat{r}(\tilde{v})]/[1 + \hat{r}(\tilde{v})]. \tag{76}$$

2. *Transmission Measurements on Free-Standing Lamellar Specimens*

The basic features of the two types of instrument used for dispersive transmission measurements are illustrated schematically in Fig. 2. Depending on the value of the absorption coefficient, either a straightforward double-pass measurement with a conventional Michelson interferometer may be suitable, as shown in Fig. 2a, or the additional complexity of a single-pass instrument, such as that shown in Fig. 2b, may be required. In the following sections we shall show how the optical constants can be related to the complex insertion loss for a number of simple optical configurations which can be described as limiting cases of Eq. (69), and this can best be done by considering first a single-pass measurement on a thick transparent specimen.

a. *Single-Pass Measurement on a Thick Transparent Specimen.* Before trying to relate the amplitude transmission coefficient $\hat{t}_{12}^{L}$ [Eq. (69)] to the complex insertion loss $\hat{L}(\tilde{v})$, it will be convenient to describe the experimental procedure in some detail, since the calculation of the refractive index spectrum $n(\tilde{v})$ of the specimen is considerably simplified if the interferograms are recorded in a suitable systematic manner (Parker *et al.*, 1978a). Let the background interferogram $I_0(x)$ be recorded with wings of equal optical path length x, symmetrically about the position of zero optical path difference, as shown in Fig. 5. The value of x is chosen in the normal way to be sufficiently large to resolve all features in the specimen spectrum, and a typical set of points at which the interferogram might be sampled experimentally is also illustrated in the figure. As a standard procedure, the sampling point nearest the zero crossing at the center of the interferogram, say the nth point on the interferogram, is located in order to balance and apodize the interferogram. In dispersive work it is also useful to designate this point as the origin of computation x_0 of the phase spectrum. Let the specimen now be inserted in the fixed arm of the interferometer, and, for the purposes of illustration, let us suppose that the dispersion in $n(\tilde{v})$ is small and that the mean value of $n(\tilde{v})$ in the spectral range of interest is $\bar{n}$. It follows from Eqs. (37) and (38) that the phase changes occurring on transmission of the partial beams described by Eq. (69) through the surfaces of the specimen will be negligible compared with those associated with the complex propagation factor $\hat{a}(\tilde{v})$ of the specimen [Eq. (66)]. If the specimen is sufficiently thick, it also follows that the signatures associated with these successive partial waves will be well separated on the specimen interferogram, and will be displaced in optical path length by amounts of approximately $(\bar{n} - 1)d$, $(3\bar{n} - 1)d$, $(5\bar{n} - 1)d$, etc., where d is the specimen thickness, from the center of the background interferogram $I_0(x)$. This, together with a typical set of experimental sampling points, is also illustrated in Fig. 5, and, under

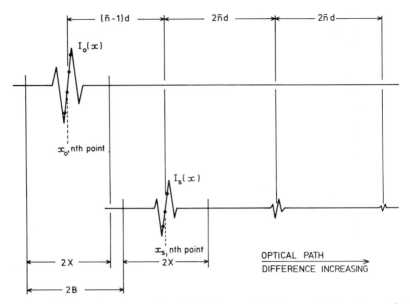

FIG. 5. A schematic representation of a background interferogram $I_0(x)$ and a specimen interferogram recorded in a single-pass transmission DFTS experiment using phase modulation. The bright interference fringe of the specimen interferogram is shifted by $(\bar{n} - 1)d$ in path difference from that of the background interferogram, where $\bar{n}$ is the mean value of the refractive index of the specimen over the measured spectral range and d its thickness. The first two interference signatures due to the rays internally reflected within the specimen are also shown.

these circumstances, the optical constants can be completely determined from a record of the first signature only. To do this, a specimen interferogram $I_s(x)$, also of optical path length $2x$, is recorded, beginning at a point shifted by an optical path length $2B$ from the initial point on $I_0(x)$, and the nth point designated the origin x_s for calculating the specimen phase spectrum. Note that B, defined in this way, corresponds to the shift on the moving mirror micrometer scale of the starting point of $I_s(x)$ from that of $I_0(x)$. It is clear that if we choose

$$2B \simeq (\bar{n} - 1)d, \tag{77}$$

then the point x_s and the zero crossing in the first signature of the specimen interferogram will nearly coincide, and leveling and apodization of $I_s(x)$ can be carried out about the point x_s.

If the experiment is performed in this way, so that only the main specimen interference signature $I_s(x)$ is recorded, the relevant complex transmission coefficient of the specimen is fully described by the first term in Eq. (69), which can be reduced, with the aid of Eqs. (23), (25), and (66), to the form

$$\hat{t}(\tilde{\nu}) = t(\tilde{\nu}) \exp i\phi(\tilde{\nu}) = \{4\hat{n}(\tilde{\nu})/[1 + \hat{n}(\tilde{\nu})]^2\}\exp i2\pi\tilde{\nu}\hat{n}(\tilde{\nu})d, \tag{78}$$

where we have again put $\hat{n}_1(\tilde{v}) = 1$ and $\hat{n}_2(\tilde{v}) = \hat{n}(\tilde{v})$, making the subscript 2 implicit. Similarly, t and ϕ are now meant to refer implicitly to the entire specimen. The complex ratio

$$\hat{L}'(\tilde{v}) = L'(\tilde{v}) \exp i\phi'_L(\tilde{v}) = FT\{I_s(x)\}/FT\{I_0(x)\} \tag{79}$$

determined from the complex transforms of the two interferograms cannot be equated to the complex insertion loss for this measurement or related to the transmission coefficient of Eq. (78) because the large phase shift $4\pi\tilde{v}B$ associated with the displacement of the starting points of the two interferograms has not yet been included. If this is done, the complex insertion loss is given by

$$\hat{L}(\tilde{v}) = L'(\tilde{v}) \exp i(\phi'_L(\tilde{v}) + 4\pi\tilde{v}B) \tag{80}$$

in terms of the experimentally determined quantities. It is also related to the optical constants of the specimen through the complex transmission coefficient of the specimen by

$$\begin{aligned}\hat{L}(\tilde{v}) &= \hat{t}(\tilde{v}) \exp(-i2\pi\tilde{v}d) \\ &= \{4\hat{n}(\tilde{v})/[1 + \hat{n}(\tilde{v})]^2\} \exp i2\pi\tilde{v}[\hat{n}(\tilde{v}) - 1]d, \end{aligned} \tag{81}$$

leading to the relationship between the experimentally determined quantities L', ϕ'_L, and B and the optical constants. The factor $\exp(-i2\pi\tilde{v}d)$ in Eq. (81) arises because the specimen replaces a length d of vacuum in the arm of the interferometer.

In Eq. (80), ϕ'_L is the principal value of the phase difference between the two complex spectra. For a highly dispersive specimen it may be necessary to replace it by

$$\phi'_L \pm 2m\pi, \tag{82}$$

where $m = 0, 1, 2$, etc., the additional term being necessary to ensure continuity in the true value of the phase difference when its computed value changes branches. Generally, additional information is required to assign a value of m to a particular branch. In the simple experiment discussed here, where we have specifically required that the dispersion in $n(\tilde{v})$ is small, this can be done by recognizing that $\phi'_L(\tilde{v})$ approaches zero at low frequencies.

Since the experiment can only be performed as described if $k(\tilde{v}) \ll n(\tilde{v})$ in Eq. (8), it follows that Eqs. (80) and (81) can be simplified to give

$$n(\tilde{v}) = 1 + \frac{1}{2\pi\tilde{v}d} [\phi'_L(\tilde{v}) + 4\pi\tilde{v}B \pm 2m\pi], \tag{83}$$

$$k(\tilde{v}) = \frac{1}{2\pi\tilde{v}d} \ln\left[\frac{4n(\tilde{v})}{[1 + n(\tilde{v})]^2} \frac{1}{L(\tilde{v})}\right], \tag{84}$$

so that $n(\tilde{v})$ and $k(\tilde{v})$ can be calculated separately in a straightforward way.

b. *Single-Pass Measurements on a Thin Absorbing Specimen.* As we have already pointed out in Section III.D.1, if a transmission measurement is out of the question, the optical constants of the specimen must be determined by measuring its front surface amplitude reflection coefficient. However, if it is feasible for a transmission measurement to be performed on a carefully thinned specimen, then a single-pass measurement is preferable to an amplitude reflection measurement because it follows from Eq. (31) that the phase change $\phi_r(\tilde{v})$ occurring on reflection will be close to π, the value for the reference mirror, as $k(\tilde{v})$ must still be comparatively small. Thus, $\phi_r(\tilde{v}) - \pi$ [Eq. (75)], which is the measured quantity, will be very small and difficult to measure accurately. In the case of a single-pass transmission measurement, the measured phase

$$\phi_L(\tilde{v}) \simeq 2\pi\tilde{v}[n(\tilde{v}) - 1]d \tag{85}$$

will be much larger and more easily measured. The interferograms can be recorded as described in Section III.D.2.a, bearing in mind that $2B$ will now be much smaller and that it may be necessary to increase significantly the length of the interferograms to resolve fully all features in the specimen spectrum. If the specimen is very thin, the signatures associated with all the transmitted partial waves described by Eq. (69) will overlap on the interferogram and the full geometric series must be included in the analysis. If we use the notation of Eqs. (79) and (80), the complex insertion loss can be expressed, with the aid of Eqs. (22)–(25), as

$$\hat{L}(\tilde{v}) = \frac{4\hat{n}}{(1 + \hat{n})^2} \frac{\exp 2\pi i\tilde{v}(\hat{n} - 1)d}{1 - [(1 - \hat{n})/(1 + \hat{n})]^2 \exp 4\pi i\tilde{v}\hat{n}d}, \tag{86}$$

with $n(\tilde{v})$ given by Eq. (8), and the complex insertion loss is again related to $\hat{t}(\tilde{v})$ by Eqs. (79)–(82).

As such measurements are usually necessary only in the neighborhood of intense absorption bands where the dispersion in $n(\tilde{v})$ is high, many branches of $\phi'_L(\tilde{v})$ may occur in Eq. (82), and it may not be possible to trace the phase spectrum continuously to zero frequency, especially if there are absorption bands at frequencies below the measured spectral range. Additional information may then be required to assign a value of m to a particular branch, and this may be obtained either from measurements on a specimen of a different thickness or from supplementary amplitude reflection measurements.

Equation (86) cannot be solved explicitly for $\hat{n}(\tilde{v})$ in terms of $\hat{L}(\tilde{v})$, but it can be used to calculate $\hat{n}(\tilde{v})$ with the desired accuracy by following an iterative procedure such as a complex secant method (Conte and de Boor, 1972). Before describing this method fully, it will be convenient to modify the notation slightly. Thus we shall refer to the experimental value of the complex insertion loss determined from Eqs. (79)–(82) as $\hat{L}_E(\tilde{v})$, and to the theoretical value, which is obtained by substituting an appropriate value of $\hat{n}(\tilde{v})$ into Eq. (86), as $\hat{L}_T(\tilde{v})$. Although for calculations at a particular frequency $\tilde{v}$ both variables occurring in Eq. (86) are two-dimensional so that the equation cannot be fully represented by a two-dimensional diagram, the principle of the secant method can be understood by referring to the diagram shown in Fig. 6 in which the curve represents schematically the relationship between L_T and $\hat{n}$ defined by Eq. (86). The significance of the quantities labeled in the diagram, which are related by the equation

$$\hat{n} = \hat{n}'' + (\hat{L}_T'' - \hat{L}_E)(\hat{n}' - \hat{n}'')/(\hat{L}_T'' - \hat{L}_T') \tag{87}$$

and the steps in the iterative procedure, are as follows. A convenient starting point at a frequency $\tilde{v}$ on the measured transmission spectrum is first chosen, and the value of $\hat{L}_E$ at this point is determined from Eqs. (79)–(82). The first trial value $\hat{n}'$ for the complex refractive index is either estimated roughly from Eq. (86) or obtained from independent information, such as a supplementary amplitude reflection measurement. A second trial value

$$\hat{n}'' = \hat{n}' + \delta\hat{n} \tag{88}$$

is obtained by adding a small increment

$$\delta\hat{n} = \delta n + i\delta k \tag{89}$$

to $\hat{n}'$. Corresponding values $\hat{L}_T'$ and $\hat{L}_T''$ of the theoretically determined complex transmission coefficient are calculated by substituting $\hat{n}'$ and $\hat{n}''$,

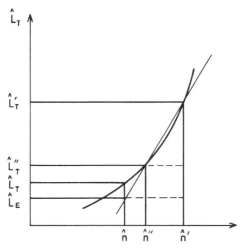

FIG. 6. Illustrating the secant method by which the complex refractive index $\hat{n}$ of a specimen is found from its measured complex insertion loss $\hat{L}$ via an iterative technique based on Eq. (86).

respectively, into Eq. (86), and a third value $\hat{n}$ of the complex refractive index is then calculated by substituting these values in Eq. (87). The value $\hat{n}$ is then used to determine an improved value $\hat{L}_T$ of the calculated transmission coefficient from Eq. (86), which can then be compared with the experimental value $\hat{L}_E$ to obtain

$$\hat{L}_T - \hat{L}_E = \delta\hat{L}. \tag{90}$$

The process is repeated, resetting $\hat{n}' = \hat{n}$ as the new starting value for the refractive index each time, until a value for $\hat{L}_T$ is found which agrees, to within the desired accuracy

$$\delta\hat{L} < \delta\hat{L}_0 \tag{91}$$

with $\hat{L}_E$. The final value for $\hat{n}(\tilde{v})$ at this frequency is then used as the first trial value $\hat{n}'(\tilde{v} + \delta\tilde{v})$ at the adjacent spectral point $\tilde{v} + \delta\tilde{v}$, and so on.

c. *Double-Pass Transmission Measurements.* In many cases, satisfactory results may be obtained by inserting the specimen in the fixed arm of a conventional Michelson interferometer so that the radiation passes through it twice, as shown in Fig. 2a. The complex double-pass insertion loss $\hat{L}_2(\tilde{v})$, which is then measured, is simply the square of the complex, single-pass, insertion loss

$$L_2(\tilde{v}) \exp i\phi_2(\tilde{v}) = [L(\tilde{v}) \exp i\phi_L(\tilde{v})]^2 \tag{92}$$

The interferograms can be recorded in the same way as before, so that the complex insertion loss is still given by the ratio of the complex Fourier transforms of the interferograms. Depending on the optical properties of the specimen, either Eqs. (83) and (84) or Eq. (86) can be used to determine $n(v)$ and $k(v)$, and this is best done by first reducing Eq. (92) to the form

$$\phi_L(\tilde{v}) = \tfrac{1}{2}\phi_2(\tilde{v}) \tag{93}$$

and

$$L(\tilde{v}) = [L_2(\tilde{v})]^{1/2}. \tag{94}$$

E. THE ORIGIN OF COMPUTATION FOR THE COMPLEX TRANSFORMATION

In order to carry out the numerical evaluation of the complex Fourier transformation of the interferogram it is common practice for it to be sampled at a finite number of equal intervals of path difference. Thus the integral equations developed in the previous section for infinite, continuous data will not be valid in general, and the limitations introduced by their application to real data, such as finite resolution and spectrum aliasing, are well understood and covered in the standard works on FTS. In this section we shall consider one aspect of the use of real data that is of particular relevance to the computation of phase spectra, the fact that, in general, the position of zero optical path difference of an interferogram does not coincide with any of the points generating the sampling comb of that interferogram. There is some confusion in the literature over the consequences of this and over the correct procedure to follow in the evaluation of phase spectra in DFTS, largely due to the unnecessary phase correction procedure of Chamberlain *et al.* (1969b). The correct procedure to follow in this situation has been outlined by Birch and Bulleid (1977) and recently considered in a more general manner by Birch and Parker (1979). In the following the problem is discussed along the lines of the latter work.

We have discussed earlier how the object of a DFTS experiment is the determination of the complex insertion loss of a specimen. It is apparent, from its definition, that this quantity is the ratio of the complex spectra obtained by the complex Fourier transformation of the specimen and reference interferograms using the same position of geometric path difference as computational origin. This position can be the position of zero path difference but need not necessarily be so. Thus in an ideal experiment the complex insertion loss would be determined from

$$\hat{L}(\tilde{v}) = \hat{S}_x^S(\tilde{v})/\hat{S}_x^R(\tilde{v}),\tag{95}$$

where the superscripts S and R to the complex spectrum $\hat{S}(\tilde{v})$ refer to the specimen and reference measurements, respectively, and the subscript x indicates the position of path difference used as the origin of computation for both complex spectra. The phase of the complex insertion loss would therefore be obtained from

$$\phi_L(\tilde{v}) = \text{ph}\{\hat{S}_x^S(\tilde{v})\} - \text{ph}\{\hat{S}_x^R(\tilde{v})\}.\tag{96}$$

In a real experiment the phase delay caused by the specimen causes the specimen interferogram to shift to positive path differences, and severe phase branching could result if the same origin of computation were used for both specimen and reference interferograms. It is generally more convenient to use a sampling point x_s near the center of the displaced bright fringe as the origin for the specimen interferogram. Similarly, in order to reduce branching in the phase of the reference spectrum, one assumes that the point x used as its origin is close to the center of the reference interferogram. Thus, in terms of the experimentally observed quantities, the Fourier transform shift theorem shows that the phase of the complex insertion loss is also given by

$$\phi_L(\tilde{v}) = \text{ph}\{\hat{S}_{x_s}^S(\tilde{v})\} - \text{ph}\{\hat{S}_x^R(\tilde{v})\} + 2\pi\tilde{v}(x_s - x)\tag{97}$$

in addition to Eq. (96). The third term of the right-hand side of Eq. (97) is called the *fringe shift* term and is readily determined from the recorded sampling combs of the two interferograms since the difference $(x_s - x)$ is a whole number of sampling intervals. This is a perfectly general result for determining the phase of the complex insertion loss of any specimen in a reflection or transmission experiment, and although we have spoken of the points x and x_s as being close to the bright fringes to reduce phase branching, this is not really necessary if the spectra extend to low wave numbers, as any branching can then be corrected by inspection.

In the analysis of Chamberlain et al. (1969b) the precise separation $\bar{x}$ of the centers of the specimen and reference interferograms is determined and used to calculate a phase spectrum

$$\phi'(\tilde{v}) = \mathrm{ph}\{\hat{S}^S_{x_s}(\tilde{v})\} - \mathrm{ph}\{\hat{S}^R_x(\tilde{v})\} + 2\pi\tilde{v}\bar{x}, \qquad (98)$$

where x_s and x are now taken, explicitly, as being the sampling points nearest to the center of each interferogram. Unless these points coincide with the centers of their respective interferograms, $\phi'(\tilde{v})$ will not be equal to the true phase of the complex insertion loss. These authors then proceed to consider the case of a double-pass transmission experiment and show how the erroneous refractive index spectrum calculated from their phase spectrum $\phi'(\tilde{v})$ may be corrected by the addition of a term obtained by estimating values of the small shifts of the computational origins of the two interferograms from their respective centers. This is an unnecessary and time-consuming process as

(i) the initial step of their procedure, the calculation of the phase spectrum $\phi'(\tilde{v})$, introduces the error that the remainder of the procedure removes, and

(ii) neither $\bar{x}$ nor the small shifts of the computational origins from the interferogram centers are whole numbers of sampling intervals. Thus, their derivation by interpolation in the interferograms will introduce systematic errors. Moreover, Chamberlain et al. (1969b) suggest finding $\bar{x}$ from a second experiment using more finely sampled interferograms. Thus, additional measurements are required which, due to instrumental instabilities, will be likely to introduce additional systematic errors.

These problems are all avoided if the interferograms are recorded in the systematic manner described in Section III.D.2 and the phase spectrum determined from Eq. (97).

IV.   Dispersive Fourier Transform Spectrometry of Gases and Vapors

Gases and vapors are, in principle, the most convenient materials to study in a DFTS experiment, even though their necessary containment in a gas cell introduces the complication of windows. The low refractive indices that they exhibit under normal experimental conditions ($\sim 1 + i0$) means that interface effects may be completely ignored, with the result that the equations relating the complex insertion loss to the complex refractive index are very simple and may be solved directly for the latter. The low absorption coefficient also means that very long path lengths may be used if required, so that large phase shifts can be produced leading to high measurement precision, even for these low refractive indices. In the strictest sense, many of the early DFTS measurements were not precise as they only measured the dispersion across an absorption line or band without ascribing a true level to it, but, as we shall see, the dispersion is an important parameter that may be used to give values for the line strength under conditions in which power transmission spectrometry cannot.

The first DFTS measurement of the refraction spectrum of a gaseous specimen was that of Chamberlain and Gebbie (1965b) on the hydrogen halides HCl, HBr, and HI between 12 and 60 cm^{-1}. They then proceeded with various coworkers to measure NH$_3$ (Chamberlain et al., 1969a) and GeCl$_4$ (Chantry et al., 1969a) at submillimeter wavelengths, while also developing a dispersive instrument for the near infrared which was used to study the fundamental vibrational band of HCl between 2500 and 3300 cm^{-1} (Chamberlain et al., 1965b). During the same period Sanderson (1967) modified an interferometer developed by Russell and Bell (1966) for the dispersive study of solids and used it to measure the refractive index spectrum of HCl between 10 and 280 cm^{-1}, and of HBr (Robinette and Sanderson, 1969). Subsequently Sanderson and Scott (1970, 1971) developed an instrument specifically designed for dispersive and nondispersive measurements on gases and vapors and studied a part of the pure rotation band of CO (Sanderson et al., 1971). More recently, Birch (1978a) has described a high-resolution interferometer that has been used for precise measurements of the refraction spectra of N$_2$, O$_2$, CO$_2$, and NH$_3$ (Birch and Bulleid, 1976, 1977) and of water, methanol, and ethanol vapors (Birch and Afsar, 1977a,b; Kemp et al., 1978).

Precise refractive index measurements on gases and vapors did not, of course, commence with the development of DFTS. They have been performed for a long time in the more accessible short-wavelength region of the spectrum, and the early refractometers, such as the Jamin, Mach–Zehnder, and Rayleigh, were all two-beam interferometers, although not Fourier transform devices. In common with multiple-beam interferometers like the Fabry–Perot, they were used with quasi-monochromatic radiation with the phase shift due to the specimen found by fringe counting as the pressure or wavelength was scanned or by an equivalent technique such as compensation of the change in path difference. Similarly, more recent instruments developed for very high-precision work which employ Michelson interferometers to determine the phase shift (see, for example, Legay, 1958; Kerl, 1974; Peck and Huang, 1977) are also operated with monochromatic radiation and are not Fourier transform instruments. In fact, one of the interferometers which shall be described later in this section has also been used to give single-frequency measurements of the refractive indices of gases (Chamberlain *et al.*, 1965a), of liquids (Gebbie *et al.*, 1965; Chamberlain *et al.*, 1967b) and of solids (Chamberlain and Gebbie, 1965a), but these were not Fourier transform determinations.

A. The Derivation of the Complex Refractive Index

The ease with which the complex refractive index of a material may be derived from dispersive Fourier transform measurements is dependent on the interactions that occur at the interfaces of the specimen, and for a gaseous specimen, this will be the interface with the containing window or windows. However, gases and vapors have the particular advantage of such low complex refractive indices that these interface effects may normally be completely ignored. There are two configurations that have been used for the containment of a gaseous specimen in one arm of an interferometer, and these are represented schematically in Fig. 7. In the upper case, used by Chamberlain *et al.* (1969a) and Birch (1978a), the gas is contained between a single window and the fixed mirror of the interferometer, while in the lower case, used by Sanderson (1967) and Sanderson and Scott (1970, 1971), it is contained in a gas cell between two windows. Each method has its own advantages: the single-window method has the lowest absorption losses, hence giving better quality spectra, while the separation of the fixed mirror mount from the gas in the two-window cell means that for high-pressure studies this method of containment is less susceptible to errors due to pressure-induced motion of this mirror. These latter errors can, however, be removed by good experimental design and, on balance, the single-window system is probably preferable. It will usually be the case that the gas cell is

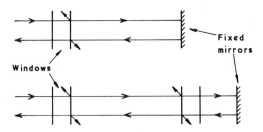

FIG. 7. The two gas cell configurations used in DFTS. In the upper, the gas is contained between a transparent window and the fixed mirror of the interferometer and in the lower between two transparent windows placed in the fixed mirror arm of the interferometer.

sufficiently long for the interference signatures from any rays reflected between the windows to fall outside the recorded range of path difference values. If this is so, and denoting the window material by the letter w, the gas by 1, and vacuum by 0, it may be shown that the complex insertion losses measured when the reference medium is vacuum are

$$\hat{L}_1(\tilde{v}) = \{(1 - \hat{r}_{w1}^2)/(1 - \hat{r}_{w0}^2)\} \exp i4\pi\tilde{v}(\hat{n}_1 - 1)d \qquad (99)$$

for the single-window system and

$$\hat{L}_2(\tilde{v}) = \{(1 - \hat{r}_{w1}^2)/(1 - \hat{r}_{w0}^2)\}^2 \exp i4\pi\tilde{v}(\hat{n}_1 - 1)d \qquad (100)$$

for the two-window system, where d is the length of the gas cell. In deriving these it is assumed that the spectral resolution in the measurement is inadequate to resolve the channel spectrum due to the windows themselves or, if it is resolved, for the system to be sufficiently stable and reproducible for it to ratio out completely in a measurement. In practice, of course, the refractive indices of most gaseous specimens are such that

$$\hat{r}_{w0} = \hat{r}_{w1} \qquad (101)$$

is a very good approximation and Eqs. (99) and (100) reduce to

$$\hat{L}_1(\tilde{v}) = \hat{L}_2(\tilde{v}) = \exp i4\pi\tilde{v}(\hat{n}_1 - 1)d. \qquad (102)$$

Thus the optical constants may be found from the measured quantity by inverting Eq. (102) to give

$$n(\tilde{v}) = 1.0 + [\mathrm{ph}\{\hat{L}(\tilde{v})\}/(4\pi\tilde{v}d)] \qquad (103)$$

and

$$k(\tilde{v}) = [1/(4\pi\tilde{v}d)] \ln(1/|\hat{L}(\tilde{v})|), \qquad (104)$$

where the subscripts have been dropped, but are implicit, and ph{ } is taken to mean the phase of the complex quantity within the brackets. If the optics of the interferometer are arranged so that the radiation in the fixed mirror arm only passes through the specimen once, as in Sanderson's original measurements (Sanderson, 1967), the factor 4 in Eqs. (103) and (104) is reduced to 2.

In a real measurement the increase in optical path difference that occurs on admission of the specimen causes the bright interference signature to be displaced to positive path differences as discussed in Section III. Under these conditions it is usual to shift the origin of computation used for the specimen interferogram by an amount $\overline{X}$ in path difference to coincide with the displaced bright signature in order to avoid excessive phase branching. The real refractive index would then be calculated from

$$n(\tilde{v}) = 1.0 + [\phi/(4\pi\tilde{v}d)] + [\overline{X}/(2d)] \qquad (105)$$

instead of Eq. (103), where ϕ represents the difference between the phase spectra computed from the reference and specimen interferograms using the different origins.

1. Possible Errors in the Refractive Index

The ability of a dispersive transmission measurement to determine the real refractive index of a gaseous specimen should, in a well-designed experiment, depend on the precisions with which the factors ϕ and $\bar{X}$ in Eq. (105) can be determined, and, ideally, the limiting factor would be random noise in ϕ, the phase difference spectrum. The gas cell should be sufficiently long that errors in the measurement of its length and departures from parallelism of the containing interfaces are negligible. Other effects, such as inadequate intensity or path difference sampling and the presence of convergent radiation (Chamberlain, 1967a; Russell and Bell, 1967a; Sanderson, 1967) are assumed to have been minimized by good experimental procedure.

The limiting effects of random phase noise may be considered with reference to the random phase noise data of Birch (1978a) and Birch and Murray (1978) for interferometers that are typical of those presently used in dispersive measurements. They found that below 50 cm^{-1}, where liquid-helium-cooled detectors would be used, it was possible to work at high resolution (~ 0.17 cm^{-1}) with an rms phase noise of about 5 mrad. This corresponds to an error in $(n - 1)$ of about 1.5×10^{-7} for a 900-mm cell at 30 cm^{-1}. At higher wave numbers, where room temperature detectors were used, a similar level of random phase noise was found for spectral resolutions of about 1 cm^{-1}.

As the effect of the shift term $\bar{X}$ is to give the overall level of the refraction spectrum, there are some applications in the measurement of dispersive spectra where errors in its determination are unimportant. If this is not so, however, one should consider two main sources of systematic error. First, $\bar{X}$ may be incorrectly estimated, although a procedure that avoids this possibility has been discussed (Birch and Bulleid, 1977; Birch and Parker, 1979). Second, backlash in the drive system of the moving mirror will cause an unknown displacement between the sampling combs of the reference and specimen interferograms, which leads to an error in $\bar{X}$. The magnitude of this depends on the type of drive system used, and for optically monitored systems, it can be totally eliminated. If, however, there is no monitoring, then for the micrometer-type drives commonly used at submillimeter wavelengths a backlash error of 0.5 μm in path difference is not uncommon. This corresponds to a systematic error in $(n - 1)$ of between 10^{-6} and 2×10^{-7} for gas cells between 0.2 and 0.9 m long.

Thus, for present state-of-the-art measurements, it should be possible to determine the real refractive index spectra of gases and vapors with a random error due to noise in the phase difference spectrum between 10^{-6} and 10^{-7}, the exact figure depending on the resolution required, the spectral region, and the length of the gas cell. This will be combined with a possible systematic error of up to 10^{-6}, again depending on the length of the gas cell. There will be, of course, other factors, such as the thermal and mechanical stability of the interferometer and line profile distortions due to the instrumental line shape, that might predominate under certain experimental conditions, but, ideally, the random phase noise should be the limiting factor. In this discussion of errors we have dealt with errors in the absolute value of the refractive index of between 10^{-6} and 10^{-7}. It should be pointed out that n is close to unity and, as $(n-1)$ is the measured quantity, typically taking values of about 10^{-3} or less, the real measurement precision is a few parts in 10^3 or 10^4.

This entire discussion is only justified if the pressure measurement is sufficiently precise. If the pressure measurement system is not accurately calibrated or is poorly used, the systematic error in the pressure measurement will dominate all other errors and limit the accuracy of the refractive index measurement. For certain gases or vapors, particularly those with small molecular dipole moments, the same will apply to the temperature measurement.

B.   The Determination of Absorption Line Strengths

Although in some applications, such as radiation propagation through a gaseous atmosphere, knowledge of the refractive index is, per se, sufficient, the greatest use made of dispersively measured refraction data has been in the determination of absorption line strengths. These may be derived from a transmission spectrum by calculating the integrated fractional absorption across the line (Benedict et al., 1956a, b; Fleming, 1976), but the methods used require line shape assumptions to be made. The refraction method due to Chamberlain (1965, 1967b) is a more general method that does not require prior knowledge of the line shape. It results from a consideration of the Kramers–Krönig integral relating the refractive index at a particular wave number $\tilde{v}_0$ to the power absorption coefficient measured over the entire spectrum:

$$n(\tilde{v}_0) - n(\infty) = [1/(2\pi^2)] \int_0^\infty [\alpha(\tilde{v})/(\tilde{v}^2 - \tilde{v}_0^2)]\, d\tilde{v}. \tag{106}$$

If this is applied to two wave numbers $\tilde{v}_1$ and $\tilde{v}_2$ on either side of an isolated absorption line, the refractive index difference between these wave numbers may be expressed as

$$n(\tilde{v}_1) - n(\tilde{v}_2) = \frac{1}{2\pi^2} \int_0^\infty \left[\frac{1}{\tilde{v}^2 - \tilde{v}_1^2} + \frac{1}{\tilde{v}_2^2 - \tilde{v}^2}\right]\alpha(\tilde{v})\, d\tilde{v}. \tag{107}$$

Chamberlain (1965) showed that when the integration in Eq. (107) is restricted to the spectral region near the feature where $\alpha(\tilde{v})$ the absorption coefficient is significantly different from zero, Eq. (107) may be written in the form

$$n_i(\tilde{v}_1) - n_i(\tilde{v}_2) = \frac{A_i}{2\pi^2}\left[\frac{1}{\tilde{v}_i^2 - \tilde{v}_1^2} + \frac{1}{\tilde{v}_2^2 - \tilde{v}_i^2}\right], \tag{108}$$

where the subscript i has been introduced to indicate that the result only describes the dispersion due to the ith line if more than one are present and A_i and $\tilde{v}_i$ are the line strength and center wave number, respectively. This only applies for wave numbers situated greater than the half-width away from the line center. Thus the original integral, Eq. (106), which would have allowed the refractive index across an absorption feature to be calculated from its measured absorption spectrum, has been effectively inverted so that the line strength may be found from the measured refractive index difference across the line. In practice, A_i is not determined from a single pair of refractive index values. Measurements are made of the refraction spectrum and pairs of points $\tilde{v}_1$ and $\tilde{v}_2$, usually equispaced about the line center, are selected. The refractive index difference for each pair is found and the value of the function in brackets on the right-hand side of Eq. (108) computed. If this equation is valid, the two quantities will be linearly related and a least-squares linear fit to the data set yields the best value of A_i.

There are two particular advantages to this method. First, the line strength is derived from the refractive index difference across the absorption line, and this does not require knowledge of the absolute level of the refraction spectrum merely its variation. This considerably relaxes the measurement precautions that have to be taken. Second, as Eq. (108) was derived without any line shape assumptions it is not necessary for the line to have been fully resolved. Thus the line strength may be found from the low-resolution refraction spectrum in the wings of an unresolved line or from measurements in the wings of a line whose center has been blacked out.

The refractive index difference described by Eq. (108) is only applicable to a single isolated feature. In real spectra, especially those of gases and vapors, the measured refractive index difference across a single line will contain contributions from adjacent lines and from any strong, shorter-wavelength dispersive mechanisms. In these circumstances the application of Eq. (108) previously described fails as the two measured quantities only become linearly related very near to the line center, the contributions from other lines being approximately constant over a sufficiently narrow spectral interval. If the strengths of the additional features are known, it is a simple matter to compute their contributions to the dispersion in the region of the unknown line and to correct the measured refractive index difference accordingly so that Eq. (108) may be used. If, however, the strengths are unknown, Chamberlain (1967b) has described an iterative procedure based on Eq. (108) that can be used to find the true line strengths of all the lines in the measured spectrum. The results of an application of this method to the pure rotation line of gaseous ammonia at 39.78 cm^{-1} are shown in Fig. 8, taken from Chamberlain *et al.* (1969a). This shows the refractive index difference plotted against f_J, the function in brackets in Eq. (108). The open circles are the results of applying Eq. (108) directly to the measured data, without allowing for the dispersion associated with the adjacent pure rotation transitions, which are approximately equispaced at 20 cm^{-1} intervals. They provide a poor fit to a straight line except for large values of f_J, which correspond to wave number pairs close to the line center. When allowance is made for the contributions of the adjacent transitions, the appearance of the data improves, and the solid circles were obtained after nine iterations of Chamberlain's procedure. These provide a much improved fit to a straight line and yield the value 3.72 cm^{-2} atm^{-1} for the integrated line strength.

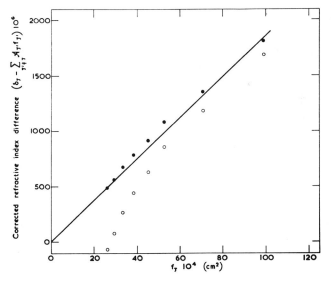

FIG. 8. The refractive index difference plotted against f_J, the function in brackets in Eq. (108) for the pure rotation line of gaseous ammonia at 39.78 cm^{-1}. The open circles are the results of applying the method of Chamberlain directly, with no allowance being made for the contribution of neighboring lines to the dispersion. They provide a poor fit to a straight line. The filled circles were obtained after nine iterations to allow for neighboring lines. These provide a good straight line fit, allowing the integrated line strength to be calculated. [From Chamberlain *et al.* (1969a). Reproduced by permission of Pergamon Press Ltd.]

Sanderson and Scott (1971) have applied this technique to the direct determination of the partial pressures of a two-gas mixture. If one of the gases has a known line spectrum and the other is essentially nondispersive, then measurement of the refractive index difference of the line spectrum in the mixture enables the partial pressures to be calculated. By applying this to HCl–inert gas mixtures Sanderson and Scott were able to obtain HCl partial pressures which were internally consistent to 1 %.

C. INTERFEROMETERS AND MEASUREMENTS

There have been five DFTS interferometers for the study of gases and vapors described in the literature (Chamberlain *et al.*, 1965b, 1969a; Sanderson, 1967; Sanderson and Scott, 1970, 1971; Birch, 1978a). In this section we shall consider the construction and use of three of these which can be taken to illustrate desirable experimental features and used to discuss practical limitations. The first DFTS measurements on gaseous specimens were those of Chamberlain and Gebbie (1965b) on the hydrogen halides HCl, HBr, and HI using an interferometer which, in a subsequent form (Chamberlain *et al.*, 1969a), is shown schematically in Fig. 9. The main feature of the instrument is its utilization of the modular concept of the NPL cube interferometer developed by Gebbie and co-workers (Gebbie and Twiss, 1966; Chantry *et al.*, 1969a). Each element of the interferometer forms its own separate module, all of which bolt together on the central "cube" containing the dielectric beam divider to form a vacuum-tight enclosure.

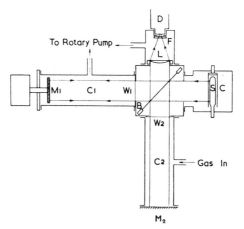

FIG. 9.   The modular interferometer used by Chamberlain and coworkers for the DFTS of gases. The abbreviations are as follows: S, high-pressure mercury arc lamp; C, 13-Hz mechanical chopper; L, polyethylene lens; B, melinex beam divider; W_1 and W_2, melinex windows; M_1, movable mirror; M_2, fixed mirror; D, Golay cell; and F, filter. The gas cell C_2 was 203 mm long. [From Chamberlain *et al.* (1969a). Reproduced by permission of Pergamon Press Ltd.]

The flexibility of this approach allows interferometers to be constructed that are specifically designed for a particular measurement, leading to an optimization of the experimental conditions. This flexibility has meant that this type of instrument has found favor with many workers and has been responsible for many of the most recent developments in the DFTS of gases, liquids, and solids.

The gaseous specimen was contained in the fixed mirror arm C_2 and was separated from the main body of the interferometer by a 25-μm-thick polyethylene terephthalate[2] window. A similar window in the moving mirror arm C_1 acted as a compensator to maintain the phase symmetry of the empty interferometer although, in another description of this interferometer (Chamberlain et al., 1969b) given in a more general article, the compensating window is not shown. The two arms of the interferometer were constructed from 50-mm-internal-diameter glass tubing, and the length of the gas cell was 203 mm. The radiation source was an uncollimated 120-W quartz-encapsulated mercury vapor arc with a coaxial cylindrical chopper to allow amplitude modulation to be used. The radiation leaving the exit aperture of the interferometer was condensed onto the detector by a polyethylene lens. The interferometer was not maintained at a constant temperature, but the low coefficient of thermal expansion of the glass arms would have minimized any problems associated with differential thermal expansion of the arms. The interferometer was generally used in the longer-wavelength region below 120 cm^{-1} at a spectral resolution of 2 cm^{-1}, although in measurements on gaseous GeCl$_4$ (Chantry et al., 1969a) the region between 300 and 600 cm^{-1} was covered.

The ultimate level of precision obtainable with this instrument would have been limited by changes in the length of the gas cell caused by bulging of the 25-μm window when the gas cell was filled with gas. This would have restricted the ability of the interferometer to provide the absolute level of a refraction spectrum. However, for the purposes of determining line strengths we have seen that this is not a restriction, and the refractive index difference spectrum of part of the pure rotation band of gaseous ammonia shown in Fig. 10 between 10 and 110 cm^{-1} is typical of the results obtained with this instrument that were used to give line strengths and, subsequently, dipole moments.

The interferometers used by Sanderson and co-workers (Sanderson, 1967; Sanderson and Scott, 1970, 1971) were completely different in concept from those of Chamberlain. They were vacuum tank instruments with all the optics mounted on base plates which were connected to the tank by three point supports in positions that were insensitive to any flexure of the tank on

---

[2] This is known by the tradenames mylar and melinex.

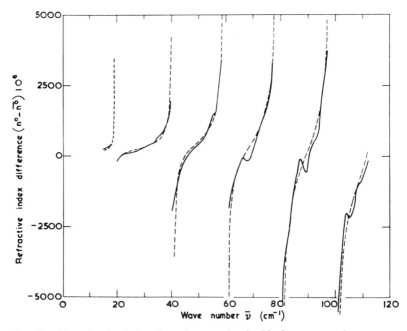

FIG. 10. The refractive index dispersion associated with the pure rotation spectrum of gaseous ammonia between 15 and 115 cm⁻¹ at 760 Torr and 300 K. The solid lines are the experimental data, the dashed lines were calculated assuming an average value of $1.426D$ for the dipole moment. [From Chamberlain *et al.* (1969a). Reproduced by permission of Pergamon Press Ltd.]

evacuation. The vacuum tank provided the additional benefit of thermal isolation from ambient conditions, which led to improved phase stability, and Sanderson and Scott (1970, 1971) further enhanced this by water-cooling parts of the optics within the tank. The optical configuration of the high resolution interferometer developed by these authors is shown in Fig. 11. The gas cell was 225 mm long by 112 mm in diameter and was sealed at either end by 1.5-mm-thick polypropylene or polyethylene windows. It was connected to a pivot under the center of the beam divider so that it could be swung into one beam for dispersive measurements or into the exit beam for nondispersive, power transmission measurements. For nondispersive measurements their gas cell was supplied with windows cut to a conical form, one convex and one concave, and assembled in the orientation that kept the length of the cell constant across its aperture. This suppressed the effects of channel spectra due to multiply reflected beams within each window but gave an appreciable loss of signal and was not used in dispersive measurements, plane parallel windows then being used instead. Sanderson (1967) investigated the consequences of pressure-induced bulging of the cell

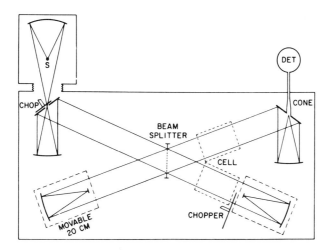

FIG. 11.   The interferometer used by Sanderson and Scott for the study of gases by dispersive and nondispersive FTS. [From Sanderson and Scott (1971). Reproduced by permission of the Optical Society of America.]

windows and found a movement of 0.2 mm atm^{-1} which, for the highest pressures that he used, corresponded to a fractional error in $(n - 1)$ of about -1.3×10^{-3}. In the interferometer shown in Fig. 11, Sanderson *et al.* (1971) found an additional problem associated with these flexible windows. When they bulged due to the introduction of gas into the cell, the channel spectrum associated with each changed its period to become significantly different from that of the reference, empty cell, measurement and caused substantial distortion of the measured phase shift spectrum. This was overcome by recording the reference measurement with an equal pressure of inert gas in the cell, the constant contribution of the refractive index of the inert gas to the measurement being unimportant in the determination of line strengths.

The radiation within this interferometer was collimated so that it was not necessary to correct for the effects of convergent radiation (Chamberlain, 1967a; Russell and Bell, 1967a), and all the collimating and condensing optics were reflecting, including cats'-eye reflectors at the end of each 700-mm-long arm. The radiation within the interferometer was amplitude-modulated, but novel positioning of a segmented chopper having reflecting blades in the gas cell arm allowed the effects of source fluctuations to be reduced and the constant term in the interferogram to be partly suppressed. The cat's-eye reflector in the moving mirror arm could be driven through 200-mm travel leading to a possible unapodized spectral resolution of 0.025 cm^{-1} for the double-sided interferograms required for DFTS. This is according to the usually applied resolution criterion which Chantry and Fleming (1976) have shown to be optimistic. The "highest" resolution that the instrument appears to have been used to in the dispersive mode was 0.4 cm^{-1} in the study of HCl from 25 to 150 cm^{-1} (Sanderson and Scott, 1971).

In the interferometers of Chamberlain and Sanderson we have seen how distortions of the flexible membranes used as the gas cell windows limit the determinations of the absolute level of the refractive index spectrum and how this may not be a problem in the study of line spectra. However, in the study of gases that do not possess a permanent dipole moment and that are largely nondispersive in the submillimeter and millimeter wavelength regions the absolute level of the refraction spectrum becomes the required quantity, and it is important to measure it precisely. Birch (1978a) has described an interferometer that has been designed to enable such precise measurements to be made, and its optical configuration is shown in Fig. 12. In common with the earlier instrument of Chamberlain *et al.* (1969a) it is based on the modular cube concept but differs from both interferometers previously described in having a rigid window to the gas cell so that it does not distort significantly when the cell is filled with gas. This window and the compensating one in the moving mirror arm were made from 5-mm-thick high-density polyethylene and was firmly clamped over the entrance to each arm. The major problem with the use of such thick windows is their increasing absorption loss above ∼50 cm^{-1} and, if a lens is used to condense the radiation onto the detector, the combined lens and window absorption losses degrade the performance of the instrument significantly. This was minimized by the use of all-reflecting Pfund optics to condense the radiation onto the detector in the higher wave number region (Dromey and Birch, 1978). The two arms of the interferometer were constructed from

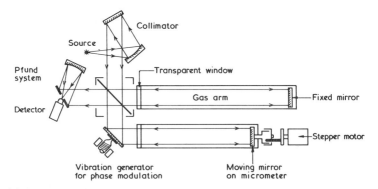

FIG. 12.   The interferometer used by Birch and coworkers for the study of gases and vapors by DFTS. [From Birch (1978a). Reproduced by permission of Pergamon Press Ltd.]

64-mm internal diameter brass tubing with several lengths available so that gas cells of between 0.1- and 0.9-m length could be used as necessary. The moving mirror arm was bent through a right angle by a vibrating plane mirror so that phase modulation (Chamberlain, 1971, Chamberlain and Gebbie, 1971a,b) could be used. This also served to make the instrument more compact and easier to stabilize. The temperature of the interferometer was maintained constant within $\pm 0.5°C$ by circulating water from a constant temperature bath through cooling coils soldered around the outside of the two arms. Changes in the length of the gas cell during measurements were monitored by electronic transducers placed at either end of it. The most significant of such changes was due to the decrease in axial pressure acting along it when it was filled to an appreciable fraction of atmospheric pressure. This caused the fixed mirror to move away from the position that it occupied in the reference, empty cell measurement leading to an error in $\bar{X}$ the fringe shift term. The change was typically 3 μm for a 760-Torr increase inside a 0.9-m cell, and the measured change was used to correct the computed refraction spectrum for the effect. The moving mirror could be step-displaced through 50-mm travel to give an unapodized spectral resolution of 0.14 cm^{-1} for a double-sided interferogram according to the most optimistic criterion of Chantry and Fleming (1976).

Figure 13 shows the refraction spectrum of water vapor measured with this interferometer at 12.8 Torr between 10 and 50 cm^{-1} (Kemp et al., 1978) and may be used to illustrate the noise levels intrinsic to the interferometer.

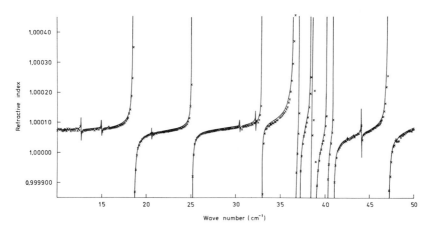

FIG. 13.   The refractive index of water vapor between 10 and 50 cm^{-1} at 12.8 Torr and 290 K. The crosses are the experimental points determined with the interferometer of Fig. 12, and the continuous lines represent the spectrum calculated for these conditions using the kinetic line shape of Zhevakin and Naumov (1963, 1967) (Kemp et al., 1978).

The measured points are shown as crosses and the spectral resolution was 0.17 cm^{-1}. The level of random noise, which is not really apparent on the scale of this spectrum, was about 5×10^{-7} in the center of the spectral range where the signal-to-noise ratio was highest (Birch and Afsar, 1977b). The level of systematic error in these measurements may be judged from the agreement between the experimental points and the continuous curves which are theoretical calculations based on the kinetic line shape of Zhevakin and Naumov (1963, 1967). The calculated spectrum is higher than the measurements by 5×10^{-6}. Such differences are easily accounted for by the possible errors in the absorption strengths used in the calculation and in the pressure measurement and illustrate the point made earlier that the pressure measurement could easily become the limiting factor in precise refractive index determinations. Differences between measurement and calculation in the region of line center were due to the omission of the instrumental line shape in the calculation of the theoretical spectra.

V. Dispersive Fourier Transform Spectrometry of Liquids

In the previous section it was demonstrated how the low complex refractive indices of gases and vapors made them the most convenient class of materials to study by DFTS techniques as interface effects could usually be completely ignored. Liquids, on the other hand, encompass much larger values of the complex refractive index so that the surface reflections become important and interface effects must be allowed for in precise measurements. This situation is, however, quite different from that of solids, as the fact that the thickness of a liquid specimen is easily variable has led to the development of several different experimental techniques for the DFTS of liquids. The first such measurement was by Chamberlain *et al.* (1967a) on symmetric tetrabromoethane, and until the recent work of Yarwood and James (1977), all of the reported DFTS measurements on liquids appear to have been by or in association with the NPL group. Many different liquids and solutions have been studied since the first measurements on tetrabromoethane, and although it is not appropriate to mention them all here, it is interesting to note that the more important technical developments have resulted from the study of highly absorbing polar liquids such as water and the primary aliphatic alcohols. Liquids are presently the phase of matter that has been most widely studied by DFTS, although the present resurgence of interest in the DFTS of solids will probably reverse this soon.

There are essentially three experimental configurations that have been used to contain liquid specimens in DFTS studies and in the following sections the subject shall be discussed in terms of these configurations.

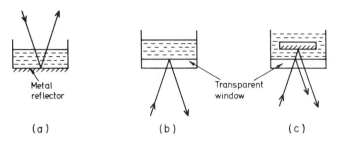

FIG. 14. The three configurations used to contain liquid specimens in DFTS studies: (a) free liquid layer; (b) window–liquid layer; (c) liquid cell.

The three methods of containment are shown schematically in Fig. 14, and as each may be used in at least two different ways, there are more than six distinct experimental techniques for the DFTS of liquids. In this figure the incident and principal reflected light beams are shown at nonnormal incidence for clarity although these configurations are only used at normal incidence. The free liquid layer method was the first used, and in it the liquid forms a gravity-held layer on a flat, highly polished, metal mirror, usually of stainless steel. The mirror forms the fixed mirror of the two-beam interferometer, and the radiation in that arm of the interferometer passes through the liquid specimen twice, once before incidence on the mirror and once after. In order to study more absorbing liquids it is necessary to use progressively thinner layers to maintain reasonable transmission levels. As surface tension effects limit the minimum thickness for which continuous, free, gravity-held films can be maintained to about 100 μm, this prohibits the application of the technique to even more absorbing liquids. Under these circumstances the window–liquid interface method shown in Fig. 14b must be used. In this the fixed arm of the interferometer is terminated by a plane parallel transparent window, the liquid forms a relatively thick layer over the upper surface of the window, and the complex reflectivity of the interface between the liquid and the window is measured. The more recent liquid cell method is shown in Fig. 14c. This combines the free liquid layer and window–liquid interface methods. As in the window–liquid interface method the liquid is poured over the upper surface of a transparent window, but the presence of a movable mirror in the liquid volume allows transmission measurements to be made if the mirror is sufficiently close to the window as well as reflection measurements when the mirror is moved further away so that the transmitted signal is strongly attenuated. In the following sections the various techniques that have been used with these three containment methods shall be considered and compared. Where it proves necessary to develop the discussion through the use of mathematical expressions this shall be done in terms of the simple notation defined in this work; much of the development of this area of DFTS in the literature has been in terms of very general approaches which, while rigorously correct, lead to different elaborate notations that can confuse someone approaching the subject for the first time.

A. THE FREE LIQUID LAYER METHOD

The interferometer used by Chamberlain *et al.* (1967a) for the first liquid DFTS measurements is shown in Fig. 15 together with a present-day, free layer instrument (Afsar *et al.*, 1976b) for comparison. In essence the two interferometers are the same, the differences between them largely arising from the use in the more recent interferometer of the modern refinements of ordinary Fourier transform spectrometry such as reflecting collimation optics, phase modulation, and polarizing beam dividers (Martin and Puplett, 1970; Costley *et al.*, 1977). In both interferometers the liquid cell is the important feature. Its polished stainless steel base (M2 in Fig. 15a, 14 in Fig. 15b) forms the fixed mirror of the two-beam interferometer, and the liquid to be studied is poured onto its reflecting surface, forming a gravity-held layer. This must be plane parallel if precise measurements are to be obtained which requires the mirror to be accurately horizontal. This is achieved by mounting the whole interferometer on an adjustable plate so that the desired orientation can be set, and the precise manner by which this may be achieved is described in the following section. In the early version of the interferometer the liquid cell was not attached to it but rested on an adjustable table below it. The thin Melinex windows W_2 and W'_2 sealed the two and the interferometer was evacuated while the air gap between W_2 and W'_2 was flushed with dry air to remove the absorption associated with atmospheric water vapor. For the empty cell reference measurement, the liquid cell would also have been flushed with dry air but was sealed off after the liquid was introduced. Thus the space above the liquid would contain dry air saturated with the liquid vapor. To compensate for the phase asymmetry introduced by W_2 and W'_2 two similar windows W_1 and W'_1 were placed in the moving mirror arm. This use of two windows in each arm leads to large transmission losses at the higher wave numbers, and in all subsequent forms of this interferometer (Chamberlain *et al.*, 1969b; Davies *et al.*, 1970) as well as the recent version in Fig. 15b, the liquid cell was directly attached to the interferometer with only one window between them and no compensating window in the moving mirror arm. (In a high-stability system the compensating window is not really necessary as the phase asymmetry due to the uncompensated window will disappear when the phase difference spectrum is formed.)

As previously mentioned, the high complex refractive indices of liquids mean that interface effects cannot be neglected so that simple relationships

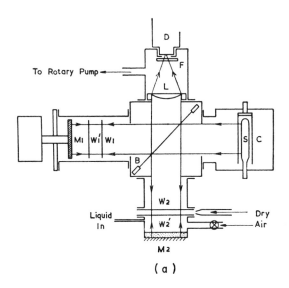

(a)

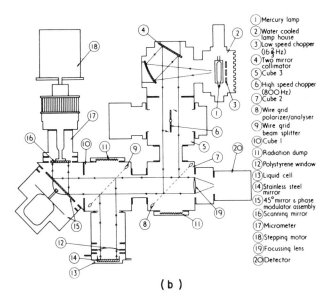

(b)

FIG. 15. Interferometers used for the DFTS of free liquid layers: (a) the first interferometer developed for such measurements. The key to its lettering is the same as for Fig. 9 [From Chamberlain *et al.* (1967a).]; (b) a recently developed interferometer shown here with polarizing grid beam dividers. [From Afsar *et al.* (1976b). Both reproduced by permission of Pergamon Press Ltd.]

between the real and imaginary parts of the complex refractive index and the measured complex insertion loss similar to those described by Eqs. (103) and (104) cannot be used, and one must either use more rigorous analyses or resort to experimental approaches that suppress the problem. In the full, general treatment of liquid layer systems, the equations relating $\hat{n}(\bar{\nu})$ to $\hat{L}(\bar{\nu})$ cannot be solved directly for the former quantity. The Leiden group (Honijk et al., 1972, 1973a,b; Passchier et al., 1975, 1976) have recently analyzed this situation and describe an iterative solution to the problem, but prior to this the determination of $\hat{n}$ was by either of two quite precise but approximate methods of treatment of the experimental observations developed by Chamberlain and co-workers (Chamberlain et al., 1967a, 1973a, 1974a,b; Davies et al., 1970). In the following sections we shall first outline the approximate methods of Chamberlain, illustrating them with some typical results, before discussing the full interferogram approach of the Leiden group, and, finally, presenting an assessment of the precision that can be achieved with current free layer techniques and of the major sources of error.

1.   The Nature of the Dispersive Interferogram

Before discussing the approximate methods of Chamberlain it is necessary to consider the nature of the interferogram obtained in a typical free layer experiment. When no liquid is in the interferometer the recorded interferogram contains one uniquely bright interference signature centered on zero path difference. With a plane parallel liquid layer on the fixed mirror the incident ray is multiply reflected within the layer, and the total reflected ray is the infinite sum of these internally reflected rays. As these all have substantially different phases the recorded interferogram will now contain many interference signatures, each corresponding to the interference of one of the internally reflected rays with the radiation from the moving mirror arm. This is illustrated by the typical free layer interferograms shown in Fig. 16. The upper interferogram (a) was obtained with no liquid in the cell, and as phase modulation was used, it is the zero crossing between the maximum and minimum intensities that corresponds to the zero path difference position. The lower interferogram was obtained with liquid in the cell. The signature labeled R at negative path differences corresponds to the ray reflected from the top surface of the liquid. The T-signature corresponds to the ray that has traveled through the layer twice, having been reflected once from the fixed mirror, while the $M(m = 1)$-signature corresponds to the internally reflected ray that has undergone two reflections at the fixed mirror. These internally reflected rays are shown schematically in Fig. 17, and although we have only discussed three signatures, there are, in general, an infinite number of the internally reflected signatures M_1, M_2, etc. In

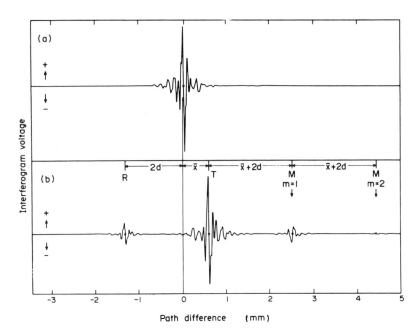

FIG. 16. Typical free layer interferograms recorded with a phase-modulated system: (a) from an empty liquid cell; (b) with liquid in the cell and showing the interference signatures due to radiation reflected (R) from the upper surface of the liquid, transmitted once (T) through the liquid layer, and internally reflected (M_1) once at the upper surface of the liquid layer. [From Afsar *et al.* (1976b). Reproduced by permission of Pergamon Press Ltd.]

practice the extra path length traveled in the liquid by the higher-order, internally reflected rays leads to their rapid attenuation and M_1 is usually the only one of practical importance. This can be seen in Fig. 16b where the signature for M_2 is indicated but is only just apparent. Although each of the signatures R, T, M_1, M_2, etc., contains sufficient information for the computation of the complex refractive index, the approximate methods of Chamberlain deal with ways of isolating the T-signature and are, therefore, essentially transmission techniques.

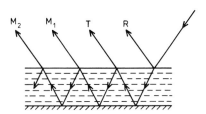

FIG. 17. The rays reflected from a free liquid layer on a mirror.

It is central to the free layer method that expressions for the path differences at which these signatures occur be derived. These are used to give the thickness of the free layer and enable the phase spectra computed from the complex insertion loss to be related to the refractive index. In the explicit presentation of the theory used by Chamberlain and coworkers (Afsar et al., 1976b,c) these path difference positions are given somewhat ambiguously, although the final equations for $\hat{n}(\tilde{v})$ in terms of the measured complex insertion loss are correct. This appears to have arisen from a confusion between path length and path difference. However, before moving on to consider this, it is appropriate to mention that Fig. 17 suggests the method that can be used for the final alignment of the interferometer. First, the mirrors of the interferometer are adjusted to provide a well-aligned empty cell interferogram as in Fig. 16a. The liquid is then placed in the liquid cell and the tilt of the entire interferometer adjusted until the R-signature is optimized. When this is achieved the liquid surface must be virtually parallel to the fixed mirror, the desired condition.

The major factor determining the path difference positions at which the various signatures occur is the thickness of the liquid layer, but it is necessary to include the contributions from the various shifts that occur on transmission through or reflection from the interfaces of the system. Thus for a liquid thickness d, the R-signature will occur at a path difference given by

$$x_a = -2d + [1/(2\pi\tilde{v})](\phi^r_{0L} - \pi) \qquad (109)$$

that is dominated by the term $-2d$ arising from the displacement of the liquid surface from the fixed mirror. In this and the following the subscripts 0, L, and M are used to indicate the vacuum (or air)–liquid (OL), the vacuum–mirror (0M), and the liquid–mirror (LM) interfaces in line with the notation of Section II. The first term in the parentheses of Eq. (109) represents the phase shift that occurs on reflection at the upper surface of the liquid, while the second term $-\pi$ represents the effect of the phase shift that occurs at the vacuum–mirror interface in the moving mirror arm. In the expressions of Afsar et al. (1976b, c) this term $-\pi$ is omitted but is meant to be understood as being included in the equivalent term to our ϕ^r_{0L} (M. N. Afsar, private communication). In a similar manner the T-signature would occur at a path difference

$$x_T = 2(\bar{n} - 1)d + [1/(2\pi\tilde{v})](\phi^t_{0L} + \phi^t_{L0} + \phi^r_{LM} - \pi) \qquad (110)$$

that is dominated by the term $2(\bar{n} - 1)d$ containing the "mean" value of the refractive index over the spectral region of the measurements, while the mth internally reflected ray signature is found at

$$x_m = 2[(m + 1)\bar{n} - 1]d + [\phi^t_{0L} + \phi^t_{L0} + (m + 1)\phi^r_{LM} + m\phi^r_{L0} - \pi][1/(2\pi\tilde{v})]. \qquad (111)$$

Although these expressions are usually dominated by the constant terms in d, the other wave number-dependent terms can become significant for very thin layers. Under these conditions the R-, T-, and M-signatures will overlap each other by significant amounts, and the simple free layer approach cannot be used.

2. The Ideal Case

In principle, for measurements on transparent liquids the liquid layer can be made sufficiently thick for the R- and M-signatures to fall outside the range of path differences involved in the recording and computation of the T-signature. Under these conditions the complex insertion loss is the ratio of the complex spectra derived from the T-signature and the empty cell interferogram, both referred to the same computational origin. This would be related to the complex refractive index through the expression

$$\hat{L}(\tilde{v}) = (1 - \hat{r}_{0L}^2)(\hat{r}_{LM}/\hat{r}_{0M}) \exp i4\pi\tilde{v}[\hat{n}(\tilde{v}) - 1]d. \tag{112}$$

It would normally be assumed that the fixed mirror was a perfect reflector so that $\hat{r}_{LM} = \hat{r}_{0M}$, which simplifies Eq. (112) to

$$\hat{L}(\tilde{v}) = (1 - \hat{r}_{0L}^2) \exp i4\pi\tilde{v}[\hat{n}(\tilde{v}) - 1]d. \tag{113}$$

Hence, the real and imaginary parts of $\hat{n}(\tilde{v})$ would be computed from the measured quantity $\hat{L}(\tilde{v})$ via the expressions

$$n(\tilde{v}) = 1.0 + [1/(4\pi\tilde{v}d)] \, \mathrm{ph}\{\hat{L}(\tilde{v})\} - [1/(4\pi\tilde{v}d)](\phi_{0L}^t + \phi_{L0}^t) \tag{114}$$

and

$$k(\tilde{v}) = [1/(4\pi\tilde{v}d)] \ln[|(1 - \hat{r}_{0L}^2)|/|\hat{L}(\tilde{v})|]. \tag{115}$$

Strictly speaking, the evaluation of these must be an iterative procedure as ϕ_{0L}^t, ϕ_{L0}^t, and $\hat{r}_{0L}$ are all functions of n and k. However, this can be simplified by recognizing that for moderately transparent liquids $\phi_{0L}^t = \phi_{L0}^t = 0$ and $n(\tilde{v}) \gg k(\tilde{v})$ so that Eqs. (114) and (115) reduce to

$$n(\tilde{v}) = 1.0 + [1/(4\pi\tilde{v}d)] \, \mathrm{ph}\{\hat{L}(\tilde{v})\} \tag{116}$$

and

$$k(\tilde{v}) = [1/(4\pi\tilde{v}d)] \ln\{[4n(\tilde{v})/(1 + n(\tilde{v}))^2][1/|\hat{L}(\tilde{v})|\}, \tag{117}$$

which are dealt with more easily. The thickness d of the layer is found by equating the path difference value at which the R-signature occurs to $-2d$. For transparent liquids of reasonable thickness this neglect of the term in $\phi_{0L}^t - \pi$ in Eq. (109) is insignificant.

3. *Editing*

The equations of the previous section were derived by assuming that the R- and M-signatures did not lie in the range of path differences about the T-signature that would be used in the Fourier transformation. There are only a limited number of transparent liquids for which this would apply, and Chamberlain developed the editing technique to use with thinner liquid layers where the R- and M-signatures came into the range of path difference values to be used but did not overlap with the structure associated with the center of the T-signature. The essentials of the method can be seen with reference to Fig. 18. The interferogram in (a) is a liquid cell interferogram, the structure labeled *p* represents the T-signature of the radiation transmitted through the liquid layer, and the interferogram is to be transformed between the path difference limits $\pm D$ within which is found the R-signature, here labeled as *q*. From the recorded interferogram, which was usually on paper-tape, Chamberlain would remove the points associated with the R-signature and replace them with an equal number taken from the noise region of the interferogram to produce the edited interferogram in (b).

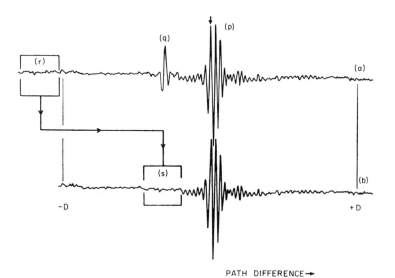

PATH DIFFERENCE→

Fig. 18. The editing procedure for dispersive free layer interferograms. The upper inter-ferogram (a) is the unedited interferogram containing the interference structure labelled *q* due to the ray reflected from the upper surface of the liquid. The editing procedure is to replace this portion of the interferogram by the portion labeled *r* taken from a region of the interferogram having no interference structure. This gives the edited interferogram (b). [From Davies *et al.* (1970). Reproduced by permission of the Chemical Society.]

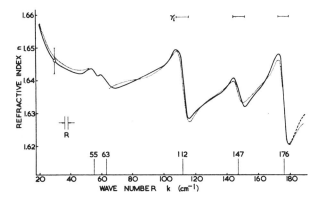

Fig. 19. A typical result obtained by the editing technique; the refraction spectrum of symmetrical tetrabromoethane at 20°C. The level of the spectrum is set by the independent laser measurement at 29.7 cm^{-1} using a Mach–Zehnder interferometer. The continuous curve is the measured spectrum, while the dotted line represents a theoretical calculation of the spectrum. R indicates the resolution of the measurement, and the horizontal lines labeled γ_i indicate the half-widths of the resonances. [From Chamberlain *et al.* (1967a). Reproduced by permission of Pergamon Press Ltd.]

An early but fairly typical result obtained by the editing technique is the refractive index of symmetrical tetrabromoethane (Chamberlain *et al.*, 1967a) shown in Fig. 19. This illustrates two important points about the manner in which the technique was used. In Section V.A.2 we described how the liquid thickness was found from the position of the *R*-signature. This only gives an approximate value so the overall level of the refraction spectrum computed from this value using Eq. (116) would be in error. In these early liquid measurements the overall level was determined in a separate experiment using a Mach–Zehnder interferometer and a submillimeter wavelength gas laser. Thus the point at 29.7 cm^{-1} in Fig. 19 was found in this manner using an HCN laser and the level of the entire spectrum determined by the broad-band method adjusted to fit this point. The second point about these early dispersive determinations is that the liquid absorption spectrum was usually not measured. At the time it was felt that a more accurate value of the absorption spectrum could be obtained from the use of conventional power transmission methods (Chamberlain *et al.*, 1968; Davies *et al.*, 1970).

4. *Subtraction*

As the liquids to be studied become more and more absorbing even thinner layers must be used and the *R*- and *M*-signatures inevitably occur at smaller path differences and eventually begin to overlap the structure of the *T*-signature. When this occurs the editing method cannot be used with any confidence and one must resort to the subtraction method originally developed for measurements on the primary aliphatic alcohols from ethanol to octanol (Chamberlain et al., 1973a, 1975).

The basis of the method is that if, in addition to recording an interferogram from the thin liquid layer, an interferogram is recorded from a layer sufficiently thick that absorption suppresses all but the R-signature, the second interferogram can in principle be subtracted from the first to isolate the T-signature. This is illustrated in Fig. 20, which shows a series of interferograms used in such a subtraction procedure. Figure 20a is the reference empty cell interferogram and Fig. 20b the thin layer interferogram for about 1 mm thickness of liquid showing the R- T-, and M-signatures as previously discussed. When the thickness of the liquid layer is increased to just over 3 mm, the R-signature remains virtually constant but is shifted to R' at about 6.5-mm path difference as shown by the interferogram of Fig. 20c. Some radiation is still transmitted through the liquid layer as shown by the displaced T-signature T', but it is much attenuated and now well removed from the R'-signature. The essence of the technique is that if the thick layer interferogram is now displaced to positive path differences so that its R'-signature occurs at the same path difference value as the R-signature of the thin layer interferogram, as shown in Fig. 20d, then the former may be subtracted from the latter to isolate the T-signature of the thin layer as shown in Fig. 20e. When this has been done Eqs. (116) and (117) can be used to derive the optical constants as discussed for the ideal case.

In practice the discrete intensity and path difference sampling of the interferograms together with the effects of intensity noise and backlash in the moving mirror drive mean that one cannot meaningfully subtract the thick and thin layer interferograms in the manner discussed, and the equivalent computation is carried out in the spectral domain using the complex spectra obtained from the reference interferogram, the full thin layer interferogram, and the R'-signature of the thick layer interferograms. This is in many ways easier to deal with than interferogram subtraction, and the algebraic formulation has been described by Afsar et al. (1976b,c) and Afsar (1977). A comparison of the editing and subtraction methods is given by Afsar et al. (1976c) and summarized by Fig. 21 which shows three computations of the refraction spectrum of chlorobenzene between 30 and 120 cm^{-1} at ambient room temperature from a free layer experiment. (Note that the wave number scales in the figure are in reciprocal meters, $100 \, m^{-1} \equiv 1 \, cm^{-1}$.) The first spectrum (a) was computed from the full free layer interferogram, similar to that of Fig. 20b with the R-signature falling into the range of path differences used in the Fourier transformation. This causes a channel spectrum-type effect in the computed phase and transmission spectra, so the refraction spectrum (a) has a large periodic contribution

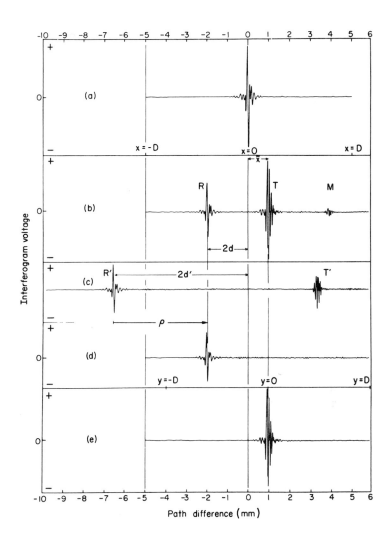

FIG. 20. Interferograms illustrating the subtraction method for free layer measurements: (a) empty cell; (b) thin liquid layer of thickness d in the cell; (c) thick liquid layer of thickness d' in the cell; (d) the thick layer interferogram shifted through ρ in path difference so that its R'-signature can be subtracted from that of the thin layer interferogram to give the single T-signature interferogram (e) [From Afsar *et al.* (1976c). Reproduced by permission of Pergamon Press Ltd.]

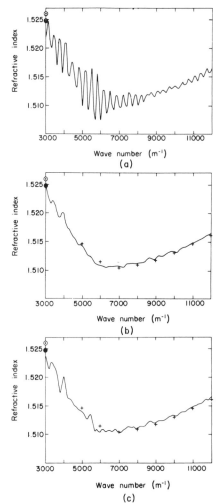

FIG. 21. The refraction spectrum of chlorobenzene computed from a free layer interferogram: (a) without any editing or subtraction; (b) after editing the upper liquid surface reflection signature, and (c) after subtraction of the upper liquid surface reflection signature. The points marked $\odot$, $\bullet$, and + are earlier measurements shown for comparison and are discussed in the text. [From Afsar *et al.* (1976c). Reproduced by permission of Pergamon Press Ltd.]

due to this superimposed upon the spectral variation of the refractive index. When the R-signature is edited out of the interferogram the periodic structure disappears, leaving behind the refraction spectrum shown in Fig. 21b. In this the points plotted as + are the results of an earlier dispersive determination of the refraction spectrum for comparison (Davies *et al.*, 1968) and the agreement between the two spectra is quite good. Figure 21c shows the same spectrum determined by the subtraction method, and it is significant to note that although the edited and subtracted spectra are in good overall agreement with each other, there is significantly more noise in the subtracted spectrum. This is a consequence of the lack of reproducibility in the measurement system which prevents the thin and thick layer R-signatures from being identical. In both the edited and the subtracted spectra the points marked ● and ⊙ at 2970 m^{-1} were the results of HCN laser Mach–Zehnder interferometer determinations of the refractive index, and the agreement of the laser and broad-band measurements to better than 0.001 (Chamberlain et al., 1974b) illustrates the significant point that the broad-band technique is now as precise as the laser measurement, if not more so. This is in contrast to the earlier work on liquids (see, for example, Chamberlain et al., 1968; Davies et al., 1968, 1970; Thomas et al., 1970; Davies and Chamberlain, 1972) in which the true level of the refraction spectrum was set by an independent laser determination. This is not now necessary.

5. Two Thicknesses

In our consideration of the ideal free layer measurement in Section V.A.2 it was stated that the thickness of the liquid layer is found from the path difference value at which the R-signature occurs. This was arrived at by ignoring the term $\phi_{0L}^{r} - \pi$ in Eq. (109), which is a good assumption for fairly thick liquid layers, but as more absorbing liquids, hence thinner layers, are studied, the effect of the neglected term is to introduce an increasingly significant error into the thickness determination. It is easily shown that this causes an underestimation of the layer thickness so that the computed refraction spectrum would be expected to be higher than the true spectrum by an amount that decreases with increasing layer thickness. This dependence on layer thickness has been observed in free layer measurements on chloroform and bromoform (Afsar et al., 1975a). One can correct for this effect by using the approximate thickness to compute a set of optical constants which are in turn used to compute a value for the phase shift on reflection at the liquid surface from Eq. (30). This is used to correct the thickness measurement with Eq. (109) and the whole process iterated until successive iterations of the optical constants converge to the required accuracy. As an alternative to this approach Afsar et al. (1976b) and Afsar (1977) have discussed a modification to the usual edit and subtraction methods that eliminates the phase shift from the measurements and so avoids the effect. The method is known as the *two-thickness method* and basically consists of making measurements on liquid layers of different thicknesses and using difference measurements so that one is concerned with thickness differences. Thus if the liquid is sufficiently transparent for the R- and M-signatures to be safely edited out, two thicknesses are used, and the refraction spectrum is found from the difference of the phase spectra computed from the two edited T-signatures. As both phase spectra contain the same contribution from the surface reflection term this vanishes when the difference is formed.

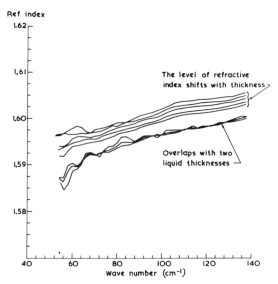

Fig. 22. The refractive index of bromoform computed from free layer interferograms. The five upper spectra are from measurements made on different thickness layers and the level of these spectra increases with decreasing layer thickness due to the increasing significance of the neglected phase shift on reflection occurring at the upper surface of the liquid. When the phase spectra obtained from measurements on two different thickness layers are subtracted, the effect of this phase shift is removed and all the refraction spectra overlap. [From Afsar *et al.* (1976b). Reproduced by permission of Pergamon Press Ltd.]

A typical result of the application of this method is shown in Fig. 22 for bromoform. The upper five spectra represent measurements made on different thickness layers, with the level of the refractive index decreasing as the thickness increases, as expected. When these are treated in pairs by the two-thickness method the resultant refraction spectra overlap each other and are independent of thickness showing the success of the method in removing the effect of the surface reflection. This simple approach fails for more absorbing liquids when the *R*- and *M*-signatures begin to overlap the *T*-signature. Measurements must then be made on an additional thick layer. This provides an isolated *R*-signature so that the two thin layer spectra may first be corrected for the effects of this signature by subtraction. These corrected spectra may then be differenced in a second subtraction procedure to remove the effects of the surface reflection in the now isolated *T*-spectra. This is known as the *double-subtraction method*.

6. *The Full Interferogram Method*

The development of the editing, subtraction, and two-thickness methods is an illustration of the gradual refinement of approximate techniques to meet ever more stringent experimental requirements. However, each signature in the full free layer interferogram contains enough information to allow the optical constants to be computed. Utilizing this fact workers at the Rijkuniversiteit, Leiden, have outlined the detailed derivation of the equations relating the complex refractive index to the complex insertion loss determined from full, unedited, free layer interferograms (Honijk *et al.*, 1972, 1973a, b; Passchier *et al.*, 1975, 1976). Their most important result comes from considering a liquid interferogram that contains all the possible interference signatures arising from the internally reflected rays of the layer. When this is combined with an empty cell interferogram the resulting complex insertion loss can be shown to be

$$\hat{L}(\tilde{v}) = [(\hat{r}_{0L} - \hat{a}_{L}^2)/(1 - \hat{a}_{L}^2 \hat{r}_{0L})] \exp(-i4\pi\tilde{v}d), \qquad (118)$$

which is the equivalent of Eq. (2.14) of Honijk *et al.* (1973b) and of Eq. (4.2) of Passchier *et al.* (1975) in the present notation. This has been derived by writing $\hat{r}_{21} = -1$ in the summation of Eq. (70) to introduce the effect of the fixed mirror and $\hat{a}_{L}$ as the complex propagation factor of the thickness d of the liquid layer.

It is apparent that there are no analytic solutions of Eq. (118) for n and k, and, therefore, iterative methods of solution must be sought. Passchier *et al.* (1975) discuss such a method but point out that it is necessary to provide initial values of the complex refractive index of the liquid for each wave number in the iterative process. This is because multiple solutions of Eq. (118) for the complex refractive index exist. These mainly affect the real refractive index for which the solutions are expected to be separated by approximately $(2\tilde{v}d)^{-1}$ at wave number $\tilde{v}$. Thus for high wave number studies or if a thick specimen is to be used, it is necessary to have a good initial estimate of the refractive index to avoid ambiguous solutions. This is a serious limitation of the method and the Leiden workers found it necessary to consider deriving these initial estimates from approximate computations on individual interference signatures in order to be able to exploit the inherent precision of the full interferogram method, i.e., from approximate methods similar to those outlined by Chamberlain and co-workers. In particular, Passchier *et al.* (1975) suggested obtaining the initial values from the isolated liquid surface R-signature, although after their error analysis of free layer measurements (Passchier *et al.*, 1976) they concluded that such reflection measurements were extremely susceptible to error and that the transmission T-signature would be preferable. Such a procedure would extend the measurement analysis considerably and this ought to lead one to consider, for each application, whether or not the extra precision theoretically implicit in the full interferogram method, compared to that obtainable from the approximate methods, warrants the extra analytical effort. In fact, Passchier *et al.*

(1975) point out that as their analysis is for ideal noise-free measurements, the practical limitations imposed by a real experimental situation may make it preferable not to attempt the full interferogram analysis but to use sub-traction-type transmission measurements instead.

7. *Measurement Precision*

In this subsection we shall discuss the precision with which the optical constants of a liquid can be determined by current free layer techniques. The level of random error in all dispersive and nondispersive measurements on materials will usually be set by the signal-to-noise ratio in the specimen interferogram. If this is maximized for the particular measurement, one is then concerned with establishing whether or not systematic errors exceed the random error. In attempting to establish measurement precision, however, one meets with the difficulty that in many of the published DFTS measurements on liquids either insufficient consideration is given to errors, which is a problem common to much of spectroscopy, or else the levels of precision that are quoted are often optimistic and occasionally appear to be at variance with the published spectra.

Measurements can be made on fairly thick layers of transparent liquids such as cyclohexane and cis- and trans-decalin, and relatively high precisions can be achieved in the refractive index although one may have to use quite considerable thicknesses to achieve good precision in the power absorption coefficient (Afsar *et al.*, 1976a). The measurements of Davies and Chamberlain (1972) on solutions of *p*-difluorobenzene in cyclohexane illustrate the accuracy that can be achieved by free layer techniques without using the two-thickness method. Between 20 and 100 cm^{-1} the total scatter of their refractive index values was ± 0.0003, which provides a measure of the high level of reproducibility of their measurements. However, it was necessary to establish the level of the refraction spectrum with an HCN laser measurement at 29.7 cm^{-1} which had a precision of only ± 0.001 and thus provided the overall limitation on their measurements. In terms of the absolute value of the refractive index this corresponded to 0.1%, although as $(n - 1)$ is actually measured the true measurement precision was about 0.25%. Recent considerations of random and systematic errors in dielectric measurements by Afsar *et al.* (1977a), in which free layer measurements on the more absorbing liquid chlorobenzene ($\alpha < 18$ Np cm^{-1}) were compared with microwave and submillimeter laser measurements, have led to what are probably more realistic estimates of present-day precision. These authors arrived at the conclusions that the absorption spectrum was no more precise than $\pm 2\%$ and that the refraction spectrum had random errors that did not exceed $\pm 0.2\%$ but also had systematic errors of as much as 0.6%.

As the liquid to be studied becomes more absorbing the precision that can be expected decreases with the decreasing transmitted power levels. In subtraction measurements on ethanol (Chamberlain *et al.*, 1973a) the refractive index between 30 and 120 cm^{-1} was determined with a scatter of about ± 0.01, which corresponds to about 1% precision. The ethanol absorption spectrum was not presented in their paper, but similar subtraction measurements on chloroform and bromoform (Afsar *et al.*, 1975a) illustrate the precision that can be achieved for fairly absorbing liquids. Chloroform and bromoform have quite narrow lines in their absorption spectra below 300 cm^{-1} with peak power absorption coefficients of between 20 and 50 Np cm^{-1}. In the region of these line centers the independent measurements shown in Afsar *et al.* (1975a) reproduce to about ± 10 Np cm^{-1} or between 10 and 20%.

Passchier *et al.* (1976) have recently presented a useful theoretical analysis of the most important sources of random and systematic error in the various free layer methods. Their general conclusion is that the subtraction technique using the two-thickness method to correct for the reflection phase shift is the most practical method. They expect that the real and imaginary parts of the complex refractive index ought to be capable of determination to within 0.01 to 0.0005 for specimen thicknesses of between 0.1 and 1.0 mm and identify the most serious systematic error in such measurements as resulting from the presence of the liquid vapor over the liquid layer.

B. REFLECTION METHODS

The ultimate limitation on the use of free layer transmission methods for the study of increasingly more absorbing liquids is provided by the difficulty of maintaining a sufficiently thin liquid layer. Below about 0.1 mm thickness the gravity-held layer is broken up by surface tension effects, and this restricts the thin layer method to liquids that have power absorption coefficients less than about 150 Np cm^{-1}. In order to study more absorbing liquids Chamberlain developed the window–liquid reflection technique in which the complex reflectivity of a plane interface between a transparent window and the liquid is measured. In principle, the R-signature of the radiation reflected from a free liquid surface could be used to give the optical constants, but, although adequate as a correction in the subtraction technique, instabilities of the free layer and the difficulties of locating its position lead to poor phase measurements so that free layer reflection is not used. The window–liquid technique was originally developed for the study of water (Chamberlain *et al.*, 1973b; Zafar *et al.*, 1973) and may be discussed with reference to Fig. 23 which shows a schematic outline of the arrangement of the apparatus used. The interferometer has its fixed mirror arm vertically upwards, terminated by an optically polished transparent

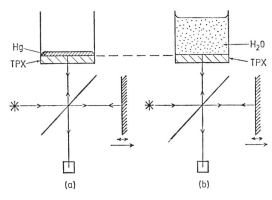

FIG. 23.   Schematic representation of the interferometer used for reflection DFTS measurements on absorbing liquids. (a) Mercury over the transparent TPX window acts as the reference reflector. (b) The mercury is replaced by an equal weight of the liquid to be studied, here shown as water. [From Chamberlain *et al.* (1973b). Reproduced by permission of Macmillan Journals Ltd.]

window, in this case the polymer TPX[3] (Chantry *et al.*, 1969b). The liquid that is to be studied is poured on top of the window in a sufficiently thick layer for no radiation to be returned after reflection at its upper free surface, and the specimen interferogram recorded. The reference measurement is provided by replacing the liquid specimen with liquid mercury. In order to obtain good-quality measurements several criteria must be satisfied. First, the reflection phase shift that one is trying to measure will be quite small so that the interferometer will have to be mechanically stable at the level of tenths of a micrometer to avoid systematic phase errors. Thus the temperature of the interferometer must be maintained at a constant level and the window must be sufficiently thick not to distort when loaded with liquid. The effects of any residual window distortion are minimized by using equal weights of liquid and mercury. Reasonable signal levels can only be obtained when there is a substantial mismatch between the window and liquid refractive indices. Therefore, the use of TPX, whose refractive index is virtually constant at 1.456 throughout the submillimeter region, allows water ($n > 2.0$) to be studied with some precision but would not be so suitable for the study of, say, aliphatic alcohols for which n is of the order of 1.5 (Chamberlain *et al.*, 1975a). With these precautions taken, the complex refractive index can be calculated from the measured complex insertion loss by using the equation

$$\hat{L}(\tilde{v}) = \{[\hat{n}_W(\tilde{v}) - \hat{n}_L(\tilde{v})]/[\hat{n}_L(\tilde{v}) + \hat{n}_W(\tilde{v})]\} \exp(-i\pi). \qquad (119)$$

The subscripts L and W refer to the liquid and window, respectively, and the phase term $\exp(-i\pi)$ arises from the assumption that the mercury is a perfect

---

[3] TPX is a polymer based on poly-4-methylpentene-1.

reflector. Use of this equation obviously requires prior knowledge of the complex refractive index of the window material, and this is obtained from a separate transmission DFTS measurement.

Afsar *et al.* (1976b) have given estimates of the precisions that can be achieved with this technique, and in measurements on water between 20 and 200 cm^{-1} in which the power absorption coefficient rose to about 1000 Np cm^{-1}, these authors found it possible to determine the phase of the window–liquid reflectivity to ± 0.04 rad and its amplitude to ± 0.02 for a TPX window. This led to precisions of $\pm 0.25 \%$ in n and a few percent in α. By ~ 170 cm^{-1} the refractive index of water begins to approach that of TPX so that higher wave number measurements become more difficult. Afsar and Hasted (1977) have avoided this problem by replacing the TPX with a silicon window. As the refractive index of silicon (~ 3.41) is considerably greater than that of most liquids, this transforms the technique into one that is also applicable to many other liquids. In addition, the reflectivity of a silicon–air interface is very much greater than that of a TPX–air interface, and Afsar and Hasted (1977) found that it was sufficient to use a silicon–air interface measurement for the reference reflection, thereby removing the complication of using liquid mercury. A further advantage of the use of a high refractive index window comes from recording the interferogram to sufficiently large path differences that the interference signature due to the reflection at the lower surface of the window is recorded in both specimen and reference interferograms. As the path difference at which this occurs is not affected by the presence of the liquid, it can be used to align the sampling combs of the two interferograms and so remove the effects of interferometer instabilities such as backlash in the moving mirror drive. Using this silicon–window method Afsar and Hasted (1977) were able to extend the measurements on H$_2$O and D$_2$O out to 450 cm^{-1}, by which wave number the power absorption coefficient of water exceeds 2000 Np cm^{-1}.

C. Variable Thickness Liquid Cells

In the previous sections we have seen how a variety of measurement techniques based on, essentially, two experimental configurations have been developed for the study of the dielectric properties of liquids and solutions ranging from the transparent to the opaque. Of these techniques the free layer methods, which have been the most widely applied, can suffer from errors due to problems of thickness determination, the presence of vapor above the liquid, and the maintenance of very thin continuous films. This last problem can be avoided by the use of the window–liquid reflection method, but this suffers from reduced measurement precision. An elegant, recent development, the use of variable-thickness liquid cells, has combined the two experimental configurations into one and overcome most, if not all, of the disadvantages of the previous methods. In these cells the liquid is contained between a transparent window and a movable mirror, which allows very thin layers down to 10 μm thickness to be used, so that very absorbing liquids can now be studied in transmission as well as reflection. Two such cells have been developed, one by Goulon at the University of Nancy that uses TPX or germanium windows and which has been partly described by Afsar *et al.* (1977b) and Stone and Chantry (1977) and one by the group at the Rijksuniversiteit, Leiden, that is generally used with silicon windows and which has been discussed in detail from both the practical (Honijk *et al.* 1977) and theoretical (Passchier *et al.*, 1975. 1977b; Honijk *et al.*, 1976) viewpoints. In this section we shall discuss the construction and operation of a liquid cell with reference to the work of the Leiden group as the less sophisticated cell of Goulon has not yet been dealt with in sufficient detail in the literature.

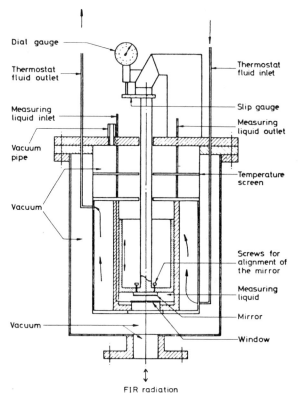

FIG. 24.   The variable thickness liquid cell of the University of Leiden. [From Stone and Chantry (1977). Reproduced by permission of Heyden and Son Ltd.]

The construction of the Leiden cell is shown schematically in Fig. 24, and although, in essence, its construction is simple, consisting of a parallel window and mirror between which the liquid specimen is placed, the practical realization of the necessary instrumental stability and measurement precision means that it is, in fact, a sophisticated instrument. The central part of the cell consists of the transparent window with an aperture just less than 40 mm in diameter and a gold-plated stainless steel mirror. The mirror is attached to the bottom of a stainless steel piston that slides within an outer stainless steel cylinder, and an axial stainless steel rod attached to the piston passes through to the outside of the cell and provides the means by which the piston is moved. The measurement volume is always connected to ambient conditions. Thus, by evacuating the space above the piston, the piston and rod arrangement can be forced by atmospheric pressure against the post supporting the dial gauge. The precise position of the mirror can then be varied continuously by adjustment of a micrometer drive above the dial gauge (not shown in the figure) or discretely by placing various slip gauges between the stainless steel rod and the dial gauge post. The whole of this assembly is mounted in a concentric jacket through which temperature-stabilized fluids may be circulated so that the specimen temperature may be varied, and this entire assembly is itself mounted inside a vacuum jacket which shares a common vacuum space with the main body of the inter-ferometer. This configuration has allowed liquids to be studied at temper-atures between 250 and 330 K. The cell is constructed with sufficiently good tolerances that, when it is assembled, the window is accurately perpendicular to the optic axis of its arm of the interferometer. The mirror can be aligned to be parallel to the window by adjustment of a push–pull screw system.

The cell window must satisfy several requirements. It should be transparent at submillimeter and millimeter wavelengths, be chemically inert to the liquids of interest, be mechanically stable, and be capable of being worked to a good optical plane parallel finish. Additionally, if window–liquid reflection measurements are to be made, it must have a refractive index that is substantially different from the liquid to be studied. Honijk et al. (1977) describe having used the cell in the region below 120 cm^{-1} with an iso-tropically cut crystal quartz window, but for most of their measurements, which were between 5 and 600 cm^{-1}, they used a high-purity, single-crystal, silicon window of about 2.5-mm thickness.

This cell may be used for transmission measurements in conjunction with methods of analysis which require a knowledge of either the absolute value of the window–mirror distance or the amount by which this distance changes between measurements on different liquid thicknesses. In the former case the absolute thickness is found from the path difference between the appropriate interference signatures in the interferogram, in a manner similar to that described for the free layer measurements. In the latter case the liquid thickness is changed by changing the slip gauge in the mirror piston assembly, and as these are precision gauges, the difference between the thicknesses of the two gauges used gives the change in liquid thickness.

1. *The Nature of the Liquid Cell Interferogram*

The interferograms obtained with this type of liquid cell are more complicated than those obtained with the free layer method outlined in Section V.A.1. This is because the liquid cell has three interfaces, interferometer–window, window–liquid, and liquid–mirror, compared to the two of the free layer method, interferometer–liquid and liquid–mirror. This leads to a more complex pattern of rays that are internally reflected within the cell and, hence, to many more interference signatures in the interferogram. The interfaces of the liquid cell and the more important of the internally reflected rays are shown in Fig. 25a with the rays labeled according to the notation adopted by Honijk *et al.* (1977). For a highly reflecting but transparent window such as silicon, the (0, 0) signature resulting from the ray reflected at the interferometer–window interface can be expected to dominate the interferogram. The intensity of the other signatures will vary in a manner which will depend on the optical constants of the window and liquid, but one would expect the (1, 0) and (1, 1) signatures to be prominent. A typical empty cell interferogram obtained with a silicon window, phase modulation, and a Golay cell detector to cover the spectral range between 20 and 120 cm^{-1} is shown in Fig. 25b. The window–mirror distance was about 1 mm, and one sees clearly the (0, 0) signature, the first six signatures of the first-order internally reflected rays, (1, 0), etc., and three of the second-order rays, (2, 0), etc. The observation of so many signatures is an indication of the quality of the parallelism of the three interfaces of the cell, and it is necessary to achieve such quality for measurements of the greatest precision. Thus the window must be worked to a good plane parallel form so that it acts in a similar manner to a high-quality Fabry–Perot etalon.

The approximate differences between the path difference values at which the (0, 0), (1, 0), and (1, 1) signatures occur have been marked in Fig. 25b. These ignore the phase shifts that occur on transmission through or reflection from the various interfaces of the cell, and although this may be reasonable for a few millimeters of high refractive index material such as silicon used with fairly transparent liquids, they will become significant for thin layers and should be included. Passchier *et al.* (1975) and Honijk *et al.* (1977) give general relations for the path difference values at which the maxima (for amplitude modulation) of these signatures occur, but again, as with the free layer case discussed in Section V.A.1, these are somewhat ambiguous as they do not explicitly include the phase term due to the effect of the reflection in

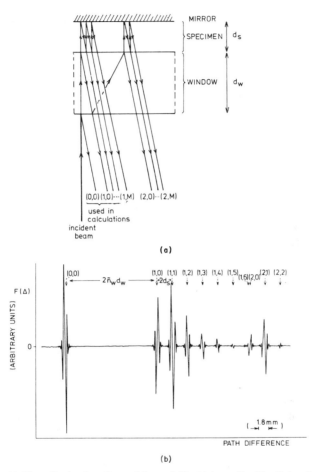

FIG. 25. (a) The reflecting interfaces of the variable thickness liquid cell showing those rays used in the computation of the refraction spectrum. (b) A typical interferogram obtained from such a system. The radiation bandwidth was from 20 to 120 cm^{-1}, the window was a 2.5-mm-thick disk of pure silicon, and the window–mirror distance was about 1 mm. [From Honijk *et al.* (1976). Reproduced by permission of Pergamon Press Ltd.]

the moving mirror arm when the partial beam in the cell is not reflected at the cell mirror. In Papers III and V of their series (Passchier *et al.*, 1975; Honijk *et al.*, 1977) the Leiden group consider in detail the problems of determining the specimen thickness from such interferograms.

2. Possible Experimental Methods

The presence of so many interference signatures in the recorded interferogram means that in common with the free layer method there are several methods of reducing liquid cell measurements to complex refractive index spectra, each based on different signatures or combinations of signatures. These have been discussed theoretically by Passchier *et al.* (1975, 1977b) and Honijk *et al.* (1976, 1977). Their general conclusion (Passchier *et al.*, 1977b) is that all of these possible methods are feasible in principle, even those using very thin layers which give rise to overlapping interference signatures and require iterative solutions. However, some of these are more susceptible to experimental error than are others, and in the following we shall discuss the various methods considered by these authors, concentrating on their preferred techniques. Most of these methods require knowledge of the complex refractive index of the window, and although this could be measured in a separate transmission DFTS experiment, Passchier *et al.* (1977a) have developed a novel technique for determining $\hat{n}(\tilde{v})$ with the window in situ in the cell. The details of this are discussed in Section VI, but basically the method consists of recording a single empty cell interferogram and using the complex spectra obtained from the $(0, 0)$ and $(1, 0)$ signatures to give $\hat{n}(\tilde{v})$ via an interative computation.

The particular liquid cell method that is used depends on the degree of opacity of the liquid to be studied, in much the same way as it does for the free layer methods. Honijk *et al.* (1977) categorize these experimental methods into three groups: transmission methods where the probing radiation propagates through the liquid and editing or subtraction techniques are used to isolate a certain interference signature; reflection methods where the reflectivity of the window–liquid interface is measured; and a quasi-full interferogram method where more than one interference signature is used in the analysis of the interferogram.

The transmission methods that Passchier *et al.* (1977b) and Honijk *et al.* (1977) consider are analogous to those of the free layer methods and similarly involve editing, subtraction, two thicknesses, and double subtraction. Transparent liquids provide the simplest transmission method that they discuss, which is labeled $\xi_2\phi_2$ in their notation. The complex insertion loss is found from a filled cell interferogram which has been edited, if necessary, to isolate the $(1, 1)$ first transmission signature and from a similar empty cell interferogram. For more absorbing liquids the signatures of the liquid interferogram begin to overlap as a result of the necessary reduction in liquid thickness. Editing then becomes impractical and the $\xi_{2M}\phi_2$ method is next considered. This is a two-thickness–subtraction method requiring three interferograms, there being an empty cell measurement, a full cell, thin layer measurement, and a full cell, thick layer measurement. The thick layer interferogram, which only contains the window–liquid reflection $(1, 0)$ signature, is subtracted from the thin layer interferogram to eliminate the reflection signature from the latter, which is then compared to that of the empty cell to give the complex insertion loss and hence $\hat{n}(\tilde{v})$. For nonopaque liquids one is not restricted to use only these filled cell–empty cell mehods. Two-thickness transmission methods are also available. The most direct of these is the $\xi_2 \xi_2'$ method which uses only the (1, 1) signatures of interferograms obtained from measurements on two different liquid thicknesses. This again requires fairly transparent liquids to avoid other signatures overlapping the (1, 1) signatures and complicating the analysis of the insertion loss.

Transmission methods can be used in these cells for liquids having power absorption coefficients up to about 1500 Np cm^{-1}, corresponding to liquid layers of about 10 μm thickness. In this high absorption regime the determination of the layer thickness becomes critical to the overall precision achieved in the calculation of the complex refractive index, and it becomes preferable to consider reflection methods. Honijk *et al.* (1977) consider two such methods. The first of these is, essentially, the window–liquid interface reflection method described in Section V.B. The mirror is removed sufficiently far from the window for the transmission (1, 1) signature to be either fully attenuated or to be outside the range of path difference values recorded. This allows the (1, 0) reflection signatures of the full and empty cell interferograms to be used to give the complex insertion loss, from which $\hat{n}(\tilde{v})$ can be found analytically if the window refractive index is known. (This is designated the $\xi_1 \phi_1$ method.) The alternative method, the $\xi_1 \phi_2$ method, utilizes the (1, 0) reflection signature of the liquid interferogram and the (1, 1) transmission signature of the empty cell interferogram. Thus it requires the thickness of the empty cell, and as this generally introduces systematic phase errors, the first $\xi_1 \phi_1$ method is preferred.

The final method that is considered by Honijk *et al.* (1977) and Passchier *et al.* (1977b) is the quasi-full interferogram method ($\xi_{1M} \phi_2$) which is meant for use with thin layers of very absorbing liquids when editing or subtraction of the overlapping signatures is either not possible or not desired. In common with the full interferogram method of the free layer configuration, this method leads to a complex insertion loss from which $\hat{n}(\tilde{v})$ can only be found by using an iterative technique.

3. *Errors*

Passchier *et al.* (1977b) have recently made a theoretical analysis of the various sources of error present in the experimental methods considered in the previous section and of the effects that each have on the value of the computed complex refractive index. When the results of this are combined with typical experimental measurements obtained with liquid cells (Honijk *et al.*, 1977) the following general conclusions emerge. The most accurate results are obtained for transparent and fairly absorbing liquids using the transmission $\xi_2 \phi_2$ (1, 1) signature method and the transmission $\xi_{2M} \phi_2$ (1, 1) signature, two-thickness–subtraction method. The former leads to values of the real refractive index which have an uncertainty of between 0.0005 and 0.001, while the latter gives rise to higher uncertainties of between 0.005 and 0.05 for the more absorbing liquids, due to the reduced signal levels and the extra noise introduced by the use of a third interferogram. The values of the absorption index, or power absorption coefficient, obtained with these two methods are found to "agree well with those produced with conventional non-dispersive techniques." Thus one anticipates that they should be accurate to within a few percent, at best.

As one moves to reflection techniques for the study of very absorbing liquids the precision that can be achieved in the determination of the optical constants becomes worse. This is primarily due to the small reflection phase shift that has to be measured. This will usually be equivalent to a path length of a few micrometers at most, while the equivalent figure for a transmission experiment can, for a sufficiently transparent liquid, exceed several millimeters. Thus, the effects of random and systematic phase errors are considerably enhanced in the reflection case. In fact Honijk et al. (1977) reported difficulty in obtaining meaningful results from the reflection techniques, which they ascribed to surface imperfections in their silicon window. Experience with similar measurements on opaque solids (Section VI) indicates that truly precise reflection measurements require tolerances that are ideally below 0.1 μm on surface finish and interferometer stability. In subsequent reflection measurements on methanol Passchier et al. (1977b) found the typical random error in both real and imaginary parts of the complex refractive index to be about 0.02, which was about a factor of 5 worse than they achieved in measurements on the same liquid by transmission liquid cell methods. In reflection measurements backlash in the moving mirror drive and general interferometer instabilities will introduce unknown displacements between the sampling combs of sequentially recorded interferograms leading to a systematic phase error. This can be avoided if each interferogram is recorded to sufficiently large values of path difference for the (0, 0) signature from the lower surface of the window to be present. The position of this signature can then be used to align the sampling combs of the interferograms to correct for any backlash or drift. In practice the alignment is usually performed in the spectral domain.

For the quasi-full interferogram method Passchier et al. (1977b) found for transparent specimens that the errors are similar to those obtained with the $\xi_2 \phi_2$ transmission method, but they do point out that as editing is readily performed for such liquids the latter technique is easier and preferable to the former. For more absorbing liquids the quasi-full interferogram method gives errors intermediate between those of the transmission and the reflection methods. This reduction in accuracy means that the full interferogram method may not be the most suitable analytic method to use for such liquids. One needs to consider whether or not greater accuracy could be achieved through use of the subtraction $\xi_{2M}\phi_2$ method (Afsar *et al.*, 1977b; Passchier *et al.*, 1977b).

D. COMPARISON OF FREE LAYER AND LIQUID CELL TECHNIQUES

The analyses of Passchier *et al.* (1976, 1977b) of the various sources of error in free layer and liquid cell techniques, together with the many liquid DFTS measurements that have been published, enable the two groups of methods to be compared, identifying the particular advantages of each and establishing whether one is generally preferable to the other. These analyses and measurements indicate that for a liquid which is sufficiently transparent for transmission measurements on it to be made, both free layer and liquid cell methods will give results of comparable accuracy provided that allowance is made for the reflection phase shift in the free layer case. This is illustrated in Fig. 26 which shows measurements of the real refractive index and the power absorption coefficient of chlorobenzene at 25°C between 10 and 220 cm^{-1} made with both the free layer and the liquid cell methods (Afsar

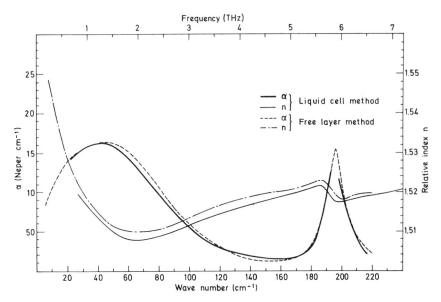

FIG. 26. The refractive index n and power absorption coefficient α of chlorobenzene at 25°C determined by both free layer and liquid cell methods. Generally, the refractive indices determined by the two methods agree to within 0.2%, while the power absorption coefficients agree within 2%. [From Afsar *et al.* (1977b). Reproduced by permission of the Institute of Electrical and Electronic Engineers, Inc.]

et al., 1977b). The free layer method used the two-thickness method to correct for the effects of the reflection phase shift at the upper surface of the liquid. Over this spectral range the refractive index values agree to within 0.2% and the power absorption coefficients to about 1 or 2%, with occasional excursions to differences of 5% or more (e.g., around 70 cm^{-1}). It has recently been suggested[4] that the agreement between the refraction spectra might, in fact, be considerably better than the 0.2% quoted above. This is because the difference between the two, which is largely systematic, could be due in part to a small difference between the free layer and liquid cell measurement temperatures not mentioned by Afsar *et al.*, (1977b).

In a similar manner it is found for reflection measurements on highly absorbing liquids that either of the two experimental configurations, the quasi-free layer or the liquid cell, will give results of comparable accuracy if similar precautions are taken over the stability of the entire interferometer. Thus, in general, we see that for transmission and reflection measurements the liquid cell methods do not presently give more accurate results than do the fully corrected free layer methods. However, there are several important reasons why the liquid cell methods are preferable:

(a)   As the liquid is fully contained there is no liquid vapor in the path of the radiation in the cell arm. Passchier *et al.*, (1976) have shown that this can provide the most serious errors in free layer measurements, especially if the liquid is particularly volatile or if the vapor has a strong line spectrum.

(b)   As the liquid is fully contained one can exercise very good control over its temperature thereby minimizing temperature-dependent errors. Additionally, the containment allows the temperature of the specimen to be varied over a much wider range than is possible with a free layer method for which, at elevated temperatures, vapor effects become exaggerated.

(c)   As the liquid cell contains all possible transmission and reflection configurations within it, in one piece of equipment, it is a versatile instrument and thus avoids the necessity of constructing several different liquid cells or interferometers if one desires a comprehensive dielectric measurement facility for liquids.

VI.   Dispersive Fourier Transform Spectrometry of Solids

We saw in Section IV how gases and vapors may be conveniently studied by DFTS, as the low values of their refractive indices usually enable inter-face effects occurring at retaining windows to be completely ignored, resulting in a simple interpretation of the complex insertion loss. In the case of liquids (Section V) retaining windows are still required, but a very wide range of

---

[4] M. N. Afsar, private communication.

values of the optical constants is encountered, and the interpretation of the complex insertion loss can become extremely complex as interface effects can no longer be ignored. From an experimental point of view, solids can be considered to fall between these two extremes, as the experimentalist has to contend with a wide range of values of the optical constants but without the complication of retaining windows. Thus dispersive transmission measurements can be made on transparent solids comparatively easily, and the limiting factors in the accuracy of optical constants determined in this way are usually the detector noise level and the geometrical quality of the specimen. In the case of dispersive reflection measurements on highly absorbing solids, however, a new difficulty arises, namely that the relative positions occupied by the specimen and reference mirror surfaces while their respective interferograms are recorded must be known with a high degree of precision. This, together with the frequent interest in solid-state physics in measurements as a function of temperature and particularly at low temperatures has provided the main source of experimental difficulties in the study of solids by DFTS. The interpretation of the complex insertion loss is usually straightforward, and, apart from one exception (Passchier *et al.*, 1977a), all reported measurements on solids by DFTS have been made with one of the three configurations illustrated in Fig. 2. Thus, apart from this case, the theoretical framework for analyzing the results has been fully developed in Section III.

Techniques for determining the optical constants of solids by DFTS were first developed simultaneously at the National Physical Laboratory by Chamberlain and coworkers, who concentrated on double-pass transmission measurements (Chamberlain *et al.*, 1963, 1969b; Chamberlain and Gebbie, 1966) and at Ohio State University by Bell and coworkers, who combined facilities for both single-pass transmission DFTS and amplitude reflection spectrometry in a single instrument (Bell, 1965, 1966; Russell and Bell, 1966, 1967a,b; Johnson and Bell, 1969). This work included measurements on crystal quartz (Chamberlain *et al.*, 1963; Russell and Bell, 1967a), sapphire (Russell and Bell 1967b), PTFE (Chamberlain and Gebbie 1966), and KCl and KBr (Johnson and Bell, 1969), which, although confined to room temperature, were in all cases sufficient to demonstrate clearly the advantages of DFTS over FTS. In later work at NPL, techniques have been extended with the development of polarizing interferometers for measurements at very low frequencies (see, for example, Afsar *et al.*, 1976b) and improved to the point that the optical constants of most specimens can now be determined at room temperature either by transmission DFTS (Afsar, 1977; Birch, 1978b) or by amplitude reflection spectrometry (Birch *et al.*, 1975, 1976; Birch and Murray, 1978) with a precision for surpassing that attainable by conventional FTS. Also, Passchier *et al.* (1977a) at the Rijksuniversiteit, Leiden, in association with the NPL group, have developed a novel technique for determining the optical constants of transparent solids with high precision, as part of their program for selecting windows for their liquid cell (Section V.C).

The past five years have also seen a rapid increase in the number of laboratories developing techniques for the study of solids by DFTS, as well as the development for the first time of techniques suitable for measurements at low temperatures. Thus the group at the Max Planck Institute for Solid State Physics in Stuttgart have developed a novel interferometer (Gast and Genzel, 1973) for amplitude reflection spectrometry of solids which was later used with a liquid helium cryostat (Gast et al., 1974) for measurements on Mg_2Si in the temperature range 80–300 K (Zwick et al., 1976) and more recently for measurements at temperatures down to 6 K on $LiTaO_3$ and sodium beta alumina (Mead and Genzel, 1978). Simultaneously, a range of techniques has been developed at Westfield College, University of London, for the study of solids by DFTS. This work began with the development of a technique for amplitude reflection spectrometry of solids which dispensed with the need for interchanging the specimen and reference mirror (Parker et al., 1974). The method was first used with an instrument with a conventional Mylar beam splitter for measurements on alkali halide crystals in the temperature range 100–300 K (Parker and Chambers, 1974, 1975, 1976). The same instrument was later used, with minor modifications, by Pai et al. (1978a,b) for similar measurements on a selection of alkali halide and ferroelectric crystals. Measurements on KH_2PO_4 and $NH_4H_2PO_4$ in the temperature range 100–300 K were later reported by Ledsham et al. (1976, 1977) using similar techniques with a polarizing interferometer. More recently, an instrument for making single-pass dispersive transmission measurements in the temperature range 100–300 K has been developed and used by Parker et al. (1977, 1978c) for measurements on alkali halide crystals, and in the latest reported work, an instrument has been constructed for use with a liquid helium cryostat and used for amplitude reflection measurements on KCl in the temperature range 7–300 K (Parker et al., 1978b). The group at the University of Cologne (Gauss et al., 1975; Gauss and Happ, 1976; Happ and Rother 1977) have described amplitude reflection measurements on KH_2PO_4 and deuterated KH_2PO_4 in the temperature range 100–300 K, but without giving detailed information on their techniques. More recently, Staal and Eldridge (1977), at the University of Vancouver, have reported measurements on KBr at room temperature using a modification of the technique of Parker et al. (1974) which enables the detector noise limit to be realized more easily in both the measured amplitude and phase spectra. Similar measurements have also been made on NaCl (Eldridge and

Staal, 1977), and the authors report[5] that measurements can now be made at temperatures down to 20 K.

A. EXPERIMENTAL ERRORS

In most dispersive work on solids the aim has been to exploit fully the potential of the technique for determining absolute values of both optical constants, rather than to acquire more limited information, such as the frequencies of particular spectral features, as is often the case in FTS. In a well-designed experiment, the uncertainty in both the amplitude $L(\tilde{v})$ and the phase $\phi_L(\tilde{v})$ of the complex insertion loss should, of course, be limited as in FTS by the detector noise level, but closer inspection shows that this constraint is far more severe for $\phi_L(\tilde{v})$ than for $L(\tilde{v})$.

In dispersive transmission spectrometry, where the measured phase angle is usually large, $\phi_L(\tilde{v})$ can often be determined with an accuracy approaching 1 part in 10^5 even though the determination of $n(\tilde{v})$ from this may be limited by the uncertainty in the crystal thickness rather than by the performance of the detector. However, in dispersive reflection spectrometry, the measured phase angle

$$0 < \phi_R - \pi \leq \pi \qquad (120)$$

is very small, and it is essential that the accuracy of the determination of ϕ_L should approach the limit set by the detector noise level. This can be very difficult to achieve experimentally as the measured phase angle corresponds, particularly at high wave numbers, to very small optical path lengths in the interferometer.

Four main sources of systematic phase errors occur in amplitude reflection spectrometry on solids (Russell and Bell, 1966; Parker et al., 1974; Staal and Eldridge, 1977; Birch and Murray, 1978) and are summarized as follows:

(a) inexact replacement of the specimen and reference mirror surfaces;
(b) differential thermal expansion between the two arms of the interferometer;
(c) backlash error in the micrometer drive of the moving mirror; and
(d) deviation of the sample geometry from the ideal shape.

The consequences of these errors can best be appreciated with reference to (a) by estimating the displacement of the specimen or reference mirror in the direction of the incident beam which would produce a change in the phase at a wave number $\tilde{v}$ equivalent to the detector noise level. If the signal-to-noise ratio at this wave number is s, so that random noise in the amplitude

---

[5] J. E. Eldridge, private communication.

and phase is approximately $1/s$ and $1/s$ rad, respectively, then the equivalent positional error is $\Delta x \simeq 1/4\pi\tilde{\nu}s$. For measurements at a wave number of 300 cm^{-1}, we find that $\Delta x \simeq 0.03$ μm, assuming that $s = 100$, which would be achieved with a fairly modest detector, such as a Golay detector. This corresponds to random noise in the phase spectrum of 0.01 rad, or about 0.6°, at 300 cm^{-1}. (With the best modern liquid helium-cooled detectors $s > 2000$, giving $\Delta x < 10^{-3}$ μm.) Thus if the systematic errors just listed cannot be reduced to this level, the limit of the detector noise level will not be reached in the phase spectrum.

The first of these errors is encountered only in amplitude reflection spectrometry, as the precise position of the specimen in the arm of the interferometer is of no consequence in dispersive transmission work, and the reduction of this error to an acceptable level has undoubtedly been the main obstacle to more widespread use of the technique. Basically, two different approaches to the problem have been followed. An instrument can be designed to carry out the replacement of the specimen and reference mirror with the desired accuracy, and instruments which do this with varying degrees of experimental sophistication have been described by Russell and Bell (1966), Gast and Genzel (1973), and Birch and Murray (1978). The instruments of Russell and Bell and of Birch and Murray are suitable for measurements at ambient temperature only, while that of Gast and Genzel has recently been used successfully at 6 K (Mead and Genzel, 1978). Alternatively, the replacement problem can be avoided by metallizing part of the specimen surface for use as a reference surface (Parker et al., 1974), and this technique has also been successfully used in the temperature range 7–300 K (Parker et al., 1978b).

The next two sources of systematic error in the list arise in all DFTS measurements but are most troublesome in amplitude reflection spectroscopy. Differential thermal expansion is easily reduced to an acceptable level by stabilizing the temperature of the interferometer, and backlash errors can usually be reduced to an acceptable level by recording the interferograms in a systematic way, so that the micrometer screw is always operated in the same direction. Parker et al. (1974) have pointed out that when part of the specimen is metallized for use as a phase reference surface, both of these errors can be eliminated by sampling both interferograms alternatively at each step on a single scan of the micrometer, and Staal and Eldridge (1977) and Parker et al. (1978b) have shown that this procedure does yield phase spectra whose accuracies are detector noise limited. Birch and Murray (1978) have described an alternative method for eliminating both of these errors by comparing the specimen and background interferograms at a point near the zero path position that depends on the accuracy of the replacement mechanism, which is $\simeq 0.01$ μm in the NPL interferometer.

Techniques have now improved to the point where the accuracy of phase measurements made with most of the instruments previously mentioned is limited not by instrumental factors but by deviations of the sample geometry from the ideal form required for the experiment. This is often the case in amplitude reflection spectroscopy since it can be difficult, especially with soft materials, to prepare samples with surfaces flat to better than ± 0.25 μm. In transmission work with thinned single crystals, where $d \simeq 100$ μm, it is often difficult to control the thickness to better than ± 2 μm (Berg and Bell, 1971; Parker *et al.*, 1978c) and even with thick specimens, i.e., $d \simeq 5$ mm, experimental uncertainties in d of the order of ± 1 μm place a limit on the accuracy of the determination of $n(v)$. However, the results of such measurements are acceptable for most applications (Russell and Bell, 1967a,b; Birch, 1978b).

When dispersive transmission spectroscopy is used to study thick transparent specimens, errors can occur in the determination of the amplitude transmission coefficient due to (a) imaging effects which are caused by thick specimens when $n(v) \gg 1$ and which are also observed in FTS, and (b) coherence effects which occur when the specimen is not perfectly plane parallel. Although these two effects are not entirely independent, the second is not observed in FTS as both beams then pass through the specimen, and its magnitude in DFTS can be estimated by considering a single-pass dispersive transmission measurement on a wedge-shaped specimen of mean thickness d, diameter D, and wedge angle α, such that the extreme thicknesses of the specimen are $d \pm \Delta d$, where $\Delta d = D\alpha/2$. If we use the same standard conditions as before, i.e., $s = 100$ at $v = 300$ cm^{-1}, then amplitude errors due to loss of coherence across the specimen will exceed the detector noise level if the difference between the two extreme values of the phase transmission coefficient [cf. Eq. (51)], $\Delta\phi \gtrsim 8°$, which is the condition that cos $\Delta\phi \gtrsim 0.01$. This occurs when

$$\Delta d \gtrsim 8/360nv = 0.4 \quad \mu m \tag{121}$$

if the refractive index $n = 2$ at $v = 300$ cm^{-1}. If $D = 25$ mm, the corresponding wedge angle, $\alpha \simeq 4$ seconds of arc and is beyond normal standards of specimen preparation. It is therefore clear that although DFTS is well suited under these circumstances to the determination of $n(v)$, large errors can arise in the value of $\alpha(v)$ determined in this way, especially with materials with very high refractive indices. Then $\alpha(v)$ can best be determined from power transmission measurements.

It is clearly advantageous when studying small specimens by DFTS to focus the radiation in the arms of the interferometer, and several instruments constructed in this way have been described in the literature. However, when dispersive transmission measurements are made with such an instrument, small systematic errors will be present in the calculated refractive index spectra if proper account is not taken of the convergence of the beam of radiation passing through the specimen. Russell and Bell (1967a) have given expressions for the convergence corrections which are appropriate for the study of isotropic samples, and for the three geometric configurations which are possible when uniaxial specimens are studied with plane polarized radiation. In the case of isotropic specimens, the true value n of the refractive index is related to the value n_{calc} determined from the complex insertion loss to a good approximation by

$$n_{\mathrm{calc}} = n[1 + \beta^2(1 - n^{-2})/4], \tag{122}$$

where β is the half-angle of the cone of radiation passing through the specimen, and the radiation is assumed to be uniformly distributed over the conical solid angle. Details of the derivation of this equation have been given by Chamberlain (1967a), and the expression is also appropriate for the ordinary ray refractive indices of (Y-cut) crystals which are measured with the optic axis parallel to the surface and perpendicular to the electric vector of an axial ray. A corresponding expression adapted from the work of Roberts and Coon (1962)

$$n_{\mathrm{e, calc}} = n_{\mathrm{e}}[1 + \beta^2(2 - n_{\mathrm{o}}^{-2} - n_{\mathrm{e}}^{-2})/8] \tag{123}$$

was used by Russell and Bell for correcting extraordinary ray refractive indices measured with Y-cut crystals with the optic axis parallel to the electric vector of an axial ray. They also pointed out that, with the subscripts o and e interchanged throughout, Eq. (123) can also be used to correct ordinary ray refractive indices measured with (Z-cut) crystals oriented with the optic axis perpendicular to the sample surface. These corrections are generally of significance only in the most accurate work, and Chamberlain (1967a) has shown, as an illustration of the magnitude of the effect, that for crystal quartz for which $n_{\mathrm{o}} \simeq 2.132$ at 110 cm^{-1} $n_{\mathrm{calc}}/n = 1.0030$ if $\beta = \frac{1}{8}$.

This discussion shall be used as a framework for the discussion of the performance of instruments in the next section. Where possible, quantitative information shall be given on the limitations of the various systematic phase errors on the performance of instruments which have been described in the literature.

B.   EARLY WORK

The first reported measurement of the refractive index spectrum of a solid by DFTS was that of Chamberlain et al. (1963). They measured the double-pass transmission spectrum of a Z-cut crystal of quartz at room temperature using a Michelson interferometer which was operated in the atmosphere despite difficulties with water vapor. We shall not describe their instrument in detail because, although it was later used for precise measurements of the optical constants of PTFE (Chamberlain and Gebbie, 1966), it was eventually superceded by more versatile instruments based on the NPL modular cube interferometer (Chantry *et al.*, 1969a). Measurements were made in the range 20–55 cm^{-1} in which $n(\tilde{v})$ is almost constant, and the results, which have already been presented in Fig. 1, were estimated to have an experimental uncertainty of about $\pm\frac{1}{2}\%$ and were in good agreement with earlier measurements by Korff and Breit (1932) and by Geick (1961). Although measurements of the absorption coefficient were not presented, the paper established for the first time the technique of broad-band refractive index measurements in the far infrared by DFTS.

Shortly afterward, Chamberlain and Gebbie (1965a) reported measurements of the refractive indices of polyethylene, Teflon, and crystal quartz at room temperature which were made using a Michelson interferometer in the asymmetric mode with a 337-μm (29.7-cm^{-1}) HCN laser source. The method was not, strictly speaking, a Fourier transform method, as neither of the Michelson mirrors was moved to produce a changing fringe pattern at the detector, specimens instead being placed in one arm of the interferometer and rotated in the beam while fringes were counted at the detector. Using this technique, the experimental uncertainty in the measured value of n was about 1.5%, and good agreement was obtained in the case of crystal quartz with the earlier measurement of Chamberlain *et al.* (1963). At the time it was considered that such laser measurements would provide a useful standard for fixing the absolute values of refractive index spectra determined by DFTS, and a similar suggestion was made by Chamberlain *et al.* (1965b). As we pointed out in the preceding section, the main limitation on the accuracy of refractive index spectra determined from dispersive transmission measurements is now the uncertainty in the specimen thickness rather than instrumental factors, so that there is no longer any need for such calibration procedures.

Chamberlain and Gebbie (1966) next used the interferometer described by Chamberlain *et al.* (1963) but evacuated to eliminate atmosphere absorption to determine both optical constants of polytetrafluorethylene at 22 °C in the frequency range 70–450 cm^{-1}. The material was chosen for a quantitative demonstration of the technique of DFTS because its absorption spectrum in the measured range contains only one feature of medium strength at 202 cm^{-1}, which is due to vibrations of neighboring CF_2 groups in the helical chains that comprise the polymer. A convergence correction was applied to their results, and the anomalous dispersion associated with the bands at 202 and 277 cm^{-1} was isolated by subtracting from the experimentally determined refractive index spectrum the dispersion due to an intense band at 516 cm^{-1}. Using a classical oscillator model for an isolated narrow absorption band of strength A_i and width $2\gamma_i$ at a wave number $\tilde{\nu}_i$ (Ditchburn, 1963; Moss, 1959), the optical constants can be written as

$$\Delta n(\tilde{\nu}) = n(\tilde{\nu}) - n(\tilde{\nu}_i) = \frac{A_i}{2\pi^2} \frac{\tilde{\nu}_i^2 - \tilde{\nu}^2}{(\tilde{\nu}_i^2 - \tilde{\nu}^2)^2 + 4\gamma_i^2\tilde{\nu}^2} \tag{124}$$

and

$$n(\tilde{\nu})\kappa(\tilde{\nu}) = \frac{A_i}{\pi^2} \frac{\gamma_i\tilde{\nu}}{(\tilde{\nu}_i^2 - \tilde{\nu}^2)^2 + 4\gamma_i^2\tilde{\nu}^2}, \tag{125}$$

where the convention $\hat{n} = n(1 - i\kappa)$ and $\alpha = 4\pi\tilde{\nu}n\kappa$, which is equivalent to Eq. (8), was used by the authors. The results which they obtained for $\Delta n(\tilde{\nu})$ and $\alpha(\tilde{\nu})$ in the wave number range from about 150 to 300 cm^{-1}, after correcting for the channel spectrum which was resolved in their measurements on a sample of only 75-μm thickness, are shown in Fig. 27, and it should be noted that the lines are drawn through the experimental points and are not derived from theory. In order to compare their results with the single oscillator model of Eqs. (124) and (125), Chamberlain and Gebbie used the result (Moss, 1959) that Eqs. (124) and (125) can be combined to give

$$[\Delta n(\tilde{\nu})]^2 + \{n(\tilde{\nu})\kappa(\tilde{\nu}) - [A_i/(8\pi^2\gamma_i\tilde{\nu}_i)]\}^2 = [A_i/(8\pi^2\gamma_i\tilde{\nu}_i)]^2. \tag{126}$$

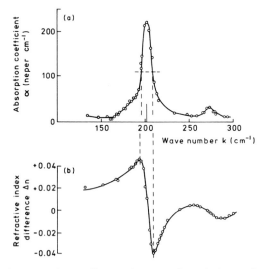

FIG. 27. (a) The absorption coefficient of polytetrafluorethylene (teflon) near 200 cm^{-1}. (b) The dispersion Δn [Eq. (124)] in the refractive index of teflon associated with the absorption feature at 202 cm^{-1}. The curves here join the experimental data points. [From Chamberlain and Gebbie (1966). Reproduced by permission of the Optical Society of America.]

Equation (126) represents a circle in the complex plane, with center at $\Delta n(\tilde{v}) = 0$ and $n(\tilde{v})\kappa(\tilde{v}) = A_i/(8\pi^2\gamma_i\tilde{v}_i)$ and with radius $A_i/(8\pi^2\gamma_i\tilde{v}_i)$, provided that

$$A_i^2/(4\gamma_i^2) \gg 4\pi^2\,\Delta n(\tilde{v}). \tag{127}$$

The results are shown plotted in this way in Fig. 28, and it can be seen that they conform closely to the expected behavior of a classical oscillator model. From the radius of the circle it was found that $A_i/\gamma_i = 657 \text{ cm}^{-1}$, giving $A_i = 4.93 \times 10^3 \text{ cm}^{-2}$, since $2\gamma_i = 15 \text{ cm}^{-1}$ from Fig. 27. However, it was concluded that a more precise value for A_i could be determined either by integrating the area under the curve of $\alpha(v)$ or from the refractive index spectrum by calculating the difference

$$n_i(\tilde{v}') - n_i(\tilde{v}'') = \frac{A_i}{2\pi^2}\left[\frac{1}{\tilde{v}_i^2 - \tilde{v}'^2 + \Gamma_i'^2} + \frac{1}{\tilde{v}''^2 - \tilde{v}_i^2 - \Gamma_i''^2}\right] \tag{128}$$

between the values of n at two wave numbers $\tilde{v}'$ and $\tilde{v}''$ situated on either side of the feature such that $\tilde{v}' < \tilde{v}_i < \tilde{v}''$, where

$$\Gamma_i^2 = (2\gamma_i\tilde{v})^2/(v_i^2 - \tilde{v}^2). \tag{129}$$

This result was obtained from Eq. (124) and was used to determine $A_i = 4.71 \times 10^{-2} \text{ cm}^{-1}$ by averaging the results from ten pairs of values of $\tilde{v}'$ and

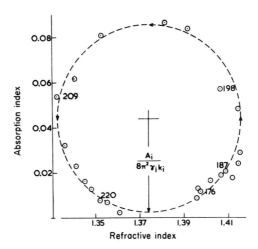

FIG. 28. Plot of the real part n of the refractive index against the imaginary part nk for the absorption in teflon at 202 cm^{-1}. The numbers near the experimental points indicate the wave number. The circular form of the plot fulfills the classical expectation. [From Chamberlain and Gebbie (1966). Reproduced by permission of the Optical Society of America.]

$\tilde{v}''$. They then fitted a classical oscillator model, using the oscillator parameters which have so far been determined to the refractive index data between 140 and 280 cm^{-1} to obtain a value for the static dielectric constant ε_0. The resulting value of $\varepsilon_0 = 1.93 \pm 0.02$ was in good agreement with the value of 1.94 ± 0.04 which was determined at 29.7 cm^{-1} by Chamberlain and Gebbie (1965a) using a laser. Thus this work gave the first complete quantitative description of an absorption band in terms of classical dispersion theory using results from DFTS.

Simultaneously with the NPL work, techniques were being developed at Ohio State University by Bell and his colleagues for studying solids by both reflection and transmission DFTS, and this work was first briefly described by Bell (1965) and, later, in more detail, by Bell (1966) and Russell and Bell (1966) in two papers describing the theoretical framework and the experimental techniques, respectively. The Michelson interferometer described by Russell and Bell is illustrated in Fig. 29, and the principal features of the instrument were as follows:

(a)   The angle of incidence of the radiation at the beam splitter was small, approximately 10°, to increase the high-frequency filtering efficiency of the

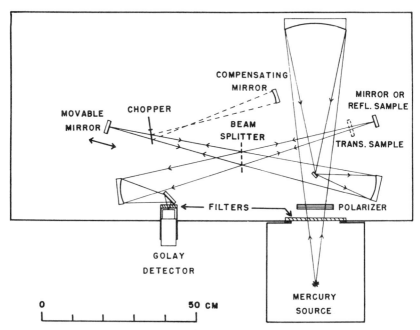

Fig. 29.   Optical plan of the vacuum far infrared interferometer designed for use in the asymmetric mode by Russell and Bell. [From Russell and Bell (1966). Reproduced by permission of Pergamon Press Ltd.]

metal mesh beam splitters which were used (Mitsuishi *et al.*, 1963; Vogel and Genzel, 1964) and to reduce the polarization of the radiation by the instrument.

(b)   The radiation was brought to a focus in each arm of the instrument to facilitate the measurement of small samples. Samples of 2-cm diameter produced an image that completely filled the detector window.

(c)   By changing the path of the radiation in the arms of the interferometer, specimens could be studied either by reflection or transmission, and Fig. 30 shows how this was achieved in the sample arm. In the reflection mode the specimen or reference mirror rested on three posts at B, and the radiation was reflected by the mirror A back along the same path to the beam splitter. In the transmission mode the specimen occupied the same position, but the radiation passed through it and was reflected back to the beam splitter by the upper mirror C so that different detector positions were required for the two measurements.

(d)   The chopper was placed in one arm of the interferometer. This is a convenient way of discriminating against source noise and reducing the demand on the dynamic range of the detector system and was suitable for use either in the reflection or the transmission mode by using a simple modification of the optics shown in Fig. 29.

To avoid indentation of soft specimens on the three posts during reflection measurements, three segments of a glass lens of large radius of curvature were used, and a replacement reproducibility of better than 0.1 μm was achieved. Errors due to thermal expansion were eliminated by stabilizing the temperature of the interferometer, and backlash errors were reduced to about 0.1 μm by scanning the interferograms in one direction only.

The instrument was suitable for measurements at room temperature only, and preliminary results which were presented in these early papers to demon-

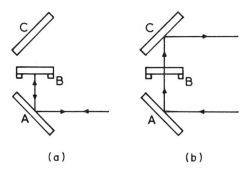

(a)   (b)

FIG. 30.   Schematic diagram of the end-mirror asembly showing the position of the sample in the sample arm of the interferometer of Russell and Bell: (a) reflection; (b) transmission.

strate the technique included the transmission spectrum of quartz (Bell, 1965), the amplitude reflectivity of KBr (Bell, 1965, 1966), and the transmission spectra of metal mesh and Mylar (Russell and Bell, 1966). As in all the early publications on DFTS, the precise distance between the positions of the grand maxima of the specimen and background interferograms was emphasized, and this no doubt provides a historical background to the computational procedure described by Chamberlain *et al.* (1969b), which, as we have already pointed out in Section III.D.2, is unnecessary (Birch and Parker, 1979).

In a subsequent paper, Russell and Bell (1967a) presented precise measurements of the optical constants of crystal quartz for both the ordinary and the extraordinary ray in the range 10–400 cm^{-1} at room temperature. They also determined dispersion parameters for two weak bands at 128.4 cm^{-1} and 263 cm^{-1} in the ordinary ray by fitting a classical oscillator model to their experimental data. Although these two bands had first been reported by Czerny (1929), they had not been included in the more quantitative analysis of Spitzer and Kleinman (1961) as their measurements only covered the range 270–2000 cm^{-1}.

For measurements below about 200 cm^{-1}, where quartz is very transparent, Russell and Bell were able to use thick specimens and record interferograms which included only the signature due to the first partial wave (Section III.D.2). For these measurements they used a Z-cut and a Y-cut specimen whose average thicknesses were determined with an error of less than 1 μm by comparison with precision gauge blocks. For measurements of the ordinary ray optical constants up to about 400 cm^{-1}, where quartz is relatively opaque, they used a very thin Y-cut specimen about 75 μm in thickness. They determined its thickness with an accuracy of 0.3 μm using Eq. (83) by comparing its measured phase at several of the channel spectrum extrema at lower frequencies with refractive index values determined with the thicker specimens. Measurements on this sample included channel spectrum effects, as signatures from several partial waves were included in the interferogram, and the optical constants were determined from the complex insertion loss using an equation similar to Eq. (86) by an iterative procedure. Their results for both rays for the wave number range 0–200 cm^{-1} are shown in Fig. 31.

They also used a pseudocoherence effect to determine the difference $n_o - n_e$ between the ordinary and extraordinary ray refractive indices at a selected number of points in the measured range. The effect has been described in more detail by Bell (1967b) and is observed as an interference effect between the ordinary and extraordinary ray as the difference $(n_o - n_e)d$ between their optical path lengths in the crystal varies with wave number when a Y-cut specimen is observed with unpolarized radiation. The order of interference is

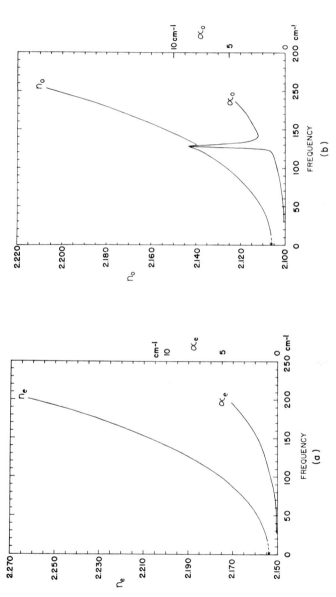

FIG. 31.   (a) Extraordinary ray refractive index and absorption coefficient of quartz for $E \perp C$. (b) Ordinary ray refractive index and absorption coefficient of quartz for $E \parallel C$. [From Russell and Bell (1967a). Reproduced by permission of the Optical Society of America.]

described by the equation

$$(n_o - n_e)d = m\lambda/2 = m/(2\tilde{v}), \tag{130}$$

where m is an integer and λ the wavelength. Thus maxima are observed when m is even and minima when m is odd, and the observed transmission spectrum is shown in Fig. 32. The differences $(n_o - n_e)$ between the two refractive indices calculated from this spectrum were in excellent agreement with the differences calculated from the separate refractive index spectra determined by DFTS. It was also pointed out that poor quality samples can lead, in DFTS, to a reduction in the height of the sample interferogram and erroneous values of $k(\tilde{v})$ and $\alpha(\tilde{v})$, due to loss of coherence, as described in Section VI.A. To avoid such problems they determined $\alpha(\tilde{v})$ from both power transmission measurements and DFTS and presented the former when discrepancies were observed between the two sets of data. Their refractive index values were estimated to have an experimental uncertainty of ± 0.001 or one part in 2000 over most of the measured range, which was considerably smaller than had previously been achieved by Roberts and Coon (1962) and Geick (1961) by measuring channel spectra. More recently, however, Loewenstein et al. (1973) have reported comparable results for quartz at room temperature and 1.5 K which were obtained from channel spectra. The corresponding errors in α varied from $\pm 100\%$ for $\alpha < 0.2 \text{ cm}^{-1}$ to $\pm 10\%$ for $\alpha > 0.8 \text{ cm}^{-1}$.

The refractive index spectra presented in this paper were more accurate than had previously been obtained for quartz in the far infrared either from channel spectra (Roberts and Coon, 1962; Geick, 1961), DFTS (Chamberlain et al., 1963), or laser measurements (Chamberlain and Gebbie, 1965a), and together with the results of Spitzer and Kleinman (1961), they provided a complete specification of the dispersion parameters for all the infrared

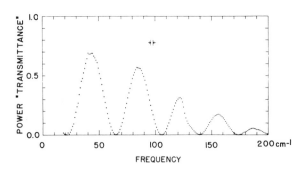

FIG. 32. Pseudocoherence between ordinary and extraordinary rays in a thin Y-cut quartz plate. The "transmittance" was observed with unpolarized radiation. [From Russell and Bell (1967a). Reproduced by permission of the Optical Society of America.]

bands of quartz. To complete the analysis, Russell and Bell were able to obtain agreement within the experimental error between the refractive index spectra calculated from classical dispersion theory and those measured experimentally by increasing the oscillator strengths determined by Spitzer and Kleinman by a few percent.

Similar measurements on sapphire were later reported by Russell and Bell (1967b). They also pointed out two additional advantages of DFTS over the channel spectrum method for determining refractive index spectra. The first is that DFTS can often be used with lower resolution than the channel spectrum method as the channel spectrum need not then be resolved, and the second results from the amplitude advantage of single-pass DFTS measurements on absorbing solids. As the square root of the power transmission coefficient is then measured, refractive index spectra can often be measured with samples that would be opaque for power transmission measurements. This is discussed later.

Johnson and Bell (1969) reported amplitude reflection measurements on single crystals of KBr and KCl at room temperature. The measurements were made at a resolution of 2 cm^{-1}, which was sufficient to resolve all features in the spectra, and their results for KCl are reproduced in Figs. 33 and 34. They also demonstrated the amplitude advantage that can be obtained by DFTS when either the power reflection coefficient $R(\tilde{v})$ or the power transmission coefficient $T(\tilde{v})$ is measured. Since

$$R(\tilde{v}) = \hat{r}(\tilde{v})\hat{r}^*(\tilde{v}) \tag{131}$$

and

$$T(\tilde{v}) = \hat{t}(\tilde{v})\hat{t}^*(\tilde{v}), \tag{132}$$

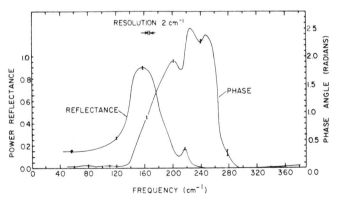

FIG. 33.   Experimentally measured power reflectance $R(\tilde{v})$ and phase spectrum $\phi_r(\tilde{v}) - \pi$ for KCl at 300 K. [From Johnson and Bell (1969). Reproduced by permission of the American Physical Society.]

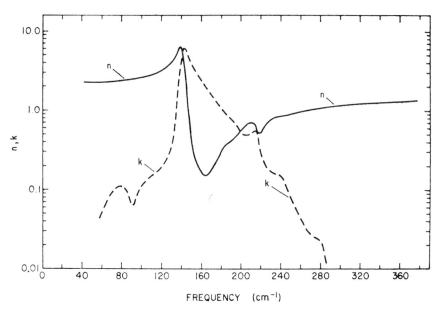

FIG. 34. Index of refraction $n(\tilde{v})$ and extinction coefficient $k(\tilde{v})$ of KCl at 300 K. [From Johnson and Bell (1969). Reproduced by permission of the American Physical Society.]

where $\hat{r}(\tilde{v})$ is the amplitude reflection coefficient and $\hat{t}(\tilde{v})$ the single pass amplitude transmission coefficient, it follows that when $R(\tilde{v})$ or $T(\tilde{v})$ is very small it can be advantageous to measure either $r(\tilde{v})$ or $t(\tilde{v})$, respectively. Johnson and Bell showed that near the minimum in the amplitude reflection–spectrum reflection coefficients as small as 0.02 could be measured with an accuracy of ± 0.005 using a Golay detector. This corresponds to determinations of power reflectivities of 4×10^{-4} with an accuracy of $\pm 2 \times 10^{-4}$, which is clearly beyond the capabilities of even the best modern instruments. Although this is not clearly visible in Fig. 33 as the power reflectivity is shown, their measurements also revealed structure in the phase spectra in the same spectral region that had not previously been observed by Kramers–Krönig analysis of power reflection spectra, and this is clearly seen in the case of KCl in Fig. 33. However, their work also revealed one of the limitations of amplitude reflection spectroscopy encountered when the specimen is fairly transparent. It can be seen from Fig. 34 that this is generally the case for KCl, as it is for all simple ionic solids, away from the *reststrahlen* band. Under these conditions, it can be seen from Fig. 33 that $(\phi_r - \pi) \to 0$, so that Eq. (76) reduces to the form

$$n(\tilde{v}) \simeq [1 + r(\tilde{v})]/[1 - r(\tilde{v})] \quad \text{and} \quad k(\tilde{v}) \simeq \{2r(\tilde{v})/[1 - r(\tilde{v})]^2\}\phi_r(\tilde{v})$$

$$(133)$$

and $n(\tilde{v}) \gg k(\tilde{v})$. Since, under these conditions, we can easily have $\phi_r(v) - \pi \lesssim 0.1°$, which cannot be satisfactorily measured, it is clear that amplitude reflection spectroscopy is not then suitable for determining $k(\tilde{v})$. To remove this limitation in the case of KBr, Johnson and Bell supplemented their amplitude reflection measurements with single-pass dispersive transmission measurements on thin ($\sim 100 \ \mu m$) single crystals. The analysis of such transmission measurements has been discussed in Section III.D.2. Although the performance of their instrument has just been described and their experimental errors are indicated in Fig. 33, the relationship between the errors in (r, ϕ_r) and those in (n, k) is not straightforward because of the complex form of Eq. (76), but this has been discussed in detail for the alkali halides by Pai et al. (1978a).

Johnson and Bell were also the first to use measurements on alkali halide crystals by DFTS for a detailed comparison of experimental values of the optical constants with theoretical descriptions of the complex dielectric response of an anharmonic crystal obtained with the techniques of thermodynamic Green's functions. Classically, the dielectric response of a simple harmonic oscillator is given by (Born and Huang, 1954)

$$\hat{\varepsilon}(\tilde{v}) = \varepsilon(\infty) + \{[\varepsilon(0) - \varepsilon(\infty)]\tilde{v}_0^2/(\tilde{v}_0^2 - \tilde{v}^2 - i\gamma\tilde{v})\}, \tag{134}$$

where $\tilde{v}_0$ is the oscillator frequency, γ a frequency-independent damping constant, $\varepsilon(0)$ and $\varepsilon(\infty)$ the static and limiting high-frequency values of the dielectric constant, respectively, and $[\varepsilon(0) - \varepsilon(\infty)]$ the oscillator strength.

This model is not capable of explaining the structure in the phase spectrum and the associated structure in the amplitude reflection spectrum which is evident in Fig. 33, because it neglects the details of the interactions between the normal modes which accompany the decay of the transverse optic phonon. Using the methods of diagrammatic perturbation theory (Maradudin and Fein, 1962), Wallis and Maradudin (1962) and Cowley (1963) have shown that for crystals with cubic symmetry the frequency-dependent dielectric susceptibility can be written as

$$\hat{\chi}_\alpha(\tilde{v}) = \chi_\alpha^E + \frac{1}{Nvh} \frac{2\tilde{v}(oj)M_\alpha^2(oj)}{\tilde{v}^2(oj) - \tilde{v}^2 + 2\tilde{v}(oj)[\Delta(oj, \tilde{v}) - i\Gamma(oj, \tilde{v})]}. \tag{135}$$

In this expression, χ_α^E is the electronic contribution to the susceptibility, $\tilde{v}(oj)$ the harmonic frequency of the transverse optic phonon at wave vector $\mathbf{q} \simeq 0$, $M_\alpha(oj)$ the α component of the leading term in the expansion of the crystal dipole moment operator in a power series of the normal mode coordinates, Nv the crystal volume, and $\Delta(oj, \tilde{v})$ and $\Gamma(oj, \tilde{v})$, respectively, the real and imaginary parts of the irreducible self-energy of the transverse optic phonon. The real part $\Delta(oj, \tilde{v})$ of the self-energy can be written as the sum of two parts:

$$\Delta(oj, \tilde{v}) = \Delta^E(oj) + \Delta^A(oj, \tilde{v}), \tag{136}$$

where $\Delta^E(oj)$ is a frequency-independent contribution which arises from the thermal expansion of the crystal and $\Delta^A(oj, \tilde{v})$ is frequency dependent and arises purely from the anharmonic interactions.

As the complex refractive index can be written as

$$n(\tilde{v})^2 = \hat{\varepsilon}(\tilde{v}) = \varepsilon(\infty) + \hat{\chi}(\tilde{v}), \tag{137}$$

Eq. (135) leads to a modified expression

$$\hat{\varepsilon}(\tilde{v}) = \varepsilon(\infty) + \frac{[\varepsilon(0) - \varepsilon(\infty)]\tilde{v}(oj)^2}{\tilde{v}(oj)^2 - \tilde{v}^2 + 2\tilde{v}(oj)[\Delta(oj, \tilde{v}) - i\Gamma(oj, \tilde{v})]} \tag{138}$$

for the dielectric response, where $\varepsilon(0)$ and $\varepsilon(\infty)$ can be expressed in terms of the coefficients given in Eq. (135). Thus the essential theoretical problem is to calculate the full frequency dependence of the functions $\Delta(oj, \tilde{v})$ and $\Gamma(oj, \tilde{v})$ which replace the constant damping coefficient γ in the classical expression given by Eq. (134).

In the model of Johnson and Bell the radiation was assumed to interact with the $\mathbf{q} \simeq 0$ transverse optic (TO) phonon mode via the first-order dipole moment and the subsequent decay of this mode by two-phonon processes with quartic and higher order anharmonic processes were neglected. The contributions to $\Delta^A(oj, \tilde{v})$ and $\Gamma(oj, \tilde{v})$ from cubic anharmonicity were calculated using phonon frequencies and eigenvectors obtained from the Karo and Hardy (1963) deformation dipole model, and the coefficient $M_\alpha(o, j)$ in Eq. (135) was evaluated using expressions given by Born and Huang (1954) and Cowley (1963). Johnson and Bell also showed that if $|\Delta(oj, \tilde{v}(oj))| \ll \tilde{v}(oj)$, then the contribution of $\Delta^E(oj)$ in Eq. (136) can be neglected, and they simplified their calculations by assuming that this inequality does hold. However, it can be shown from later work that this is not a reasonable approximation. In the case of KCl, for instance, at 300 K, $\Delta^E(oj) \simeq -7$ cm^{-1} (Lowndes and Rastogi, 1976) and $\Delta^A(oj, \tilde{v}(oj)) \simeq -3$ cm^{-1} (Rastogi et al., 1977), so that $\Delta(oj, \tilde{v}(oj)) \simeq -10$ cm^{-1}, compared with $\tilde{v}(oj) \simeq 151$ cm^{-1} (Lowndes and Martin, 1969). In recent work, which shall be discussed later, Pai et al. (1977) and Rastogi et al. (1977) have shown that both of the contributions to the self-energy in Eq. (136) can be fully determined by a combination of DFTS and high-pressure, far infrared spectroscopy (Lowndes and Rastogi, 1976). Johnson and Bell next used the values of $\Gamma(oj, \tilde{v})$ and $\Delta(oj, \tilde{v})$ and the coefficients $M_\alpha(o, j)$ and χ^E to calculate the complex susceptibility using Eq. (135), which was then used to calculate the refractive index and extinction coefficient from Eq. (137).

Their calculated and measured spectra for the extinction coefficient are shown in Fig. 35, and it can be seen that they obtained fair qualitative agreement between the two results, particularly above the TO frequency, where the three weak experimental features are reproduced quite well in the calculated

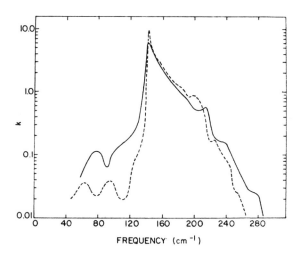

FREQUENCY (cm⁻¹)

Fig. 35. Experimental (solid curve) and calculated (dashed curve) extinction coefficient for KCl at 300 K. [From Johnson and Bell (1969). Reproduced by permission of the American Physical Society.]

curve. Inaccuracies in the calculated frequencies of these features were attributed to poor values of the phonon frequencies in the model. The more serious discrepancies between the theoretical and experimental curves on the low-frequency side of the *reststrahlen* band were attributed in part to inaccuracies in the experimental values of $k(\tilde{v})$, which were obtained for KCl from amplitude reflection measurements only and in part to the neglect of quartic contributions to the anharmonic self-energy.

In a later paper, Berg and Bell (1971) determined the optical constants of KI in the vicinity of the transverse optic phonon mode by a combination of amplitude reflection and single-pass dispersive transmission measurements. The transmission measurements were made on single crystals of KI approximately 100 μm in thickness, and the complex refractive index was computed by an iterative procedure from an expression similar to Eq. (86) which included contributions from the first two transmitted partial waves only. The values of the optical constants determined from the transmission measurements on either side of the region of intense absorption near the lattice resonance were used to fix the absolute value of the phase in the reflection measurements to avoid the need for precise replacement of the sample and reference mirror surfaces. Actually, this procedure is not strictly necessary when alkali halide crystals are studied, for the following reason. Most workers have preferred to develop techniques for determining the absolute value of the phase reflection coefficient with high precision without recourse to physical arguments, so that a wide range of materials can be studied. However, when specimens are studied that are transparent over some part of the measured spectral range, it follows from Eq. (30) that in these regions, $\phi_r - \pi \simeq 0$. Thus if this argument is used in the case of alkali halide crystals, the (incorrectly) measured phase $[\phi_r(v) - \pi + 4\pi\tilde{v}\,\Delta x]$, where Δx is an (unknown) replacement error, can be set to zero at some point of sufficiently high wave number, and the appropriate phase correction $-4\pi\tilde{v}\,\Delta x$ can be applied to the whole spectrum to remove the (now known) replacement error. This does require, however, that the value of k at this wave number be sufficiently small for the real phase delay to be less than the random phase noise in the measurement.

As in the earlier work (Johnson and Bell, 1969), structure was revealed in the phase that had not previously been observed by Kramers–Krönig analysis of power reflection spectra of comparable resolution (Hadni et al., 1968). The measured values of the optical constants were again compared with a theoretical determination based on the thermodynamic Green's functions which included only cubic anharmonicity, but the phonon dispersion curves used in the calculation were generated from a shell model with parameters selected by Dolling et al. (1966) to give phonon frequencies in good agreement with values determined from neutron diffraction experiments. Good agreement was obtained between experiment and theory in the spectral range where three-phonon interactions are allowed by energy conservation to contribute to the self-energy, but it was again concluded that at higher frequencies quartic anharmonicity should have been included.

C. RECENT DEVELOPMENTS

The early work on amplitude reflection spectroscopy at NPL was carried out in collaboration with Queen Mary College, University of London, using a simple replacement method suitable for use at room temperature only and an amplitude modulated cube interferometer (Parker et al., 1970). Measurements were presented on KH_2PO_4 (KDP) and $NH_4H_2PO_4$ (ADP), and the accuracy of the results was severely limited by random noise from the source. Progress on the development of this instrument was briefly reviewed by Chamberlain et al. (1974b), who presented measurements on KBr for comparison with the results of other groups (Johnson and Bell, 1969; Parker et al., 1974). Phase modulation was later added and the instrument was subsequently used for measurements on a wide selection of solids. This included intrinsic InSb (Afsar et al., 1975b) and ordinary soda–lime–silica glass (Birch et al., 1975). This latter work formed part of a wider spectroscopic study of the optical constants of glass over the range 0.29–4000 cm^{-1}. Dispersive measurements in the range 50–350 cm^{-1} were presented and were in excellent agreement with conventional measurements. Birch et al. (1976) also measured the a- and c-axis complex reflectivity of KDP and ADP in the range 5–350 cm^{-1}, and their measurements clearly revealed the low-frequency ferroelectric mode in KDP at a frequency of about 40 cm^{-1}.

The present form of the NPL interferometer for dispersive reflection measurements on solids at ambient temperature has been described in detail by Birch and Murray (1978), and the instrument is illustrated in Figs. 36 and 37. The instrument is oriented with the fixed mirror assembly positioned vertically upward, and the specimen or reference mirror can be placed in turn on three equilaterally spaced stainless steel balls. Indentation of soft specimens is avoided by using three additional balls which, as shown in Fig. 37, complete the points of a regular hexagon. The three additional supports are each attached to an arm that is free to rock in a vertical plane on a knife-edge fulcrum with an adjustable counterbalancing weight on the other end. The weights can be adjusted to take nearly all the weight of the specimen, thus removing it from the three locating balls. The fixed supports are mounted on adjustable gimbal rings so that the reflecting plane can be aligned normal to

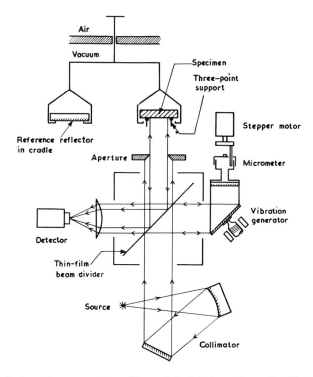

FIG. 36. A schematic representation of the modular interferometer used by Birch and Murray for dispersive reflection measurements on highly absorbing solids. [From Birch and Murray (1978). Reproduced by permission of Pergamon Press Ltd.]

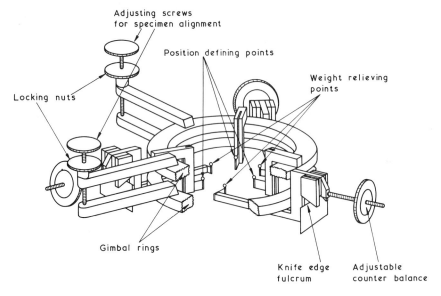

FIG. 37.   The specimen support assembly and counterbalancing arms of the modular interferometer used by Birch and Murray for dispersive reflection measurements on highly absorbing
solids. The gimball rings have been partly cut away for clarity. [From Birch and Murray (1978).
Reproduced by permission of Pergamon Press Ltd.]

the beam and the specimen and reference mirror can be lifted on and off in
cradles operated from outside the evacuated interferometer as shown in
Fig. 36. The aperture is defined by a circular stop sufficiently far removed from
the specimen that the signal reflected from it falls outside the usable path
difference range of the instrument.

Over most of the useful spectral range, which extended from 5 to 450 cm^{-1},
the level of random noise was equivalent to a displacement error of 0.008 μm,
and the precision of replacement of the specimen and reference mirror surfaces was about 0.01 μm, which corresponds to a phase error of about 0.6 m
rad, or 0.03°, at 100 cm^{-1}. As the replacement error was so close to the random
noise level, the following procedure could be used to eliminate systematic
phase errors due to backlash and differential thermal expansion. The background interferogram was first recorded, and the moving mirror returned to a
point on the steepest part of the interferogram near the position of zero optical
path difference, which, for phase modulated interferograms, lies between the
two grand maxima. The interferogram intensity I_0 at this point was recorded
and the reference mirror was replaced by the specimen and the new interferogram intensity I_s recorded without moving the moving mirror. Finally,
the specimen interferogram was recorded. As the two reference intensities

I_0 and I_s were recorded at the same position of optical path difference, they could be used to correct any inadvertent shift Δx between the sampling combs of the two recorded interferograms. This was done by locating the two intensities I_0 and I_s by interpolation near the positions of zero path difference on the two interferograms and determining the shift Δx between their recorded positions on the optical path difference scale. A correction $-2\pi\tilde{\nu}\,\Delta x$ was then added to the computed phase spectrum of the specimen. Although the procedure is not suitable for use in dispersive transmission spectrometry because the grand maxima of the two interferograms are then widely separated in optical path difference, the very small corrections entailed are then usually insignificant as the measured phase is large.

The performance of the instrument was illustrated with measurements on a crystal of CdTe recorded at a resolution of 1 cm^{-1} using a Golay detector. The experimental results are shown in Fig. 38, and over most of the measured range the standard deviation of the reflection spectrum was 0.005 and of the phase spectrum was 10 mrad ($\simeq 0.6°$). The real and imaginary parts of the complex relative permittivity obtained from these spectra and shown by the crosses in Fig. 39, were fitted with a model of a classical damped harmonic oscillator [Eq. (134)]. As there was no evidence of plasma effects in their results, no term was included to account for the response of the charge carriers. They were able to obtain good agreement between the model and their experimental results by fitting to either ε' or ε'' separately, as shown by the solid lines in Fig. 39, but in each case the corresponding fit to the other part, ε'' or ε', respectively, of the permittivity was comparatively poor, as shown by

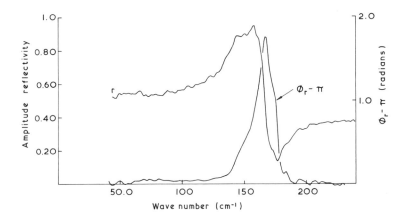

FIG. 38. The measured amplitude $r(\tilde{\nu})$ and phase $\phi_r(\tilde{\nu}) - \pi$ of the complex reflectivity of cadmium telluride measured at 290 K with a spectral resolution of 1 cm^{-1}. [From Birch and Murray (1978). Reproduced by permission of Pergamon Press Ltd.]

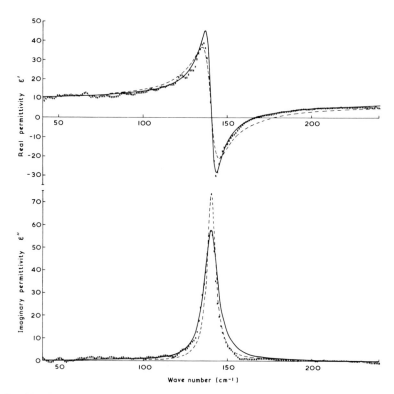

FIG. 39.   The real and imaginary parts of the complex relative permittivity $\varepsilon' + i\varepsilon''$ of cadmium telluride at 290 K. The crosses were computed from the complex reflectivity shown in Fig. 38. In each part the continuous curve is the best fit of the classical damped harmonic oscillator model to that particular data set. It not being possible to find a unique set of parameters capable of adequately describing both ε' and ε''. In the ε' plot the dotted curve was calculated from the best fit to the ε'' data and vice versa in the ε'' plot. [From Birch and Murray (1978). Reproduced by permission of Pergamon Press Ltd.]

the dotted lines in Fig. 39. The oscillator parameters that they determined by fitting to ε', which are dominated by the refractive index, were in good agreement with parameters determined by Randall and Rawcliffe (1968) and Danielewicz and Coleman (1974) mainly by fits to the refractive index spectrum, obtained in each case from channel spectrum measurements on thin specimens.

During the past few years the optical constants of many transparent solids have been determined at room temperature at NPL from dispersive transmission measurements. In one of the earlier investigations, Birch and Jones (1970) used transmission DFTS to compare the optical properties of two spinel ferrites as part of a program to find a suitable material for a Faraday rotation modulator for use from 50 cm^{-1} into the microwave region. The motivation for much of the later work has been the need for well-characterized materials for use as optical components such as windows in the work on liquids described in Section V, and the measurements were made by inserting the specimens in the fixed arm of modular cube interferometers constructed in either the conventional mode (Chantry et al., 1969a) or the polarizing mode (Afsar et al., 1976b) and illustrated in Fig. 15. All of the measurements were made in the double-pass configuration (Fig. 2a) on comparatively thick samples, and the analysis of the results is fully described in Section III.D.2. Among the materials for which optical constants have been reported are polyethylene (Chantry and Chamberlain, 1972; Afsar et al., 1976a), polypropylene (Chantry and Chamberlain, 1972; Afsar and Chantry, 1977), polymethylmethacrylate (Perspex) (Chamberlain and Gebbie, 1971a; Chantry and Chamberlain, 1972), TPX (Chantry and Chamberlain, 1972; Afsar et al., 1976a), and PTFE (Chamberlain and Gebbie, 1966; Chantry and Chamberlain, 1972). Afsar (1977) has reviewed the NPL dispersive transmission work and presented measurements of the optical constants of crystal quartz for both the ordinary and the extraordinary ray made with a polarizing interferometer (Fig. 15b) (Afsar et al., 1976b), which are not reported elsewhere, and they have reported satisfactory agreement with the results of Russell and Bell (1967), Loewenstein et al. (1973), and Passchier et al. (1977a).

Passchier et al. (1977a), at the Rijksuniversiteit, Leiden, have described a novel method for determining the complex refractive indices of transparent solids by DFTS, which they have demonstrated with measurements at room temperature on specimens of pure silicon and crystal quartz. The technique was developed for the determination of the optical constants of materials for use as windows in their liquid cell (Honijk et al., 1977), which is described in Section V.C. The principle is as follows: A plane parallel specimen is inserted normal to the beam in the fixed arm of the interferometer in the double-pass transmission arrangement of Fig. 2a, and the signatures associated with the first two partial waves, labeled $\hat{r}_{12}$ and $\hat{t}_{12}\hat{a}_2^2\hat{r}_{21}\hat{t}_{21}$ in Fig. 4 reflected from the front and back surfaces, respectively, are recorded. The specimen should be thick enough to separate fully these signatures, and no signatures arising from reflection at the fixed mirror are required in the measurement. The ratio of the complex Fourier transforms of the two recorded interferograms does not in this case correspond to the complex insertion loss defined in Section III.A, as neither of the interferograms is independent of the optical constants of the specimen. However, the ratio can be described as a complex relative frequency response,

$$\hat{\gamma}(\tilde{\nu}) = FT\{I_B(x)\}/FT\{I_F(x)\}, \tag{139}$$

where $I_F(x)$ and $I_B(x)$ are, respectively, the interferograms associated with the front and back surface signatures, and the optical constants of the specimen are related to $\hat{\gamma}(\tilde{v})$ by

$$\hat{\gamma}(\tilde{v}) = \hat{t}_{12}\hat{a}_2^2\hat{r}_{21}\hat{t}_{21}/\hat{r}_{12} = -\{4\hat{n}(\tilde{v})/[1 + \hat{n}(\tilde{v})]^2\}\exp 4\pi i\tilde{v}\hat{n}(\tilde{v})d, \quad (140)$$

where we have used Eqs. (22)–(25) and (68) and simplified the notation by putting $\hat{n}_1 = 1$ and $\hat{n}_2 = \hat{n}$, as before. Passchier et al. (1975) have described an iterative procedure for calculating $\hat{n}$ from $\hat{\gamma}$, but it should be noted that the experiment can only be performed as described if $n(\tilde{v}) \gg k(\tilde{v})$, so that it will usually be possible to simplify Eq. (140), as in the case of Eqs. (81)–(84), and solve it explicitly for $n(\tilde{v})$ and $k(v)$ in terms of the amplitude and phase of $\hat{\gamma}(\tilde{v})$. Although the technique permits both interferograms to be recorded in a single scan so that micrometer backlash errors can be avoided, such precautions are seldom justified as backlash errors are generally significantly smaller than the uncertainty in the specimen thickness. Similar remarks also apply to the other transmission techniques described in Section III.D. The authors point out that for the cancellation of the front surface $\hat{r}_{12}$ and back surface $\hat{r}_{21} = -\hat{r}_{12}$ amplitude reflectivities in Eq. (140) to be rigorously justified; the magnitudes of the two reflectivities should be identical. This is another manifestation of the coherence problems arising in transmission DFTS, which are discussed in Section VI.A., and again results in stringent conditions on the flatness and plane parallelism of the specimen surfaces. Apart from the fact that the authors already had the necessary experimental arrangement readily available in their liquid cell, the method has no obvious advantages over the conventional double-pass arrangement and has the disadvantage that each specimen must be accurately aligned normal to the incident beam.

Passchier et al. reported measurements using this technique of the optical constants of pure silicon and of the ordinary ray optical constants of crystal quartz, both at room temperature. In each case the real refractive index was determined with an accuracy of a few parts in 10^4, which was limited by systematic errors such as the uncertainty in the specimen thickness, and the uncertainty in the absorption coefficient was estimated as less than 0.3 Np cm^{-1}, with $\alpha(\tilde{v}) \lesssim 6$ Np cm^{-1} over most of the measured range for both crystals. The results for crystal quartz were in excellent agreement with those of Russell and Bell (1967a), differing systematically from the results of Loewenstein et al. (1973) in the same way as did the results of Russell and Bell. In the case of silicon, good agreement was obtained with the real refractive index measurements of Loewenstein et al. (1973) and Randall and Rawcliffe (1967), but values of the absorption coefficient were consistently lower than those reported by Loewenstein et al. The discrepancy was attributed by Passchier et al. to systematic errors in their interferometric system, the sources of which were not specified. The difficulty of measuring the absorption coefficients of very transparent materials by transmission DFTS has been discussed in Section VI.A.

Conventionally, amplitude or power reflectivities in the far infrared are measured by comparing the reflectivity of the specimen with that of a reference mirror, which is generally assumed to be a perfect reflector with a complex reflectivity

$$\hat{r} = 1.0 \exp i\pi, \tag{141}$$

although Staal and Eldridge (1977) have preferred to calculate $\hat{r}$ for their aluminized surface from the Drude free electron theory. Reference mirrors are usually made by vacuum-evaporating thin metal films on optically polished glass substrates, and their reflectivities may depart significantly from unity as a result of skin depth effects (Carli, 1977) or surface imperfections. Birch (1978b) has shown that the refractive index spectra of suitable thick transparent specimens, for which $n(\tilde{v}) \gg k(\tilde{v})$, determined by dispersive transmission spectroscopy, can be used to provide an absolute determination of the complex reflectivities of the specimens with a precision far beyond that normally achieved in either amplitude or power reflection spectroscopy. The refractive index determination is regarded as absolute as it requires [Eq. (83)] a knowledge only of the defined refractive index of vacuum

$$\hat{n} = 1.0 + i0.0 \tag{142}$$

and the length standard, which is required to sample the interferogram and measure the crystal thickness. The amplitude and phase reflection coefficients of the specimen can then be calculated with a high degree of precision from Eqs. (29) and (30) and the specimen subsequently used as the reference mirror in amplitude reflection measurements.

A sample of pure silicon was chosen as the reference standard, as silicon is transparent and contains no significant structure in the far infrared due to two-phonon bands. The complex refractive index was measured by transmission DFTS in the range 5–120 cm^{-1} and the refractive index spectrum determined with an accuracy of about 1 part in 4000, which was limited by the uncertainty in the sample thickness. This result was used to establish the complex amplitude reflectivity of the specimen with a precision of about 10^{-4}. Birch also measured the temperature dependence of the refractive index spectrum of silicon in the range 18–$24°C$, and found that his results were consistent with the slightly higher values of Loewenstein et al. (1973) measured at $27°C$. A similar comparison with the work of Randall and Rawcliffe (1967) and Passchier et al. (1977a) was not possible as these authors did not specify the temperature precisely.

The silicon crystal was used as the reference reflector in a determination of the complex reflectivities of four aluminum-on-glass mirrors using the instrument described by Birch and Murray (1978). The mirrors were made by vacuum evaporating films of aluminum with thicknesses in the range 0.017–0.20 μm on optically polished glass blanks. These thicknesses are comparable to the skin depth in bulk aluminum, which is about 0.15μm at 10 cm^{-1}, and the dependence of reflectivity on film thickness was clearly revealed in the results with a pronounced drop in amplitude reflectivity occurring at lower frequencies, particularly in the thinner films. It was concluded that the complex reflectivity scale of the reference reflector had been transferred to the film measurements with an accuracy of about 10^{-3} in the amplitude and 10 mrad in the phase, both being limited by the reflectivity comparison. The observed behavior was in reasonable agreement with a

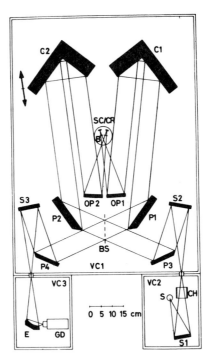

Fig. 40.   Optical plan of the asymmetric vacuum Michelson interferometer developed by Genzel and coworkers for the study of solids. The abbreviations are: S1–S3, spherical mirrors; P1–P4, plane mirrors; OP1, OP2, off-axis paraboloids; C1, C2, cube-corner mirrors; BS, beam splitter; E, off-axis ellipsoid; S, source; CH, chopper; GD, Golay detector; SC, sample chamber; CR, cryostat; B, beam reflected upwards; VC1–VC3, vacuum chambers. [From Gast *et al.* (1974). Reproduced by permission of the Institute of Electrical and Electronic Engineers, Inc.]

model that accounted for all multiple reflections in the vacuum–metal–glass–vacuum structure (Rouard, 1937), as well as for the size effect that occurs when the thickness of the metal films is much smaller than the mean free path of the free electrons (Fuchs, 1938; Chopra, 1969).

The group at the Max Planck Institute in Stuttgart have described an interferometer (Gast and Genzel, 1973), used with a liquid helium cryostat (Gast *et al.*, 1974), for making amplitude reflection measurements on solids from 4 to 300 K. The instrument has been designed with a number of novel features and is illustrated in Fig. 40. The radiation is incident on the beam splitter, which can be constructed with either metal mesh or Mylar films at the small angle of incidence of 23° to minimize unwanted polarization, the measured polarization of the beam being less than 10%. The two partial beams are reflected by plane mirrors P1 and P2 onto cube corner mirrors C1 and C2 and focused by off-axis parabolic mirrors OP1 and OP2 onto the specimen and reference mirror inside the sample chamber and cryostat SC/CR. Thus each arm of the interferometer is terminated inside the cryostat at a focus, and samples as small as 6 mm in diameter can be studied. The interferometer is scanned by moving the cube corner reflector C2, as indicated, so that the change in optical path difference is four times the displacement of the mirror. The cryostat and sample chamber have been described briefly by Gast *et al.* (1974), and further details of the sample chamber have been given

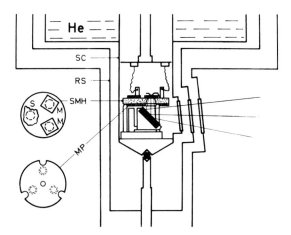

FIG. 41. Sample holder in the cryostat used with the interferometer shown in Fig. 40 (section parallel to the plane of the beam splitter). The abbreviations are: SC, sample chamber; RS, liquid-nitrogen-cooled radiation shield; SMH, top view of the sample–mirror holder; MP, metal plate mounted beneath the sample–mirror holder with three notches 120° apart for determining the three combinations, namely mirror–mirror, sample–mirror, and mirror–sample. [From Zwick *et al.* (1976). Reproduced by permission of Pergamon Press, Ltd.]

by Zwick *et al.* (1976). The latest forms of the cryostat and sample chamber have been described in detail by Mead (1978) and (1979), respectively. The cryostat of Zwick *et al.* is illustrated in Fig. 41, and a more recent sample chamber described by Mead (1979) is illustrated in Fig. 42. The beams from the parabolic mirrors are reflected upward through an angle of 90° from two plane mirrors toward the sample and reference mirror which are wrung onto an optically flat quartz plate used to define the reference plane. Measurements are made with two mirrors and a sample, as shown in Fig. 42, and the quartz plate can be rotated manually from outside the vacuum chamber in 120° steps so that the two beams can sample the combinations mirror–mirror, mirror–sample, and sample–mirror in turn.

The liquid helium and liquid nitrogen cans of the cryostat are suspended from the top by their filler tubes in the conventional way. A stout stainless steel screw is threaded through the radiation shield and clamped between two hardened balls located in the bottom of the copper tail of the sample

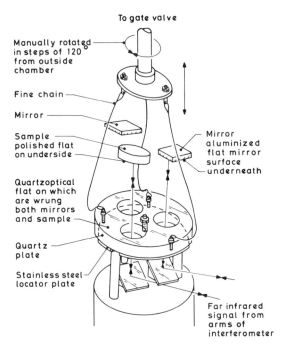

FIG. 42.   Illustration of the method by which the sample and two mirrors are rotated in 120° steps so that the two beams can sample the combinations mirror–mirror, mirror–sample, and sample–mirror for measurements at room temperature. A modification of this arrangement (Mead, 1978) has been developed for measurements in the temperature range 4.2–300 K. [From Mead (1979). Reproduced by permission of Pergamon Press, Ltd.]

mount and in the base of the cryostat. The balls act as pivots and reduce the thermal input to the specimen chamber, and the stainless steel bolt prevents movement of the sample holder during measurements at low temperatures. The radiation shield contains a number of radial slots to enable it to accommodate the strain imposed during thermal contraction.

Low-temperature measurements were made by using helium exchange gas to cool the specimen in the sample chamber, which was sealed with a cold polyethylene window. A second polyethylene window isolated the high vacuum ($< 10^{-5}$ Torr) in the cryostat from the interferometer chamber, but the cryostat could not be pumped during measurements because of the disturbance of the sample position by pump vibrations. The two partial beams were arranged to cross each other at the vacuum isolation window to provide better compensation for any dispersion arising in the window, and both windows were inclined at an angle of greater than $5°$ to the two incident partial beams to prevent any coherently reflected components from reaching the detector.

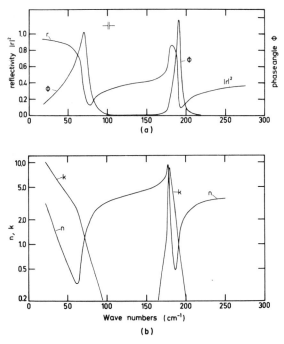

FIG. 43. (a) The power reflectivity r and phase reflection coefficient $\phi_r - \pi$ of InSb measured at room temperature; (b) the corresponding real and imaginary parts of the complex refractive index (n, k). [From Gast and Genzel (1973). Reproduced by permission of North-Holland Publishing Company.]

The performance of the instrument was first demonstrated by Gast and Genzel (1973) with measurements on InSb and InAs at room temperature. A replacement accuracy of better than 0.3 μm was reported, and their results for InSb are reproduced in Fig. 43. The measured range includes both the fundamental lattice resonance at about 180 cm^{-1} and the onset of free carrier absorption at lower frequencies. The results were in good agreement with the earlier data of Yoshinaga and Oetjen (1956) and Bell (1967a). From the known relationship between the position of the minimum in the reflectivity associated with the plasma edge and the effective mass and concentration of the free charge carriers (Yoshinaga and Oetjen 1956), the authors calculated a value for the charge carrier concentration which was in almost perfect agreement with the manufacturer's specification for the crystal.

The first measurements with the cryostat were described by Zwick *et al.* (1976), who reported measurements on Mg$_2$Si at 80 and 300 K in the range 20–600 cm^{-1}, which are reproduced in Figs. 44 and 45. The measured range included the single infrared active lattice mode at about 300 cm^{-1} and the effects of free carrier absorption at longer wavelengths. The transverse optic mode frequencies $\tilde{\nu}_{\text{TO}} = 271$ cm^{-1} at 300 K and 278 cm^{-1} at 80 K were

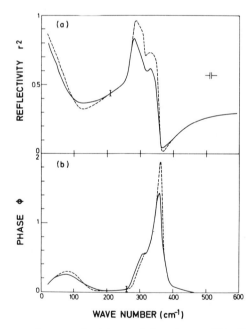

FIG. 44.   The measured power reflectivity r and phase $\phi_r - \pi$ of Mg$_2$Si at 300 K (solid curve and 80 K (dashed curve). [From Zwick *et al.* (1976). Reproduced by permission of Pergamon Press, Ltd.]

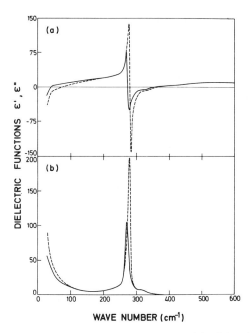

FIG. 45. (a) The real part ε' and (b) the imaginary part ε'' of the dielectric function of Mg_2Si at 300 K (solid curve) and 80 K (dashed curve). [From Zwick *et al.* (1976). Reproduced by permission of Pergamon Press Ltd.]

determined from the positions of the resonances in the dielectric functions, and the longitudinal optic (LO) mode frequencies $\tilde{\nu}_{LO} = 350$ cm^{-1} at 300 K and 354 cm^{-1} at 80 K were derived from the positions of the zero crossings in the real part ε' of the dielectric function, after subtracting out the Drude term due to the contribution of free carriers. The room-temperature value of $\tilde{\nu}_{TO}$ was in good agreement with that of McWilliams and Lynch (1963), who fitted a damped harmonic oscillator [Eq. (134)] to the measured power reflectivity, and the values of $\tilde{\nu}_{LO}$ were in good agreement with those determined by Anastassakis and Burstein (1971) from resonant Raman scattering measurements.

In the latest reported work with this instrument (Mead and Genzel, 1978) the typical positional accuracy was given as 0.5 μm, corresponding to a phase error at 200 cm^{-1} of about 7°, and measurements were presented on GaTe at 300 K, LiTaO$_3$ at 6 K, and sodium beta alumina at 6 and 300 K. The amplitude reflectivity of a small single crystal of LiTaO$_3$, which is a uniaxial ferroelectric material, was measured with the electric vector perpendicular to the c-axis, and the results are shown in Figs. 46 and 47. The authors were prevented by the shape of their specimen from making measurements with the

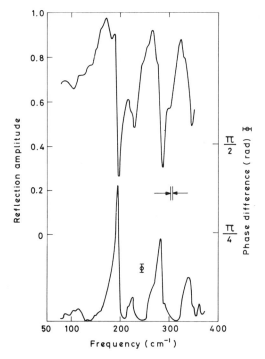

FIG. 46.   The far infrared complex reflection spectrum of LiTaO$_3$ measured at 6 K with
E ⊥ **C**. [From Mead and Genzel (1978). Reproduced by permission of Pergamon Press, Ltd.]

electric vector parallel to the c-axis and relied for mode assignments on the
work of Barker et al. (1970). Mode frequencies were determined from the
poles in the dielectric functions in the usual way, but although there was
insufficient information for an unambiguous identification of the modes
observed at 6 K with those assigned by Barker et al., measurements on
LiTaO$_3$ had not previously been reported below room temperature.

A wide range of techniques has been developed at Westfield College,
University of London, for measuring the optical constants of solids by DFTS
in the temperature range 4–300 K using instruments based on the NPL
modular cube interferometer. As the application of these techniques has been
largely confined to studying the dielectric response of alkali halide and ferroelectric crystals, which are highly absorbing over most of the spectral range
of interest, the emphasis has been mainly on amplitude reflection spectroscopy. Thus techniques were first developed for dispersive reflection spectroscopy in the temperature range 100–300 K using instruments equipped with
either dielectric, thin-film beam splitters and operated in the conventional
mode (Parker et al., 1974; Parker and Chambers, 1974, 1975, 1976) or with

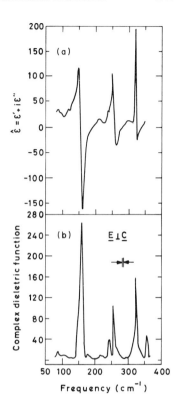

FIG. 47. (a) The real part ε' and (b) the imaginary part ε'' of the dielectric function of LiTaO$_3$ at 6 K for $E \perp C$. [From Mead and Genzel (1978). Reproduced by permission of Pergamon Press, Ltd.]

free-standing, wire grid beam splitters and operated in the polarizing mode (Parker *et al.*, 1976; Ledsham *et al.*, 1976, 1977). More recently, however, an instrument has been developed for making single-pass dispersive transmission measurements on solids in the temperature range 100–300 K by Parker *et al.* (1977, 1978c) to remove the limitation that occurs in dispersive reflection spectroscopy when $k(\tilde{\nu})$ is small and $\phi_r - \pi \to 0$. In the latest work, a liquid helium cryostat and interferometer have been described by Parker *et al.* (1978b) for dispersive reflection spectroscopy in the temperature range 4–300 K.

The common feature of all the instruments developed for amplitude reflection spectroscopy has been the elimination of the need for the precise mechanical replacement of the reference mirror by the specimen. The technique was originally demonstrated by Parker *et al.* (1974) with measurements at room temperature on a KBr crystal using an amplitude-modulated interferometer. This arrangement proved excessively noisy because of the large dc load carried on the interferogram, but a later form of the instrument with phase modulation proved satisfactory (Parker and

Chambers, 1974) and is illustrated in Fig. 48a. The basic idea is that the optically flat specimen can be inserted as a fixed reflector in the fixed arm of the interferometer, with part of its surface metallized for use as a reference mirror. In the original arrangement, shown in Fig. 48b, areas B and C of the specimen were aluminized and area A was left exposed. Interferograms could be recorded by reflection of the partial beam from areas A, B, or C by moving two screens S1 and S2 in front of the specimen and, by careful design of the screens, cross talk between the three areas was avoided.

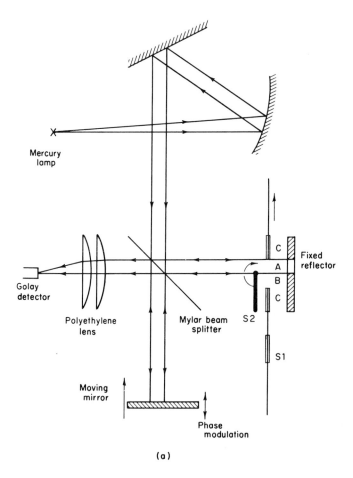

(a)

FIG. 48.   (a) Schematic diagram of the Michelson interferometer developed by Parker and coworkers for dispersive reflection measurements on solids, showing the two movable screens, S1 and S2, in front of the fixed reflector and the paths of the coherent beams of radiation using part A.

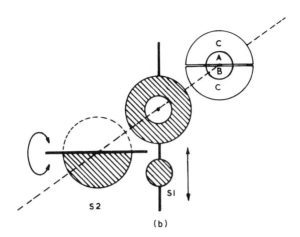

FIG. 48.   (b) The geometry of the two screens and the division of the field of view at the fixed reflector. [From Parker and Chambers (1974). Reproduced by permission of the Institute of Electrical and Electronic Engineers, Inc.]

The specimen was first aligned using area C. A specimen interferogram S was recorded with part A exposed, followed by a background interferogram B′ with part B exposed. Then their complex Fourier transforms $\rho_S \exp i\phi_S$ and $\rho_{B'} \exp i\phi_{B'}$, respectively, computed. In these expressions, ρ represents the modulus and ϕ the phase of the computed spectrum in the notation of Section III.B. In an ideal, perfectly aligned, interferometer the ratio of these two complex spectra would give directly the complex insertion loss $r \exp i(\phi_r - \pi)$ of the specimen [Eq. (75)]. However, although this condition could be approached fairly closely by carefully aligning the interferometer, small systematic differences remained between the background spectra $\rho_A \exp i\phi_A$ and $\rho_B \exp i\phi_B$, which were recorded from areas A and B, respectively, by the same procedure after replacing the specimen by a perfect mirror. As these systematic differences were accurately reproducible (Parker et al., 1974, 1975) each time the reference reflector was installed, they could be eliminated from the final result by computing the amplitude and phase reflection coefficients of the specimen from

$$r = (\rho_S/\rho_{B'}/(\rho_A/\rho_B)) \qquad \text{and} \qquad \phi_r - \pi = (\phi_S - \phi_{B'}) - (\phi_A - \phi_B). \quad (143)$$

[In the original articles the π was omitted, it being understood that the phase spectrum is always measured relative to that of a reference mirror (Parker and Chambers, 1976).]

The accuracy of the phase spectrum determined in this way is limited by the detector noise level and by systematic errors due to backlash in the moving mirror drive, but tests showed that as backlash errors were less than about 0.1 μm in magnitude systematic errors in the phase spectrum were comparable to the detector noise level (Parker et al., 1976). However, the technique can be extended to eliminate systematic errors due to screw backlash and differential thermal expansion by programming the screen assembly and micrometer drive so that the specimen and background interferograms are sampled alternately at each step on a single scan of the moving mirror. Thus all systematic phase errors apart from those associated with the sample geometry can be eliminated. The technique also has the advantage that as the specimen need not be mechanically isolated as in most replacement methods it can be placed in good mechanical and thermal contact with a suitable reservoir for measurements at low temperatures. There are, however, several disadvantages that are as follows. First, if the method is used as previously described, four interferograms are required to determine the specimen spectra from Eq. (143). In more recent work, to be discussed later in more detail, Staal and Eldridge (1977) have shown that if the screen geometry is changed to make better use of the symmetry of the beam, the sample spectra can be determined satisfactorily from two interferograms recorded in a single scan. They also found that all systematic instrumental errors were eliminated in this way and that the constraint on the sample geometry was considerably relaxed. Second, comparatively large crystals are required as the specimen is placed in a collimated beam, and part of its surface is sacrificed for the reference measurement. In practice, it has been found that, with minor modifications to the screen geometry (Ledsham et al., 1976), samples as small as about 18 mm in diameter can be conveniently measured with a Golay detector. The technique could clearly be extended to accommodate still smaller samples by focusing the radiation in the arms of the interferometer, as in the instruments of Russell and Bell (1966) and Gast and Genzel (1973). A third possible disadvantage, namely that the technique is destructive of the metallized part of the surface, has not been found to be of any significance in any of the reported applications.

For measurements at temperatures between 100 and 300 K the sample mount shown in Fig. 49 was used (Parker and Chambers, 1974). The specimen was mounted on a copper block attached at three points by nylon bolts and stainless steel springs to a sensitive alignment mechanism, and the back of the block was attached by thick copper braid to a liquid nitrogen cold finger. The sample mount was thermally isolated from the interferometer by the nylon bolts, and the sample could be accurately aligned normal to the beam from outside the interferometer after it had been cooled. Preliminary measurements were made on a KBr crystal at 5 cm^{-1} resolution at 200 and 300 K, and it was reported that there was no loss of phase accuracy on cooling the specimen. In this work the thickness (~ 0.08 μm) of the metal film on the crystal surface was neglected in the calculation of the phase spectrum, but in

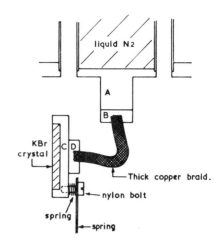

FIG. 49. Schematic diagram of the specimen mount and cold finger used with the interferometer shown in Fig. 48 for dispersive reflection measurements on solids in the temperature range 100–300 K, with pressure approximate 5×10^{-6} torr. Parts A, B, C, and D are copper. [From Parker and Chambers (1974). Reproduced by permission of The Institute of Electrical and Electronic Engineers, Inc.]

all subsequent work thicker films were used (~ 0.25 μm) to avoid skin depth effects (Carli, 1977) and their thicknesses measured optically so that an appropriate phase correction could be applied. After minor modifications to the sample mount, the instrument was used by Parker and Chambers (1975, 1976) to determine the complex amplitude reflectivity of KBr at 100 and 300 K, and the results are shown in Fig. 50. The room-temperature measurements were in good agreement with those of Johnson and Bell (1969), and the results at 100 K were the first reported measurements by DFTS below room temperature, apart from the preliminary, resolution limited, measurements on KBr at 200 K already mentioned (Parker and Chambers, 1974).

The results shown in Fig. 50 were used with the aid of Eqs. (10) and (76) to calculate the real and imaginary parts of the anharmonic self-energy of the transverse optic phonon from an expression similar to Eq. 138 but with the harmonic frequency renormalized to include the effects of thermal expansion (Bruce, 1973). The experimentally measured values of the transverse optic mode frequency at 90 and 290 K reported by Lowndes and Martin (1969) were used as the quasi-harmonic mode frequencies, and values of $\varepsilon(0)$ and $\varepsilon(\infty)$ were also taken from their paper, but, as in the work of Johnson and Bell (1969), the contribution of the thermal strain contribution $\Delta^{E}(oj)$, to $\Delta(oj, \tilde{v})$ [Eq. (136)] was neglected. The values of the self-energy functions determined in this way at room temperature were found to be in reasonable agreement, both qualitatively and quantitatively, with those calculated by Hisano et al. (1972) neglecting quartic anharmonicity and in slightly better agreement with the later calculations of Bruce (1973) which included quartic anharmonicity. The temperature dependence of the results was in reasonable agreement with the early calculations of Cowley (1963) but no recent calculations had been reported at 100 K.

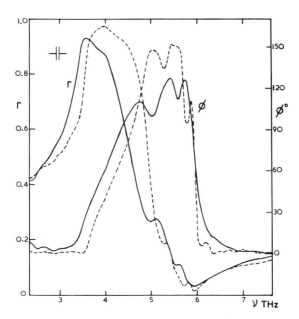

FIG. 50. The amplitude and phase reflection spectra of KBr measured at 100 K (dashed curve) and 300 K (solid curve). [From Parker and Chambers (1975). Reproduced by permission of the Institution of Electrical Engineers.]

An apodizing procedure used by Parker and Chambers (1976) is of particular interest in DFTS since it takes account of the intrinsic asymmetry which is present in dispersive interferograms. Generally, filtering in a Fourier spectrometer is designed to give a smooth background interferogram containing no sharp features in the spectral range of interest, and it follows that the associated interferogram rapidly decays to zero at relatively small optical path differences $\pm l$. Although the specimen spectrum may contain sharp features, all spectral components reflected from the specimen are delayed in phase by between π and 2π rad [Eq. (30)], so that the specimen interferogram can be recorded from the same starting point as the background interferogram and extended to a positive optical path difference $l' > l$, which is sufficiently large to include all the structure associated with the response of the specimen. The lengths of the wings of the four interferograms were chosen in this way by Parker and Chambers, and an apodizing function was used on each wing which was flat for two-thirds of the wing and rolled off to zero as $(1 - \dot{x}^2)^2$ over the remaining third. This procedure was found to be a useful way of discriminating against additional noise which would be included if all four interferograms were recorded with wings of equal length l' and was found to approximately halve the time taken to record the data with a simultaneous reduction in the final random noise level by a factor of approximately $\sqrt{2}$. A careful comparison of results obtained in this way and from interferograms recorded in the usual way, with all wings of length l', confirmed that the expected reduction in the noise level was realized with no visible effect on the spectra themselves. It should be noted, however, that when a very asymmetric, apodizing function is used, random noise fluctuations in the real and imaginary parts of the complex reflectivity can appear to be related by the Kramers–Krönig relations (Cardona, 1969), so that normal precautions should be taken when using this procedure to reduce the noise to an acceptable level.

For all later work the screen geometry shown in Fig. 48b was modified so that area C could be used to record the background interferograms, with both areas A and B left unmetallized for recording the specimen interferograms (Ledsham et al., 1976). With this arrangement, smaller samples could be studied, and similar measurements were later reported on NaCl at room temperature (Parker et al., 1976). The real and imaginary parts of the anharmonic self-energy of the transverse optic mode were calculated as described above and again found to be in reasonable agreement with theoretical results reported by Hisano et al. (1972). However, a sharp feature at 234 cm^{-1} in the experimental results was not present in the results of Hisano et al. but has since been confirmed both experimentally and theoretically by Eldridge and Staal (1977) and identified as a feature associated with the summation of two acoustic branches of the dispersion curves.

This instrument was later used by Pai et al. (1978a), with minor modifications, to determine the optical constants of the alkali halide crystals NaF, NaCl, KCl, KI, RbCl, RbBr, and RbI at 100 and 300 K from measurements of their amplitude and phase reflection spectra and to extend to lower frequencies the previous measurements on KBr (Parker and Chambers, 1975, 1976). Similar measurements were made on the three pseudodisplacive ferroelectric crystals $KTaO_3$, $SrTiO_3$, and TlBr in the same temperature range (Pai et al., 1978b). From a statistical analysis of the results, the errors in the measured complex amplitude were estimated to have an rms magnitude of less than ± 0.005 in all cases, corresponding to amplitude errors of about $\pm \frac{1}{2}\%$ and phase errors of about $\pm 0.3°$ in spectral regions where the amplitude reflectivity is close to unity, with a corresponding reduction in accuracy elsewhere, and tabulated data were given in all cases of the optical constants $n(\tilde{v})$ and $k(\tilde{v})$ and their associated errors.

In the case of the alkali halide crystals (Pai et al., 1978a), good agreement was obtained with the previous room-temperature measurements of Johnson and Bell (1969) on KBr and KCl and with the results of Berg and Bell (1971) for KI. In most cases structure associated with anharmonicity was revealed that had not previously been observed by Kramers–Krönig analysis of power reflection spectra. Although it was not possible because of this to make a detailed comparison with earlier experimental results, the frequencies of the transverse optic modes determined from the positions of the peak values of the conductivity $\sigma(\tilde{\nu}) = \tilde{\nu}\varepsilon''(\tilde{\nu})/2$ (Seitz, 1940) were in all cases in good agreement with values reported by Jones et al. (1961), Wilkinson (1963), and Lowndes and Martin (1969). It was also pointed out that, although the technique is suitable for the determination of $n(\tilde{\nu})$ when $\phi_r - \pi \to 0$, $k(\tilde{\nu})$ can then be best obtained from transmission measurements.

It follows from Eqs. (137) and (138) (Rastogi, 1975) that the real and imaginary parts of the anharmonic self-energy of the transverse optic mode can be written as

$$\Delta^A(oj, \tilde{\nu}) = \tfrac{1}{2}\{\tilde{\nu}(oj)[(\varepsilon(0) - \varepsilon(\infty))\eta'(\tilde{\nu}) - 1] + [\tilde{\nu}^2/\tilde{\nu}(oj)]\} - \Delta^E(oj), \qquad (144)$$

$$\Gamma(oj, \tilde{\nu}) = -\tfrac{1}{2}\tilde{\nu}(oj)[\varepsilon(0) - \varepsilon(\infty)]\eta''(\tilde{\nu}) \qquad (145)$$

where

$$\eta'(\tilde{\nu}) = [\varepsilon'(\tilde{\nu}) - \varepsilon(\infty)]/\{[\varepsilon'(\tilde{\nu}) - \varepsilon(\infty)]^2 + \varepsilon''(\tilde{\nu})^2\}, \qquad (146)$$

$$\eta''(\tilde{\nu}) = -\varepsilon''(\tilde{\nu})/\{[\varepsilon'(\tilde{\nu}) - \varepsilon(\infty)]^2 + \varepsilon''(\tilde{\nu})^2\} \qquad (147)$$

and can be calculated from the real and imaginary parts ε' and ε'', respectively, of the complex dielectric response determined from the amplitude reflection measurements. Experimental values of the transverse optic mode frequency determined at low temperatures (Lowndes and Rastogi, 1976) were used as a close approximation for the harmonic frequency $\tilde{\nu}(oj)$ and recent values of the thermal strain contribution $\Delta^E(oj)$ determined from high-pressure, far infrared spectroscopy (Lowndes and Rastogi, 1976), together with published data for $\varepsilon(0)$ and $\varepsilon(\infty)$ (Lowndes and Martin, 1969), were used to provide a complete determination of all terms in Eqs. (144) and (145). Thus a complete determination of the real and imaginary parts of the anharmonic self-energy was achieved for the first time using methods that took proper account of the thermal strain contribution $\Delta^E(oj)$. Results determined in this way have been presented for NaCl (Pai et al., 1977), KCl (Rastogi et al., 1977), and RbBr (Rastogi et al., 1978). In each case fairly good qualitative and quantitative agreement was obtained with theoretical results, which included contributions from quartic anharmonicity (Rastogi et al., 1974; Rastogi, 1975; Lowndes and Rastogi, 1976). The results for RbBr at 300 K are shown in Fig. 51. The excellent agreement between experiment and theory which is evident in these figures was among the best obtained for any of the eight alkali halides whose optical constants were measured by Pai et al. (1978a), and it was concluded that current theories of anharmonicity may now account extremely well for the behavior of weakly anharmonic systems at temperatures up to their Debye temperatures.

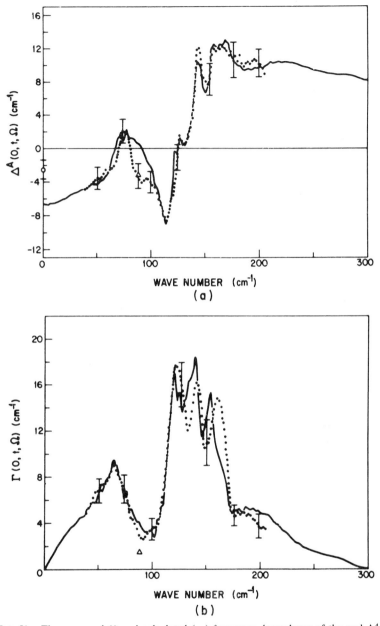

FIG. 51. The measured (·) and calculated (—) frequency dependence of the real Δ^A and imaginary Γ parts of the anharmonic self-energy of the transverse optic mode of RbBr at 300 K. In each case Δ represents an experimental value determined from far infrared transmission measurements by Lowndes and Rastogi (1976), and 0 represents an experimental value determined from dielectric constant measurements (Lowndes and Martin, 1970). [From Rastogi *et al.* (1978). Reproduced by permission of Flammarion Sciences.]

Of the three pseudodisplacive ferroelectrics studied by Pai *et al.* (1978b), TlBr crystallizes in the body-centered-cubic structure and has a single infrared-active mode whose frequency decreases slightly with decreasing temperature. Apart from this and the comparatively large values of its static and high-frequency dielectric constants, the far infrared properties of TlBr are similar to those of the alkali halides. The values of the transverse optic and longitudinal optic mode frequencies determined for TlBr by Pai *et al.* from the poles in the dielectric response were in agreement, within the limits of the experimental errors, with values determined by Cowley and Okazaki (1967) from inelastic neutron scattering experiments and by Lowndes (1972) from far infrared power spectroscopic measurements.

On the other hand, $KTaO_3$ and $SrTiO_3$ are perovskite crystals that crystallize with a TiO_6 or TaO_6 oxygen octahedral structure (Last, 1957). These crystals typically have high reflectivities in the far infrared associated with a strong ferroelectric "soft" mode, which accounts for a large part of the static dielectric constant $\varepsilon(0)$. The temperature dependence of the soft mode frequency has been shown (Cochran, 1960; Anderson, 1960) to be related to the characteristic Curie–Weiss behavior of $\varepsilon(0)$ in the paraelectric phase. A number of difficulties are encountered when studying the temperature

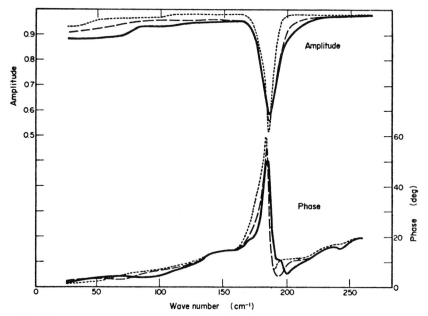

FIG. 52.   Experimentally measured amplitude reflectivity $r(v)$ and phase spectrum $\phi_r(v) - \pi$ for $KTaO_3$ at 300 K (solid curve), 200 K (dashed curve), and 100 K (dotted curve). [From Pai *et al.* (1978b). Reproduced by permission of Pergamon Press, Ltd.]

dependence of the soft mode in such materials by Kramers–Krönig analysis of power reflection spectra. These stem in part from the fact that the soft mode occurs at very low frequencies, so that difficulties can arise in truncating the Kramers–Krönig integral, and in part from the fact that power reflection measurements must be made over a very wide frequency range to include all features associated with the three infrared active lattice modes that are allowed to occur by the crystal symmetry. These problems can be avoided by DFTS since the dielectric functions can then be determined from information which is in principle complete in the measured range and limited only by experimental errors, although it should be noted that as $r \simeq 1$ and $\phi_r - \pi \simeq 0$ in the vicinity of the soft mode, large changes in the dielectric response are associated with comparatively small changes in the measured quantities, so that very accurate measurements are required. This can be clearly seen in the results of Pai *et al.* for $KTaO_3$, which are reproduced in Figs. 52 and 53. The main features of these results are in good agreement with the work of Miller and

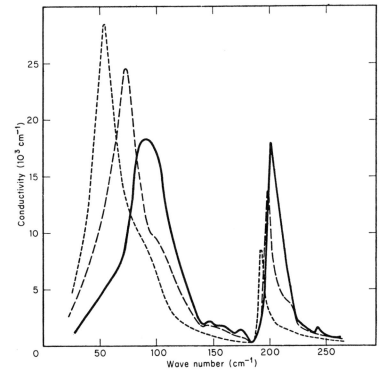

FIG. 53.   The conductivity $\sigma(v)$ of $KTaO_3$ at 300 K (solid curve), 200 K (dashed curve), and 100 K (dotted curve), calculated from the amplitude and phase reflection spectra shown in Fig. 52. [From Pai *et al.* (1978b). Reproduced by permission of Pergamon Press, Ltd.]

Spitzer (1963) and Perry and McNelly (1967), and the temperature dependence of the frequency of the soft mode taken from Fig. 53 is in excellent agreement with the results of Perry and McNelly, although the structure evident in the conductivity spectrum in Fig. 53 had not been observed before. These results have been used by Pai et al. (1978c) and Rastogi et al. (1978) to determine the temperature dependence of the real and imaginary parts of the anharmonic self-energy of the ferroelectric soft mode in $KTaO_3$ using the techniques previously described (Pai et al. 1977; Rastogi et al., 1977, 1978). Both self-energy components were found to have very large magnitudes compared to those of the transverse optic modes of the alkali halides, as expected, with a strong temperature dependence over the measured frequency range. Good agreement was obtained with earlier values determined by Lowndes and Rastogi (1973) at zero frequency from measurements of the static dielectric constant, and the results were in reasonable agreement with the approximate calculations of Cowley (1965).

In later work at Westfield College, a polarizing interferometer was developed by Ledsham et al. (1976) for studying the dielectric response of uniaxial crystals such as ferroelectrics as a function of temperature. The instrument has the same basic design as that described by Afsar et al. (1976b), which is illustrated in Fig. 15b, and was fitted with a development of the screen assembly (Parker et al., 1974) and cold finger arrangement (Parker and Chambers, 1974) needed for amplitude reflection spectroscopy at low temperatures. A polarizing interferometer is well suited to the study of optically anisotropic materials by DFTS as the radiation in each partial beam is highly polarized at wavelengths that significantly exceed the wire grid spacing (Costley et al., 1977), and the advantages of such an instrument over a conventional Michelson interferometer at low frequencies (Martin and Puplett, 1970) are also particularly valuable when studying "soft mode" behavior in ferroelectrics. However, in the work of Ledsham et al. (1976, 1977), the full potential of the instrument at low frequencies was not realized because of the lack of a suitable liquid-helium-cooled bolometer. A simple analysis of the mode of operation of the polarizing interferometer in DFTS, which neglects diffraction effects, has been given by Parker et al. (1978d).

The performance of the instrument was first demonstrated with measurements on CsI (Parker et al., 1976, Ledsham et al., 1976) and KDP and ADP (Ledsham et al., 1976) at room temperature. The transverse and longitudinal optic mode frequencies of CsI determined from the poles in the dielectric response were in good agreement with earlier data (Lowndes and Martin, 1969), and the real and imaginary parts of the anharmonic self-energy were calculated as previously described (Parker and Chambers, 1976), but no theoretical results were available for comparison. The measurements on KDP and ADP, although confined to frequencies above 40 cm^{-1}, were in good agreement with the earlier measurements of Birch *et al.* (1976) by DFTS, as well as with the work of Barker and Tinkham (1963), Kawamura *et al.* (1970), and Onyango *et al.* (1975) in which the dielectric functions were determined by Kramers–Krönig analysis of power reflection spectra.

The measurements of the *c*-axis amplitude reflectivity of KDP were later extended to about 20 cm^{-1} in frequency and down to 150 K in temperature by Ledsham *et al.* (1977), and their results are shown in Fig. 54. Over most of the measured frequency range the amplitude spectra were determined with an accuracy of better than $\pm 1\%$ and the phase spectra to better than $\pm\frac{1}{2}°$. The real and imaginary parts of the dielectric constant calculated from these

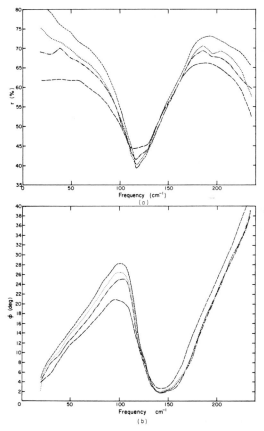

FIG. 54.   The amplitude *r* and phase $\phi_r - \pi$ reflection spectra, respectively, of the *c*-axis of KDP. The resolution is 5 cm^{-1}. The values given are: 300 K ($-\,-$), 200 K ($-\cdot-$), 175 K ($\cdots$), and 150 K ($---$). [From Ledsham *et al.* (1977). Reproduced by permission of Pergamon Press, Ltd.]

results are indicated by the dotted lines in Fig. 55, and the overdamped low frequency mode, first reported by Barker and Tinkham (1963), which is associated with the ferroelectric transition, is clearly resolved in most of these curves.

Kaminow and Damen (1968) reported the first quantitative study of the temperature dependence of the ferroelectric mode in KDP with Raman scattering measurements, and they fitted their data with a model of a simple, damped, harmonic oscillator. However, later Raman scattering experiments on KDP by She *et al.* (1972) and Lagakos and Cummins (1974) revealed that the ferroelectric soft mode is strongly coupled to an optic phonon mode of the same symmetry near 180 cm^{-1}, and She *et al.* analyzed their spectra in terms of a simple classical oscillator model proposed by Barker and Hopfield (1964) in which an additional parameter was added to describe coupling between two optic modes. In this way, Barker and Hopfield were able to explain the infrared dispersion of $BaTiO_3$ and other perovskite materials which could not be satisfactorily fitted by any choice of parameters using uncoupled oscillators. Although Ledsham *et al.* (1977) were similarly unable to explain their results for KDP (Fig. 55) with a superposition of independent oscillators, they were able to obtain good agreement over most of the measured frequency range by using the coupled oscillator model of Barker and Hopfield. In the notation of Barker and Hopfield, if two particles of unit mass with effective charges e_1 and e_2 are coupled mutually by a spring of

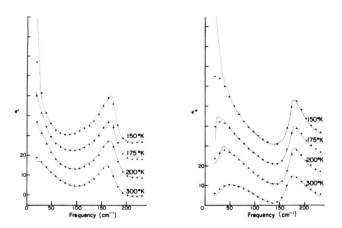

FIG. 55.   Theoretical fits to the dielectric functions (ε', ε'') of KDP. The dotted lines are the experimental curves calculated from the amplitude and phase reflection spectra shown in Fig. 54. The 200, 175, and 150 K curves are displaced by 10, 20, and 30 units, respectively, up the ε axes, from the 300 K curve, for clarity. The crosses are points from theoretical curves calculated using the reactive coupling model described in the text. [From Ledsham *et al.* (1977). Reproduced by permission of Pergamon Press, Ltd.]

constant k_{12}, and to fixed points by springs of constants k_1 and k_2, the dielectric response is given by

$$\hat{\varepsilon}(\tilde{v}) = \varepsilon(\infty) + [(e_1^2 G_1 + e_2^2 G_2 + 2e_1 e_2 G_1 G_2)/(1 - k_{12}^2 G_1 G_2)], \quad (148)$$

where

$$G_j = 1/(k_j + k_{12} - \tilde{v}^2 + i\tilde{v}\Gamma_j), \quad j = 1 \text{ or } 2 \quad (149)$$

is the response of the uncoupled mode j with damping constant Γ_j. Equation (148) was fitted to the experimental results for ε' and ε'' using $(k_1 + k_{12})$, $(k_2 + k_{12})$, k_{12}, Γ_1, Γ_2, e_1, and $-e_2$ as adjustable parameters and taking $\varepsilon(\infty) = 3.2$ from the paper of Barker and Tinkham (1963). The results, which are indicated by the crosses in Fig. 55, were in good agreement with the experimental measurements over most of the measured frequency range. This model describes the case of purely reactive coupling between the two modes and was preferred to the equivalent models with purely resistive, or complex, coupling, as the unphysical occurrence of a negative resistance in the high-frequency mode, which can then arise, is avoided. The significance of this negative resistance, which corresponds to the high-frequency mode giving power to the electric field, has been discussed by Barker and Hopfield (1964), and the need for more information to provide a physical justification for the choice of parameters in the model has been discussed by Wehner and Steigmeier (1975). Although the frequency of the ferroelectric mode determined by Ledsham et al. from models with either reactive or resistive coupling exhibited the expected softening, the mode frequencies were in each case consistently higher than those calculated by She et al. (1972), or by Lagakos and Cummins (1974) and Gauss et al. (1975), respectively, using similar models.

In more recent work, Parker et al. (1978c) have described a Fourier spectrometer for determining the optical constants of transparent solids in the far infrared from 77 to 300 K from single-pass, dispersive, transmission measurements. A single-pass configuration was chosen in preference to a double-pass configuration to obtain the full advantage of amplitude spectroscopy [Eq. (132)] when investigating detailed structure close to strong absorption bands, such as that occurring near the lattice resonance in alkali halide crystals, and measurements made with the instrument complement those obtained by amplitude reflection spectroscopy, which cannot be used to obtain a satisfactory determination of $k(\tilde{v})$ away from the reststrahlen band, where $\phi_r - \pi \to 0$ [Eq. (30)]. The instrument has been constructed in a configuration very similar to that shown in Fig. 2b, which is basically that of the Martin–Puplett polarizing interferometer (Martin and Puplett, 1970), so it can be easily converted to the polarizing mode for measurements at very low frequencies by changing to a wire grid beam splitter and adding a suitable polarizer and analyzer. Measurements are made at low temperatures by placing the specimen in a movable copper mount attached to the base of a liquid nitrogen cold finger.

Measurements at 90 and 300 K were reported on single crystals of KBr thinned to about 140 μm using techniques similar to those described by Berg and Bell (1971). Interferograms were recorded as described in Section III.D.2.b to facilitate the calculation of the refractive index spectrum, and an asymmetric apodizing function similar to that described by Parker and Chambers (1976) was used on the specimen interferograms. As the signatures associated with the different transmitted partial waves could not be separated on the specimen interferograms, channel spectrum effects were observed on the computed transmission spectra, so values of the optical constants were calculated by an iterative procedure from Eq. (86) which includes all terms in the geometric series [Eq. (69)] contributing to the channel spectrum. The correct branch indices [Eq. (82)] for the phase spectrum above the *reststrahlen* band were found by comparing the phase on one of the branches with values calculated from the refractive index determined from earlier amplitude reflection measurements (Parker and Chambers, 1976), and starting values for $n(\tilde{v})$ and $k(\tilde{v})$ for the iterative calculation were obtained from the same paper.

The values of $n(\tilde{v})$ and $k(\tilde{v})$ determined in this way at 300 K were in excellent agreement with those of Johnson and Bell (1969), and the results at 90 K,

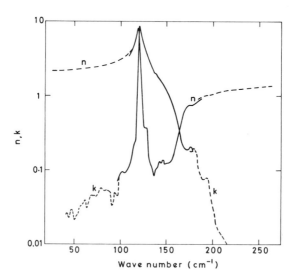

FIG. 56.   The refractive index $n(v)$ and extinction coefficient $k(v)$ of KBr determined at 90 K by dispersive transmission spectroscopy (dotted lines) and at 100 K by dispersive reflection spectroscopy (solid lines). [From Parker *et al.* (1978c). Reproduced by permission of Pergamon Press, Ltd.]

which are reproduced in Fig. 56, are the first reported measurements of optical constants at low temperatures by dispersive transmission spectroscopy. Also shown in the figure are the optical constants in the intermediate frequency range, where the crystal is opaque, determined at 100 K by Parker and Chambers (1976) by amplitude reflection spectroscopy, and it can be seen that, in regions of overlap, good agreement was obtained between values obtained by reflection and transmission. The accuracy of the values obtained from transmission measurements was limited by the detector noise level and by the error in the determination of the crystal thickness, the uncertainty in the resulting values of $n(\tilde{v})$ and $k(\tilde{v})$ being a few percent, at most.

In the latest work at Westfield College, Parker *et al.* (1978b) have developed a Fourier spectrometer and liquid helium cryostat based on the NPL modular cube interferometer and illustrated in Figs. 57 and 58, respectively, for dispersive reflection measurements on solids in the temperature range 4–300 K. The cryostat is completely demountable, and has a number of novel features designed to enable phase spectra to be measured with an accuracy limited only by the detector noise level. One of the most important of these

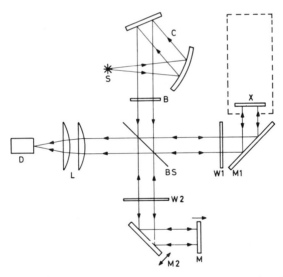

FIG. 57. Schematic diagram of the interferometer and cryostat developed by Parker and coworkers for amplitude reflection measurements on solids in the temperature range 4.2–300 K, with components as follows: S, mercury lamp source; C, collimator; B, black polyethylene filter; BS, beam splitter; W1, white polyethylene vacuum window; W2, compensating window; M1, second beam splitter module; x, specimen; M2, phase modulation mirror; and M, moving mirror. The specimen is mounted on the base plate of the liquid helium can of a cryostat which is indicated by the dotted lines. [From Parker *et al.* (1978b). Reproduced by permission of Pergamon Press, Ltd.]

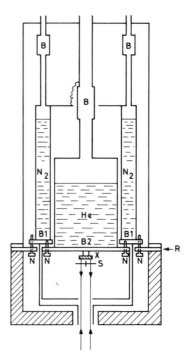

FIG. 58.   Schematic diagram of the liquid helium cryostat used with the interferometer shown
in Fig. 57 with components as follows: R, metal reference ring; B1 and B2, liquid nitrogen and
liquid helium can base plates respectively; N, nylon bolts; B, stainless steel bellows; x, specimen;
and S, screen assembly. [From Parker *et al.* (1978b). Reproduced by permission of Pergamon
Press, Ltd.]

is that thermally insulating nylon bolts *N* (Fig. 58) are used to fix the base
plates B1 and B2 of the liquid nitrogen and liquid helium cans, respectively,
rigidly in position in a metal reference plate R in the fixed arm of the interferometer. Thermal contraction of the liquid nitrogen and liquid helium cans
that occurs as they are filled and later as the liquids boil off is accommodated
by flexible stainless steel bellows B welded in the filler tubes. In this way,
mechanical drift of the base plate B2, which would be continuously present in
a conventional cryostat with the cans suspended from the top by their filler
tubes, is completely eliminated, so long as liquids are present in the two cans.
The specimen X occupies a stable position in good mechanical and thermal
contact with the base B2. The cryostat can be evacuated to a pressure 10^{-6}
Torr, and the rest of the interferometer, which is separated from it by a polyethylene vacuum window W1, can be evacuated to a pressure of about
10^{-2} Torr, which is sufficient to reduce atmospheric absorption to an acceptable level. The window W1 is compensated by a similar window W2 in

the other arm of the interferometer, and a black polyethylene filter B together with the white polyethylene window W1, removes all radiation of frequencies greater than about 450 cm^{-1} from the collimated beam transmitted to the crystal.

For amplitude reflection measurements the specimen was partially aluminized using a screen geometry similar to that described by Staal and Eldridge (1977), which shall be described later, and the screen assembly S in Fig. 58 was mounted just in front of the specimen, in good thermal contact with the liquid helium can base plate, and could be satisfactorily operated with a simple electromagnetic system at all temperatures between 7 and 300 K, the operating range of the cryostat. Complete background and specimen interferograms could be recorded in turn in the usual way (conventional mode), and the screen mechanism and moving mirror drive were also programmed to permit the two interferograms to be sampled alternately on a single scan of the moving mirror (switching mode). Tests made in the switching mode with a perfect mirror in place of the specimen established that all systematic errors were eliminated and that the amplitude and phase reflection spectra of the specimen could be determined (with an accuracy limited only by the detector noise level) from just one background and one specimen interferogram recorded on the same scan, as reported by Staal and Eldridge (1977). Staal and Eldridge have also pointed out that a fast detector is needed to operate conveniently in the switching mode. Thus, because of the slow response time of their Golay detector, the measurements of Parker et al. on KCl which are described below were made in the conventional mode. Although it was still possible to determine the amplitude and phase spectra from pairs of interferograms rather than from sets of four [cf., Eq. (143)], the accuracy of the phase spectra was consequently limited by systematic errors due to micrometer backlash ($\simeq 0.1 \ \mu$m), rather than by the detector noise level.

Measurements were made at 7, 80, and 300 K on a single crystal of KCl. A collimated beam of 30-mm diameter, defined by the size of the aperture in the screen S, was used in the sample arm of the interferometer. Although slight misalignment of the specimen did occur on cooling, alignment of the interferometer was easily restored by adjusting the phase modulation mirror M2, and the specimen position became very stable as soon as liquids began to collect in the two cans. The results which were obtained at 300 K were in good agreement with those of Johnson and Bell (1969), and the transverse and longitudinal optic mode frequencies $\tilde{\nu}_T$ and $\tilde{\nu}_L$, respectively, determined at 300 and 80 K were in good agreement with the results of Pai et al. (1978a), which were also obtained by dispersive reflection spectroscopy. The amplitude and phase reflection spectra of KCl measured by Parker et al. at 7 K are shown in Fig. 59. Although no previous measurements by DFTS are available

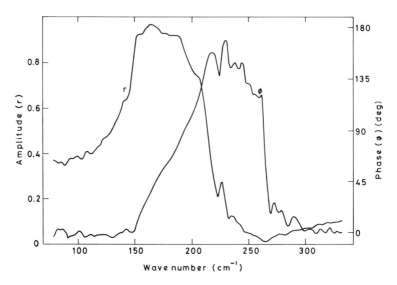

FIG. 59.   The measured amplitude and phase reflection spectra of KCl at 7 K. [From Parker *et al.* (1978b). Reproduced by permission of Pergamon Press, Ltd.]

for comparison, the value of $\tilde{\nu}_T$ obtained from these measurements was in good agreement with the results of Lowndes and Martin (1969) at 2 K.

The group at the University of Cologne (Gauss *et al.*, 1975; Gauss and Happ, 1976; Happ and Rother, 1977) have recently studied the dielectric response of KDP and deuterated KDP (DKDP) in the temperature range 100–300 K by DFTS.

In the earliest reported work, Gauss *et al.* (1975) used an asymmetric Michelson interferometer with a CSF Carcinotron source operating at a wavelength of 2.16 mm to measure the complex amplitude reflectivity of KDP with $\mathbf{E} \parallel C$ and $\mathbf{E} \perp C$ between 100 and 300 K. They also measured the far infrared spectra with $\mathbf{E} \parallel C$ and $\mathbf{E} \perp C$ at room temperature with an asymmetric Fourier spectrometer. Apart from the statement that a replacement technique was used in each instrument, no details were given of the dispersive aspects or of the phase accuracy of either instrument. However, their measurements showed almost zero phase at the minimum near 140 cm^{-1} in the room-temperature, *c*-axis spectrum, compared with a value of $3.2 \pm 0.5°$ reported by Ledsham *et al.* (1977), which was in good agreement with the result of Birch *et al.* (1976). A value of $3.0 \pm 0.5°$ was also obtained by Gledhill *et al.* (1977) from power transmission measurements on the narrow *c*-axis transmission window which occurs at 136 cm^{-1} in KDP, so the low value of Gauss *et al.* may result from systematic phase errors associated with a small replacement error of approximately 0.5 μm. Although they were unable to describe the c-axis spectrum satisfactorily with independent oscillators, they obtained a good fit using a coupled oscillator model with parameters in good agreement with those of Lagakos and Cummins (1974). After obtaining a similar set of parameters at 130 K by fitting the coupled oscillator model to the far infrared reflection data of Sugawara and Nakamura (1970), they used the model to deduce the temperature dependence of the soft mode frequency $\tilde{\nu}$ between room temperature and the ferroelectric transition temperature from their millimeter wave amplitude reflection data. Their final values of $\tilde{\nu}_1^2$ were within 12% of those obtained in a more direct way by Lagakos and Cummins from light scattering measurements. They were able to explain their results for $\mathbf{E} \perp C$ in terms of a heavily damped oscillator at a frequency below 80 cm^{-1} with weak temperature dependence, consistent with a proton E-mode (Havlin et al., 1975).

The complex amplitude reflectivity of DKDP was later measured between 140 and 300 K at 4.62 cm^{-1} by Gauss and Happ (1976) and between 200 and 300 K with the asymmetric Fourier spectrometer by Happ and Rother (1977). Results were obtained that were quite different at low frequencies from those of KDP. The c-axis measurements below 30 cm^{-1} were satisfactorily described by a Debye relaxational mechanism with a characteristic frequency near 3 cm^{-1} at room temperature, and two additional oscillators were required to explain the measurements up to 300 cm^{-1}. The low-frequency a-axis mode, which occurs at 80 cm^{-1} in KDP at room temperature, was shifted to about 10 cm^{-1} in DKDP and could be satisfactorily fitted with an overdamped oscillator which was attributed to E-mode deuteron motion, as described by Kaminow (1965).

Staal and Eldridge (1977) have described simple modifications to a commercial RIIC FS 720 Fourier spectrometer based on the technique developed by Parker et al. (1974), which convert it to an amplitude reflection spectrometer suitable for measurements on specimens greater than about 2 cm in diameter. The specimen is partially aluminized in the pattern illustrated in Fig. 60 and placed with the switching mask, also illustrated in Fig. 60, in the fixed arm of the interferometer. The background and specimen interferograms are sampled alternately at each micrometer step on a single scan of the micrometer by rotating the mask backward and forward between the two adjustable stops with the solenoid. Because of the need for rapid switching between the two interferograms, a Golay detector with a comparatively long response time was found to be unsuitable, so a liquid-helium-cooled bolometer with a fast response time was used. This mask design is more compatible with the symmetry of the beam than those used by either Parker et al. (1974) or Ledsham et al. (1976), and it was found that systematic errors due to differences between the two parts of the field of view were completely eliminated in this way. Thus spectra that were limited in accuracy only by

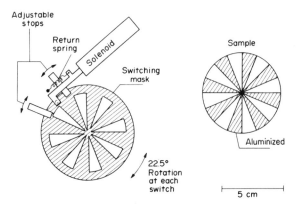

FIG. 60.   A diagram of the aluminized sample and eight-section switching mask developed by Staal and Eldridge. [From Staal and Eldridge (1977). Reproduced by permission of Pergamon Press, Ltd.]

the detector noise fluctuations could be obtained from just two interferograms recorded in a single scan, obviating the need for reratioing with four interferograms [Eq. (143)]. An additional advantage was that phase errors and amplitude errors due to loss of coherence, both of which derive from lack of flatness of the specimen surface, were considerably reduced in magnitude because of the more equal averaging of surface irregularities between the exposed and aluminized segments. Consequently, less stringent limits are placed on the degree of flatness of the specimen surface, which is a considerable advantage, especially when studying soft materials. The amplitude and phase reflection spectra of KBr at room temperature presented by Staal and Eldridge were in agreement, within the limits of the detector noise level, with those of Johnson and Bell (1969).

The instrument was later used by Eldridge and Staal (1977) to measure the amplitude and phase reflection spectra of NaCl from 25 to 500 cm^{-1} at room temperature, and the dispersive measurements were supplemented with power transmission measurements in spectral regions where the phase was small. The results were used to calculate the optical constants and the real and imaginary parts $\Delta(oj, \tilde{v})$ and $\Gamma(oj, \tilde{v})$ of the anharmonic self-energy for comparison with theoretical calculations which included contributions from cubic and quartic anharmonicity. The damping coefficient $\Gamma(oj, \tilde{v})$ was calculated from dispersion curves generated by the shell model of Schmunk and Winder (1970) and fitted to inelastic neutron scattering data, and the real part $\Delta'(oj, \tilde{v})$, which neglects an unknown constant, was calculated from $\Gamma(oj, \tilde{v})$ using the appropriate Kramers–Krönig dispersion relation (Eldridge and Howard, 1973). By renormalizing the harmonic frequency $\tilde{v}(oj)$ in Eq. (135) to include the effects of both thermal expansion and anharmonicity

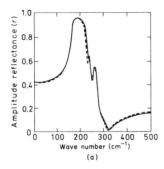
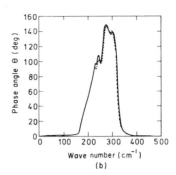

FIG. 61. (a) Calculated (solid curve) and measured (dashed curve) amplitude reflection spectra of NaCl at 290 K. (b) Calculated (solid curve) and measured (dashed curve) reflectance phase angle of NaCl at 290 K. [From Eldridge and Staal (1977). Reproduced by permission of the American Physical Society.]

(Berg and Bell, 1971; Eldridge and Howard, 1973), the susceptibility was rewritten as

$$\hat{\chi}_\alpha(\tilde{v}) = \chi_\alpha^E + \frac{1}{Nvh} \frac{\tilde{v}_0 M_\alpha^2(oj)}{\tilde{v}_0^2 - \tilde{v}^2 + 2\tilde{v}_0[\Delta'(oj, \tilde{v}) - \Delta'(oj, \tilde{v}_0) - i\Gamma(oj, \tilde{v})]}, \quad (150)$$

where $\tilde{v}_0$ is the measured resonant frequency. The need to determine the unknown constant in $\Delta'(oj, \tilde{v})$ and, hence, the thermal strain contribution Eq. (136) was thus avoided, since the total frequency shift $\Delta'(oj, \tilde{v}) - \Delta'(oj, \tilde{v}_0)$ was required to vanish at $\tilde{v}_0$. Equation (150) was used as the basis for the calculation of the optical constants and the amplitude and phase reflection spectra. Excellent overall agreement was obtained between the measured and calculated spectra for NaCl at room temperature. This is illustrated for the amplitude and phase reflection spectra in Figs. 61a and 61b, respectively. This work is to be extended to low temperatures, and it has recently been reported[6] that measurements have been successfully made on NaCl at 20 K.

VII. Conclusion

The development of dispersive Fourier transform spectrometry and its application to the study of gases, liquids, and solids has been reviewed from its inception in 1963 up to the latest reported work still in press at the time of writing. During this period, it has been clearly established that the technique enables a more accurate determination of the optical constants of specimens taken from all three material phases than would otherwise have been possible, and, in many cases, this has led to a better understanding of

---

[6] J. E. Eldridge, private communication.

the microscopic behavior of the system. In much of the published work, the fundamental detector noise limit has either been reached or closely approached in the experimentally determined values of the optical constants, and it is hoped that as the realization of this goal is facilitated by improvements in technique dispersive Fourier transform spectrometry will be more widely used as a spectroscopic tool. This review was completed in July 1978.

ACKNOWLEDGMENTS

We wish to thank Dr. T. G. Blaney, Dr. W. G. Chambers, and Dr. J. E. Gibbs for their comments on this chapter and, particularly, Miss P. Hultum for typing the manuscript.

REFERENCES

Afsar, M. N. (1977). *National Physical Laboratory Rep.* DES 42, June.
Afsar, M. N., and Chantry, G. W. (1977). *IEEE Trans.* **MTT-25**, 509–511.
Afsar, M. N., and Hasted, J. B. (1977), *J. Opt. Soc. Am.* **67**, 902–904.
Afsar, M. N., Hasted, J. B., Zafar, M. S., and Chamberlain, J. (1975a). *Chem. Phys. Lett.* **36**, 69–72.
Afsar, M. N., Birch, J. R., and Chamberlain, J. (1975b). *IEE Conf. Publ. No.* 129, pp. 131–134.
Afsar, M. N., Chamberlain, J., and Chantry, G. W. (1976a). *IEEE Trans. Instrum. Meas.* **IM-25**, 290–294.
Afsar, M. N., Hasted, J. B., and Chamberlain, J. (1976b). *Infrared Phys.* **16**, 301–310.
Afsar, M. N., Chamberlain, J., and Hasted, J. B. (1976c). *Infrared Phys.* **16**, 587–599.
Afsar, M. N., Chamberlain, J., Chantry, G. W., Finsy, R., and Van Loon, R. (1977a). *Proc. IEE* **124**, 575–577.
Afsar, M. N., Honijk, D. D., Passchier, W. F., and Goulon, J. (1977b). *IEEE Trans. Microwave Theory Tech.* **MTT-25**, 505–508.
Anastassakis, E., and Burstein, E. (1971). *In* "Light Scattering in Solids" (M. Balkanski, ed.), p. 52. Flammarion Sciences, Paris.
Anderson, P. W. (1960). *in* "Fizika Dielektrikov" (G. I. Skanavi, ed.), P. N. Lebedev Physical Institute, USSR Academy of Sciences, Moscow.
Avery, D. G. (1952). *Proc. Phys. Soc. London Sect. B* **65**, 425–428.
Barker, A. S., and Hopfield, J. J. (1964). *Phys. Rev.* **135A**, 1732–1737.
Barker, A. S., and Tinkham, M. (1963). *J. Chem. Phys.* **38**, 2257–2264.
Barker, A. S., Ballman, A. A., and Ditzenberger, J. A. (1970). *Phys. Rev. B* **2**, 4233–4239.
Bell, E. E. (1965). *Jpn. J. Appl. Phys.* (*Suppl. I*) **4**, 412–416.
Bell, E. E. (1966). *Infrared Phys.* **6**, 57–74.
Bell, E. E. (1967a). *Handb. Phys.* **25**, 1–58.
Bell, E. E. (1967b). *J. Phys.* (*Suppl. 3–4*) **28**, C2-18–C2-25.
Bell, E. E. (1970). *Proc. Aspen Int. Conf. Fourier Spectrosc.* (G. A. Vanasse, A. T. Stair Jr., and D. J. Baker, eds.). AFCRL-71-0019, Special Rep. No. 114, pp. 71–82 (January 5, 1971).
Bell, R. J. (1972). "Introductory Fourier Transform Spectroscopy," Chapter 8, Academic Press, New York.
Benedict, W. S., Herman, R., Moore, G. E., and Silverman, S. (1956a). *Can. J. Phys.* **34**, 830–849.
Benedict, W. S., Herman, R., Moore, G. E., and Silverman, S. (1956b). *Can. J. Phys.* **34**, 850–875.
Berg, J. I., and Bell, E. E. (1971). *Phys. Rev.* **4B**, 3572–3580.

Berreman, D. W. (1967). *Appl. Opt.* **6**, 1519–1521.

Birch, J. R. (1978a). *Infrared Phys.* **18**, 275–282.

Birch, J. R. (1978b). *Infrared Phys.* **18**, 613–620.

Birch, J. R., and Afsar, M. N. (1977a). *Conf. Fourier Transform Infrared Spectrosc. Columbia, South Carolina.*

Birch, J. R., and Afsar, M. N. (1977b). *Conf. Precise Elec. Meas., Sussex, England.* pp. 67–69. IEE Conf. Publ. 152.

Birch, J. R., and Bulleid, C. E. (1976). *Conf. Precision Electromagn. Meas., Boulder, Colorado* pp. 48–59. Conf. Digest IEEE Cat. No. 76CH1099-1 IM.

Birch, J. R., and Bulleid, C. E. (1977). *Infrared Phys.* **17**, 279–282.

Birch, J. R., and Jones, R. G. (1970). *Infrared Phys.* **10**, 217–224.

Birch, J. R., and Murray, D. K. (1978). *Infrared Phys.* **18**, 283–291.

Birch, J. R., and Parker, T. J. (1979). *Infrared Phys.* **19**, 109–113.

Birch, J. R., Cook, R. J., Harding, A. F., Jones, R. G., and Price, G. D. (1975). *J. Phys. D* **8**, 1353–1358.

Birch, J. R., Price, G. D., and Chamberlain, J. (1976). *Infrared Phys.* **16**, 311–315.

Bloor, D. (1970). *Infrared Phys.* **10**, 1–55.

Born, M., and Huang, K. (1954). "Dynamical Theory of Crystal Lattices." Oxford Univ. Press, London and New York.

Born, M., and Wolf, E. (1965). "Principles of Optics." Pergamon, Oxford.

Bruce, A. D. (1973). *J. Phys. C* **6**, 174–188.

Cardona, M. (1969). *In* "Optical Properties of Solids" (S. Nudelman and S. S. Mitra, eds). Plenum Press, New York.

Carli, B. (1977). *J. Opt. Soc. Am.* **67**, 908–910.

Chamberlain, J. E. (1965). *Infrared Phys.* **5**, 175–178

Chamberlain, J. E. (1967a). *Appl. Opt.* **6**, 980–981.

Chamberlain, J. E. (1967b). *J. Quant. Spectrosc. Radiat. Trans.* **7**, 151–168.

Chamberlain, J. E. (1971). *Infrared Phys.* **11**, 25–55.

Chamberlain, J. E. (1972a). *Infrared Phys.* **12**, 145–164.

Chamberlain, J. E. (1972b). *IEEE Trans. Instrum. Meas.* **IM-21**, 438–442.

Chamberlain, J., and Chantry, G. W. (1973). "High Frequency Dielectric Measurement." IPC Press, Guildford.

Chamberlain, J. E., and Gebbie, H. A. (1965a). *Nature (London)*, **206**, 602–603.

Chamberlain, J. E., and Gebbie, H. A. (1965b). *Nature (London)*, **208**, 480–481.

Chamberlain, J. E., and Gebbie, H. A. (1966). *Appl. Opt.* **5**, 393–396.

Chamberlain, J. E., and Gebbie, H. A. (1971a). *Infrared Phys.* **11**, 57–73.

Chamberlain, J. E., and Gebbie, H. A. (1971b). *Appl. Opt.* **10**, 1184–1185.

Chamberlain, J. E., Gibbs, J. E., and Gebbie, H. A. (1963). *Nature (London)*, **198**, 874–875.

Chamberlain, J. E., Findlay, F. D., and Gebbie, H. A. (1965a). *Nature (London)*, **206**, 886–887.

Chamberlain, J. E., Findlay, F. D., and Gebbie, H. A. (1965b). *Appl. Opt.* **4**, 1382–1385.

Chamberlain, J. E., Costley, A. E., and Gebbie, H. A. (1967a). *Spectrochim. Acta.* **23A**, 2255–2260.

Chamberlain, J. E., Werner, E. B. C., Gebbie, H. A., and Slough, W. (1967b). *Trans. Faraday Soc.* **63**, 2605–2609.

Chamberlain, J. E., Gebbie, H. A., Pardoe, G. W. F., and Davies, M. (1968). *Chem. Phys. Lett.* **1**, 523–525.

Chamberlain, J. E., Costley, A. E., and Gebbie, H. A. (1969a). *Spectrochim. Acta* **25A**, 9–18.

Chamberlain, J. E., Gibbs, J. E., and Gebbie, H. A. (1969b). *Infrared Phys.* **9**, 185–209.

Chamberlain, J. E., Afsar, M. N., and Hasted, J. B. (1973a). *Nature (London) Phys. Sci.* **245**, 28–30.

Chamberlain, J., Zafar, M. S., and Hasted, J. B. (1973b). *Nature (London) Phys. Sci.* **243**, 116–117.
Chamberlain, J. E., Afsar, M. N., Davies, G. J., Hasted, J. B., and Zafar, M. S. (1974a). *IEEE Trans.* **MTT-22**, 1028–1032.
Chamberlain, J. E., Afsar, M. N., Murray, D. K., Price, G. D., and Zafar, M. S. (1974b). *IEEE Trans. Instrum. Meas.* **IM-23**, 483–488.
Chamberlain, J. E., Afsar, M. N., Hasted, J. B., Zafar, M. S., and Davies, G. J. (1975). *Nature (London)*, **255**, 319–321.
Chambers, W. G. (1975). *Infrared Phys.* **15**, 139–141.
Chantry, G. W. (1971). "Submillimeter Spectroscopy." Academic Press, New York.
Chantry, G. W., and Chamberlain, J. E. (1972). *In* "Polymer Science" (A. D. Jenkins, ed.), pp. 1330–1381. North-Holland Publ.
Chantry, G. W., and Fleming, J. W. (1976). *Infrared Phys.* **16**, 655–660.
Chantry, G. W., Evans H. M., Chamberlain, J., and Gebbie, H. A. (1969a). *Infrared Phys.* **9**, 85–93.
Chantry, G. W., Evans, H. M., Fleming, J. W., and Gebbie, H. A. (1969b). *Infrared Phys.* **9**, 31–33.
Chopra, K. L. (1969). "Thin Film Phenomena." McGraw-Hill, New York.
Cochran, W. (1960). *Adv. Phys.* **9**, 387–423.
Conte, S. D., and de Boor, C. (1972). "Elementary Numerical Analysis," p. 32. McGraw-Hill, New York and Kogakusha, Tokyo.
Costley, A. E., Hursey, K. H., Neill, G. F., and Ward, J. M. (1977). *J. Opt. Soc. Am.* **67**, 979–981.
Cowley, R. A. (1963). *Adv. Phys.* **12**, 421–480.
Cowley, R. A. (1965). *Phil. Mag.* **11**, 673–706.
Cowley, E. R., and Okazaki, A. (1976). *Proc. R. Soc. London Ser. A* **300**, 45–59.
Czerny, M. (1929). *Z. Phys.* **53**, 317–325.
Danielewicz, E. J., and Coleman, P. D. (1974). *Appl. Opt.* **13**, 1164–1170.
Davies, G. J., and Chamberlain, J. (1972). *J. Phys. A* **5**, 767–772.
Davies, M., Pardoe, G. W. F., Chamberlain, J., and Gebbie, H. A. (1968). *Trans. Faraday Soc.* **64**, 847–860.
Davies, M., Pardoe, G. W. F., Chamberlain, J., and Gebbie, H. A. (1970). *Trans. Faraday Soc.* **66**, 273–292.
Ditchburn, R. W. (1973). "Light," Blackie, London. pp. 569–570.
Dolling, G., Cowley, R. A., Schittenhelm, C., and Thorson, I. M. (1966). *Phys. Rev.* **147**, 577–582.
Dromey, J. D., and Birch, J. R. (1978). *Infrared Phys.* **18**, 243–245.
Eldridge, J. E., and Howard, R. (1973). *Phys. Rev.* **7B**, 4652–4665.
Eldridge, J. E., and Staal, P. R. (1977). *Phys. Rev.* **16B**, 4608–4618.
Fellgett, P. (1951). Thesis, Univ. of Cambridge.
Fellgett, P. (1958). *J. Phys. Radium* **19**, 187–191.
Fleming, J. W. (1976). *J. Quant. Spectrosc. Radiat. Trans.* **16**, 63–68.
Fuchs, K. (1938). *Proc. Cambridge Phil. Soc.* **34**, 100–108.
Gast, J., and Genzel, L. (1973). *Opt. Commun.* **8**, 26–30.
Gast, J., Genzel, L., and Zwick, U. (1974). *IEEE Trans.* **MTT-22**, 1026–1027.
Gauss, K. E., and Happ, H. (1976). *Phys. Status Solidi (b)* **78**, 133–138.
Gauss, K. E., Happ, H., and Rother, G. (1975). *Phys. Status Solidi (b)* **72**, 623–630.
Gebbie, H. A., and Twiss, R. Q. (1966). *Rep. Prog. Phys.* **29**, 729–756.
Gebbie, H. A., and Vanasse, G. A. (1956). *Nature (London)*, **178**, 432.
Gebbie, H. A., Stone, N. W. B., Findlay, F. D., and Pyatt, E. C. (1965). *Nature (London)*, **205**, 377–378.
Geick, R. (1961). *Z. Phys.* **161**, 116–122.

Gledhill, G. A., Angress, J. F., Martin, R. W., and Chambers, W. G. (1977). *J. Phys. C* **9**, L617–L618.

Hadni, A., Claudel, J., Mariot, G., and Strimer, P. (1968). *Appl. Opt.* **7**, 161–165.

Happ, H., and Rother, G. (1977). *Phys. Status Solidi (b)* **79**, 473–477.

Harrick, N. J. (1967). "Internal Reflection Spectroscopy." Wiley (Interscience), New York.

Havlin, S., Litov, E., and Sompolinsky, H. (1975). *Phys. Lett.* **A53**, 41–42.

Hisano, K., Placido, F., Bruce, A. D., and Holah, G. D. (1972). *J. Phys. C* **5**, 2511–2522.

Honijk, D. D., Passchier, W. F., and Mandel, M. (1972). *Physica*, **59**, 536–540.

Honijk, D. D., Passchier, W. F., and Mandel, M. (1973a). *Physica*, **64**, 171–188.

Honijk, D. D., Passchier, W. F., and Mandel, M. (1973b). *Physica*, **68**, 457–474.

Honijk, D. D., Passchier, W. F., Mandel, M., and Afsar, M. N. (1976). *Infrared Phys.* **16**, 257–262.

Honijk, D. D., Passchier, W. F., Mandel, M., and Afsar, M. N. (1977). *Infrared Phys.* **17**, 9–24.

Jacquinot, P. (1954). *J. Opt. Soc. Am.* **44**, 761–765.

Jacquinot, P., and Dufor, C. J. (1948). *J. Rech. CNRS*, **6**, 91–103.

Johnson, K. W., and Bell, E. E. (1969). *Phys. Rev.* **187**, 1044–1052.

Jones, G. O., Martin, D. H., Mawer, P. A., and Perry, C. H. (1961). *Proc. R. Soc. London Ser. A* **261**, 10–27.

Kaminow, I. P. (1965). *Phys. Rev. A* **138**, 1539–1543.

Kaminow, I. P., and Damen, T. C. (1968). *Phys. Rev. Lett.* **20**, 1105–1108.

Karo, A. M., and Hardy, J. R. (1963). *Phys. Rev.* **129**, 2024–2036.

Kawamura, T., Mitsuishi, A., and Yoshinaga, H. (1970). *J. Phys. Soc. Jpn. Suppl.* **28**, 227–229.

Kemp, A. J., Birch, J. R., and Afsar, M. N. (1978). *Infrared Phys.* **18**, 827–833.

Kerl, K. (1974). *Ber. Bunsengellsch. Phys. Chem.* **78**, 1209–1214.

Korff, S. A., and Breit, G. (1932). *Rev. Mod. Phys.* **4**, 471–503.

Lagakos, N., and Cummins, H. Z. (1974). *Phys. Rev. B* **10**, 1063–1069.

Last, J. T. (1957). *Phys. Rev.* **105**, 1740–1750.

Ledsham, D. A., Chambers, W. G., and Parker, T. J. (1976). *Infrared Phys.* **16**, 515–522.

Ledsham, D. A., Chambers, W. G., and Parker, T. J. (1977). *Infrared Phys.* **17**, 165–172.

Legay, F. (1958). *Rev. Opt.* **37**, 11–17.

Lindquist, R. E., and Ewald, A. W. (1963). *J. Opt. Soc. Am.* **53**, 247–249.

Loewenstein, E. V., and Smith, D. R. (1971). *Appl. Opt.* **10**, 577–583.

Loewenstein, E. V., Smith, D. R., and Morgan, R. L. (1973). *Appl. Opt.* **12**, 398–406.

Lowndes, R. P. (1972). *Phys. Rev.* **6B**, 1490–1498.

Lowndes, R. P., and Martin, D. H. (1969). *Proc. R. Soc. London Ser. A* **308**, 473–496.

Lowndes, R. P., and Martin, D. H. (1970). *Proc. R. Soc. London Ser. A* **316**, 351–375.

Lowndes, R. P., and Rastogi, A. (1973). *J. Phys. C* **6**, 932–944.

Lowndes, R. P., and Rastogi, A. (1976). *Phys. Rev. B* **14**, 3598–3620.

Maradudin, A. A., and Fein, A. E. (1962). *Phys. Rev.* **128**, 2589–2608.

Martin, D. H., and Puplett, E. (1970). *Infrared Phys.* **10**, 105–109.

McWilliams, D., and Lynch, D. W. (1963). *Phys. Rev.* **130**, 2248–2252.

Mead, D. G. (1978). *Infrared Phys.* **18**, 257–258.

Mead, D. G. (1979). *Infrared Phys.* **19**, 19–25.

Mead, D. G., and Genzel, L. (1978). *Infrared Phys.* **18**, 555–564.

Michelson, A. A. (1881). *Am. J. Sci. Ser. 3* **22**, 120–129.

Miller, R. C., and Spitzer, W. G. (1963). *Phys. Rev.* **129**, 94–98.

Mitsuishi, A., Otsuka, Y., Fujita, S., and Yoshinaga, H. (1963). *Jpn. J. Appl. Phys.* **2**, 574–577.

Moss, T. S. (1959). "Optical Properties of Semiconductors." Butterworths, London.

Onyango, F., Smith, W., and Angress, J. F. (1975). *J. Phys. Chem. Solids* **36**, 309–313.

Pai, K. F., Rastogi, A., Tornberg, N. E., Parker, T. J., and Lowndes, R. P. (1977). *J. Opt. Soc. Am.* **67**, 914–917.

Pai, K. F., Parker, T. J., Tornberg, N. E., Lowndes, R. P., and Chambers, W. G. (1978a). *Infrared Phys.* **18**, 199–214.

Pai, K. F., Parker, T. J., Tornberg, N. E., Lowndes, R. P., and Chambers, W. G. (1978b). *Infrared Phys.* **18**, 327–336.

Pai, K. F., Parker, T. J., and Lowndes, R. P. (1978c). *J. Opt. Soc. Am.* **68**, 1322–1325.

Palik, E. D. (1960). *J. Opt. Soc. Am.* **50**, 1329–1336.

Palik, E. D. (1963). U.S. Naval Research Laboratory Bibliography No. 21.

Parker, T. J., and Chambers, W. G. (1974). *IEEE Trans.* **MTT-22**, 1032–1036.

Parker, T. J., and Chambers, W. G. (1975). *IEE Conf. Publ.* No. 129, 169–172.

Parker, T. J., and Chambers, W. G. (1976). *Infrared Phys.* **16**, 349–354.

Parker, T. J., Chamberlain, J., and Burfoot, J. C. (1970). *J. Phys. Soc. Jpn. Suppl.* **28**, 230–232.

Parker, T. J., Chambers, W. G., and Angress, J. F. (1974). *Infrared Phys.* **14**, 207–215.

Parker, T. J., Ledsham, D. A., and Chambers, W. G. (1976). *Infrared Phys.* **16**, 293–297.

Parker, T. J., Ledsham, D. A., Chambers, W. G., and Ford, J. E. (1977). *Conf. Fourier Transform Infrared Spectrosc., Columbia, South Carolina.*

Parker, T. J., Ford, J. E., and Chambers, W. G. (1978a). *Infrared Phys.* **18**, 215–219.

Parker, T. J., Lowndes, R. P., and Mok, C. L. (1978b). *Infrared Phys.* **18**, 565–570.

Parker, T. J., Chambers, W. G., Ford, J. E., and Mok, C. L. (1978c). *Infrared Phys.* **18**, 571–576.

Parker, T. J., Ledsham, D. A., and Chambers, W. G. (1978d). *Infrared Phys.* **18**, 179–183.

Passchier, W. F., Honijk, D. D., and Mandel, M. (1975). *Infrared Phys.* **15**, 95–109.

Passchier, W. F., Honijk, D. D., and Mandel, M. (1976). *Infrared Phys.* **16**, 389–401.

Passchier, W. F., Honijk, D. D., Mandel, M., and Afsar, M. N. (1977a). *J. Phys. D* **10**, 509–517.

Passchier, W. F., Honijk, D. D., Mandel, M., and Afsar, M. N. (1977b). *Infrared Phys.* **17**, 381–391.

Peck, E. R., and Huang, S. (1977). *J. Opt. Soc. Am.* **67**, 1550–1554.

Perry, C. H., and McNelly, T. F. (1967). *Phys. Rev.* **154**, 456–458.

Randall, C. M., and Rawcliffe, R. D. (1967). *Appl. Opt.* **6**, 1889–1895.

Randall, C. M., and Rawcliffe, R. D. (1968). *Appl. Opt.* **7**, 213.

Rastogi, A. (1975). PhD. Thesis, Northeastern Univ. (unpublished).

Rastogi, A., Hawranek, J. P., and Lowndes, R. P. (1974). *Phys. Rev. B* **9**, 1938–1950.

Rastogi, A., Pai, K. F., Tornberg, N. E., Parker, T. J., and Lowndes, R. P. (1977). *Phys. Lett.* **62A**, 239–241.

Rastogi, A., Pai, K. F., Parker, T. J., and Lowndes, R. P. (1978). *In* "Lattice Dynamics" (M. Balkanski, ed.). pp. 142–143. Flammarion Sciences, Paris.

Roberts, D., and Coon, D. D. (1962). *J. Opt. Soc. Am.* **52**, 1023–1029.

Robinette, W. H., and Sanderson, R. B. (1969). *Appl. Opt.* **8**, 711–712.

Robinson, T. S. (1952). *Proc. Phys. Soc. London Sect. B* **65**, 910–911.

Robinson, T. S., and Price, W. C. (1953). *Proc. Phys. Soc. London Sect B* **66**, 969–974.

Rouard, P. (1937). *Ann. Phys.* **7**, 291–384.

Rubens, H. (1889a). *Wiedemann Ann. Phys. Chem.* **37**, 249–268.

Rubens, H. (1889b). *Wiedemann Ann. Phys. Chem.* **37**, 522–523.

Russell, E. E., and Bell, E. E. (1966). *Infrared Phys.* **6**, 75–84.

Russell, E. E., and Bell, E. E. (1967a). *J. Opt. Soc. Am.* **57**, 341–348.

Russell, E. E., and Bell, E. E. (1967b). *J. Opt. Soc. Am.* **57**, 543–544.

Sanderson, R. B. (1967). *Appl. Opt.* **6**, 1527–1530.

Sanderson, R. B., and Scott, H. E. (1970). *Aspen Int. Conf. Fourier Spectrosc.* pp. 167–174. AFCRL Special Rep. No. 114.

Sanderson, R. B., and Scott, H. E. (1971). *Appl. Opt.* **10**, 1097–1102.

Sanderson, R. B., Scott, H. E., and White, J. E. (1971). *J. Mol. Spectrosc.* **38**, 252–256.
Schmunk, R. E., and Winder, D. R. (1970). *J. Phys. Chem. Solids*, **31**, 131–141.
Seitz, F. (1940). "Modern Theory of Solids," Chapter XVII. McGraw-Hill, New York.
She, C. Y., Broberg, T. W., Wall, L. S., and Edwards, D. F. (1972). *Phys. Rev. B* **6**, 1847–1850.
Simon, I. (1951). *J. Opt. Soc. Am.* **43**, 336–345.
Slater, J. C., and Frank, N. H. (1947). "Electromagnetism." McGraw-Hill, New York.
Spitzer, W. G., and Kleinman, D. A. (1961). *Phys. Rev.* **121**, 1324–1335.
Staal, P. R., and Eldridge, J. E. (1977). *Infrared Phys.* **17**, 299–303.
Stone, N. W. B., and Chantry, G. W. (1977). *Adv. Infrared Raman Spectrosc.* **3**, 43–86.
Strong, J. (1957). *J. Opt. Soc. Am.* **47**, 354–357.
Sugawara, F., and Nakamura, T. (1970). *J. Phys. Soc. Jpn.* **28**, 158–160.
Thomas, T. E., Orville-Thomas, W. J., Chamberlain, J., and Gebbie, H. A. (1970). *Trans. Faraday Soc.* **66**, 2710–2719.
Vogel, P., and Genzel, L. (1964). *Infrared Phys.* **4**, 257–262.
Wallis, R. A., and Maradudin, A. A. (1962). *Phys. Rev.* **125**, 1277–1282.
Wehner, R. K., and Steigmeier, E. F. (1975). *RCA Rev.* **36**, 70–88.
Wilkinson, G. R. (1963). *In* "Infrared Spectroscopy and Molecular Structure" (M. Davies, ed.), pp. 85–110. Elsevier, Amsterdam.
Yarwood, J., and James, P. L. (1977). *Conf. Fourier Transform Infrared Spectros. Columbia, South Carolina.*
Yoshinaga, H., and Oetjen, R. A. (1956). *Phys. Rev.* **101**, 526–531.
Young, R. H. (1977). *J. Opt. Soc. Am.* **67**, 520–523.
Zafar, M. S., Hasted, J. B., and Chamberlain, J. (1973). *Nature (London) Phys. Sci.* **243**, 106–109.
Zhevakin, S. A., and Naumov, A. P. (1963). *Radiofizika*, **6**, 674–694.
Zhevakin, S. A., and Naumov, A. P. (1967). *Radio Eng. Electron. Phys.* **12**, 1067–1076.
Zwerdling, S. (1970). *J. Opt. Soc. Am.* **60**, 787–790.
Zwick, U., Irslinger, C., and Genzel, L. (1976). *Infrared Phys.* **16**, 263–267.

CHAPTER 4

Far Infrared Submillimeter Spectroscopy with an Optically Pumped Laser

B. L. Bean   *and*   *S. Perkowitz*

I. Introduction

Most portions of the electromagnetic spectrum have proven useful for various forms of spectroscopy, and the far infrared/submillimeter (FIR/submm) region 10–200 cm^{-1} is no exception. Many physical processes have characteristic energies corresponding to this range and are thus good candidates for submillimeter spectroscopy. Examples include lattice and free carrier effects in crystalline solids, superconducting behavior, and the continuum absorption of water.

For such spectroscopy to be practicable, certain instrumentation requirements must be met. A spectroscopic system must include a source of wide wavelength coverage with sufficient intensity to give good signal-to-noise ratios even for highly absorbing samples, and sufficient stability to allow accurate spectroscopic measurement. In some wavelength ranges, for instance, in the visible, such requirements are easily met by blackbodies. These sources provide continuous, broadband coverage at sufficient intensities for the most accurate work. In other ranges, although it may not be possible to produce continuous sources, it is at least possible to produce fixed wavelength sources of high intensity. An example is the use of microwave klystrons.

Neither approach, however, is suitable for the submillimeter region. Only a small fraction of the radiation from a hot blackbody lies in the FIR range. The construction of fixed wavelength devices analogous to klystrons has proven difficult. In fact, no source combining the desirable features of continuous coverage and moderate to high intensity has existed for the submillimeter region. As a result various clever stratagems have been adapted.

One approach has been to use conventional grating spectrometers with additional filtering to discriminate against undesired middle infrared radiation from the blackbody source. Such instruments have been used successfully, but anomalous results due to unfiltered radiation have occurred in spite of the most stringent care. The FIR power level is so low that really excellent low-temperature detectors are often necessary. A second approach is to maximize the signal-to-noise ratio available from a blackbody source by using a Fourier transform spectrometer, which is basically a Michelson interferometer. The so-called Jacquinot and Fellgett advantages of this system—one pertaining to the absence of conventional slits, the other relating to the simultaneous observation of many spectral elements—give it a significant advantage over the conventional grating spectrometer. Such Fourier instruments, which became useful fifteen years ago with the development of high-speed computers, have proven the most successful FIR devices to date. They have allowed accurate spectroscopy between 5 and 200 cm^{-1} with good signal-to-noise ratios, although the most sensitive work also often requires the best available cooled detectors.

With the advent of a new submillimeter source, the optically pumped laser, the possibility of a new type of FIR spectroscopy has developed. These devices provide moderate to high levels of power for spectroscopy with sufficient stability for precise work. They are not continuously tunable, but instead produce a large array of individually pumped laser lines, sufficient to cover the submillimeter range in detail. However, unlike Fourier systems, where different designs are preferred for the middle and extreme FIR ranges, a single laser can cover the whole span. The high power of the laser, which is orders of magnitude beyond what is available from a Fourier system, often eliminates the need for a cooled detector allowing great flexibility and economical operation.

This chapter shall describe in detail an operating laser spectrometer now functioning at our laboratory at Emory University. The intent of the chapter is to provide a practical guide to the design and construction of such a system, and to show a wide range of its applications so that the interested reader can judge for himself where it may or may not prove useful. Following an introduction to the theory of operation of the submillimeter pumped laser, we shall discuss the design and construction of our laser spectrometer, illustrate its strengths and weaknesses with various spectroscopic applications, and make projections about the future usefulness of such devices.

II. Pumped Laser Theory of Operation

An optically pumped FIR laser is relatively simple in design. A gaseous medium inside the FIR laser cavity is optically pumped with an infrared beam, usually from a CO_2 laser. The CO_2 laser works well for this purpose because it has high output power, sufficient for good optical pumping, on over 130 emission lines in the 9- to 11-μm region. Molecules with rich absorption spectra overlapping the CO_2 laser emission lines are good possibilities for FIR laser media. The molecules must also have a permanent dipole moment for laser action to be obtained on a rotational transition (Chang, 1974). Polar molecules such as methanol, formic acid, and the methyl halides satisfy these conditions and make very good laser media.

Figure 1 is a partial energy level diagram of a symmetric top molecule that can be used to illustrate the pumping and lasing transitions of a submillimeter laser. The infrared pump radiation excites molecules from a certain rotational level in a lower vibrational state to a rotational level (J, K) in the upper vibrational state. The rotational levels in the upper vibrational state are not heavily populated so a population inversion can be easily achieved between the (J, K) state and the (J − 1, K) state. FIR laser action occurs by stimulating the transition from the (J, K) to (J − 1, K) state. As this process proceeds a population inversion may also be achieved between the (J − 1, K) and (J − 2, K) states. Thus a cascading effect can develop in which two or more lines may lase with excitation from only one pump line. The starting level may also be depleted (as indicated in the figure) such that a population inversion exists between it and an upper rotational state. Both of these effects have been observed to cause multiple FIR outputs associated with many of the CO_2 pumping transitions (Chang and Bridges, 1970).

Optically pumped FIR lasers were first constructed according to a Fabry–Perot design with one plane and one spherical mirror (Chang and Bridges, 1970). A later design consisted of an oversized cylindrical metallic waveguide with flat mirrors close to or inside the ends of the tube (Hodges and Hartwick, 1973). We chose the latter design for our system because of the advantages it

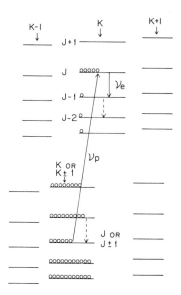

Fig. 1. Partial energy level diagram for a symmetric top molecule showing transitions in an optically pumped submm laser (adapted from Chang, 1974).

offered. Waveguide lasers can be more compact than Fabry–Perot designs while still providing efficient pump absorption and low diffraction losses. They also allow efficient collisional deexcitation at the walls, minimum pump losses other than by the gas absorption, convenience in fabricating more efficient output couplers (Danielewicz and Coleman, 1976), and low-loss FIR modes that are compatible with free space propagating modes (Hodges *et al.*, 1977).

The power output from FIR lasers has been below the theoretical value obtained from the Manley–Rowe relation. A typical photon efficiency for a laser with a cylindrical, metallic waveguide and hole-coupled metal mirrors would be 1% or less. Some recent work to optimize the FIR output has reported conversion efficiencies approaching 20% (Hodges *et al.*, 1977). However, even efficiencies of 1% provide large powers relative to those from conventional sources.

III.   Availability of Submillimeter Lines and Media

Recent sources have reported over 800 relatively powerful FIR lines from more than 35 different lasing media (Rosenbluh *et al.*, 1976; Yamanaka *et al.*, 1976; Gallagher *et al.*, 1977). These numbers are increasing constantly as further investigations are made and as improved techniques are employed to search for additional lines in new and old media (Busse and Thurmaier, 1977; Mathieu and Izatt, 1978). All but a few of these lines are obtained by optically

TABLE I

SUBMILLIMETER LASER LINES OBTAINED FROM FOUR PUMPED MEDIA[a]

f (cm^{-1})	λ (μm)	Pumped medium	Pumping line	Pol.	P_{rel}
8.2	1217	A	9P(16)	‖	1
9.8	1020	D	10P(14)		1
11.2	890	D	10P(22)	‖	3
13.1	764.1	D	10P(10)	⊥	2
14.3	699.5	A	9P(34)	⊥	3
15.1	663.3	D	10P(24)	‖	2
17.5	570.5	A	9P(16)	‖	8
18.1	554.4	D	10P(14)	⊥	2
21.2	471	A	10R(38)		4
24.0	417.1	B	9P(6)	⊥	2
24.6	406.4	D	10P(14)		1
24.6	406	C	10R(12)	⊥	1
25.5	392.3	A	9P(36)	⊥	4
26.8	372.7	D	10P(12)	‖	3
30.3	330	B	9R(4)	⊥	3
31.3	320	B	9P(30)	‖	2
32.8	305	B	9R(8)	⊥	2
33.4	299	C	10R(24)	⊥	1
34.2	292.6	B	9R(8)	⊥	2
39.4	254	A	10R(38)		2
42.9	232.9	A	9R(10)	‖	2
43.6	229.1	B	9P(6)	‖	7
53.8	185.9	A	9R(18)	‖	1
54.3	184	C	10R(24)	⊥	1
58.6	170.6	A	9P(36)	‖	2
61.3	163	A	10R(38)	‖	9
68.5	146	B	9P(30)	⊥	2
74.2	134.7	B	9P(6)	⊥	3
84.2	118.8	A	9P(36)	⊥	10
97.1	103	B	9P(30)	⊥	2
103.6	96.5	A	9R(10)	‖	9
141.6	70.6	A	9P(34)	⊥	1
175.4	57	B	9R(8)	⊥	6
214.1	46.7	B	9R(8)		6
243.9	41	C	10R(18)	⊥	2

[a] Frequency f, wavelength λ, pumped medium, pumping line, polarization Pol., and relative power P_{rel} for the 35-submillimeter lines reported here. The media are denoted as A for CH_3OH, B for CH_3OD, C for CD_3OD, and D for CH_2CF_2. In the P_{rel} column, the strong line at 84.2 cm^{-1} is arbitrarily assigned a power level of 10 (reproduced from Bean and Perkowitz, 1977a).

pumping with either a cw or pulsed CO_2 laser. Methanol, formic acid, the methyl halides, and the deuterated forms of these materials account for a large percentage of the observed lines. Output powers for cw pumped FIR lines range from a few microwatts to hundreds of milliwatts (Hodges et al., 1977).

It can be rather complicated and troublesome to consider using over 35 different media for FIR spectroscopic work. The task can be made easier by selectively choosing appropriate media with many lines in the spectral region of interest. The number of different materials required is then dependent on the resolution required and the spectral coverage needed. An example of this is given in Table I where four substances—methyl alcohol (CH_3OH), two deuterated methyl alcohols (CH_3OD and CD_3OD), and 1,1-difluoroethylene (CH_2CF_2)—were chosen which give a good spectral coverage from 8 to 244 cm^{-1} (Bean and Perkowitz, 1977a). This coverage gives an average line spacing of just under 7 cm^{-1} which decreases to a little more than 3 cm^{-1} for the frequencies less than 100 cm^{-1}. These four media are safe and simple to use.

IV. The Spectroscopic System

This section gives a full description of our laser spectrometer. The details of the pump laser, the FIR laser, and the system electronics are given in turn.

A. THE PUMP LASER

The laser used to optically pump a submillimeter laser should lend itself to ease of operation. In practice, it is not difficult to obtain 85–90 of the 130 emission lines at powers of 6 to 30 W—sufficient power for useful pumping—in most commercial and homemade CO_2 lasers. Another important factor is that the pump must be well stabilized, since any frequency instability in it causes amplitude instability in the FIR output. Methods for stabilizing CO_2 lasers through a phase-sensitive piezoelectric feedback system are well established. These considerations make the CO_2 laser an excellent practical choice for pumping.

An illustration of our CO_2 and FIR lasers is given in Fig. 2. The CO_2 laser cavity is 183 cm long with a discharge region of about 152 cm. The tube has an 8-mm bore diameter with a water jacket surrounding it for cooling purposes and has ZnSe Brewster's angle windows on the ends. The cavity reflectors are external and consist of a plane grating with 150 lines/mm and a ZnSe meniscus lens with a 65 % reflective coating on the inner surface. The grating is blazed at 10 μm and is set so the polarized beam is perpendicular to the grooves. The meniscus lens output coupler has a 20-m radius of curvature on the inner concave surface and a 10-m radius of curvature on the convex surface.

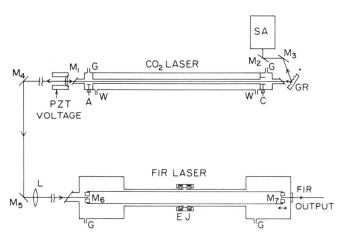

FIG. 2.   Block diagram of the CO_2 and FIR laser system: M, mirrors; GR, grating; SA, spectrum analyzer; G, gas inlet and outlet; W, cooling water inlet and outlet; C, cathode; A, anode; L, lens; and EJ, expansion joint.

The gas is a mixture of about 12% CO_2, 15% N_2, and 73% He which is flowed continuously through the cell and maintained at about 13-torr pressure. The laser operates with about 30 mA of current through the discharge tube and a voltage of 13 kV across the series combination of the tube and a 100-kΩ resistor.

The output of the laser is greater than 20 W for the strongest lines and about 5 to 10 W for the weakest transitions that will lase. This output passes through the meniscus lens and goes to the FIR laser. There is another low power output which comes off the grating as the zeroth order and which is directed into a commercial spectrum analyzer in order to monitor the wavelength. Each end of the laser tube is mounted in an x–y positioner for adjustment in the directions perpendicular to the axis of the tube. This adjustment often improves the beam quality and power output.

The entire CO_2 laser is mounted on a 15-cm-wide and 7.5-cm-thick marble slab. The slab's large mass and low coefficient of expansion make it a good laser base.

B.   The Far Infrared Laser

The FIR laser consists of a vacuum chamber containing the mirrors and waveguide. Several variations are reported in the literature, but the vacuum chamber for our system consists of two rectangular boxes which house the mirror mounts and a tube connecting them through which the cylindrical waveguide passes. It was found that a slip or expansion joint was required in the tube because the heat generated in the absorbing gas would cause the tube to expand and push the end boxes apart. This would displace the end mirrors, producing a drift in the FIR output power.

The FIR cavity is formed by a gold-plated metallic waveguide with internal mirrors. The waveguide is 91 cm long and 2.54 cm in diameter. The gold-coated copper mirrors have the same diameter with a 2-mm center hole which admits the pump radiation and transmits the FIR radiation. The CO_2 radiation enters the vacuum cavity through a NaCl window set at Brewster's angle and the FIR output exits through a crystal quartz window. The output window is quite transparent to the FIR radiation, but it absorbs the CO_2 radiation that may have passed through the output coupling hole. This feature helps eliminate unwanted CO_2 radiation in the FIR output.

The cavity mirrors fit just inside the waveguide tube and one of them is mounted on a translation stage that can be used to adjust the cavity length. The drive for the differential screw on the translation stage passes through the vacuum wall and can be operated manually or by a stepping motor. The total drive on the differential screw is 1.3 mm. It is only necessary to be able to translate a distance which is more than half the length of the longest wavelength expected from the laser. It was found that the fine control given by the differential screw was much more important than the longer distance of travel when using a regular micrometer screw.

The gases that compose the lasing media are flowed continuously through the system and exhausted outside the laboratory building. The laser is operated at pressures ranging from 0.1 to 0.3 torr depending upon the optimum lasing for the particular output frequency.

The CO_2 pump beam is focused into the FIR cavity through a large $f/\#$ lens in order to have the beam fill only half the surface area of the output mirror on the first pass and all of the input mirror after the return pass. Observations by Hodges et al. (1977) indicate that there is a high pump loss at the waveguide walls, so maximum pumping by the CO_2 radiation before it strikes the walls is important. It has also been noted by the same investigators that the FIR laser performance degrades noticeably when the CO_2 laser beam deviates from a Gaussian profile.

The rest of the FIR optical train includes two crystal quartz beam splitters, a mechanical chopper, and a Fabry–Perot interferometer, as shown in Fig. 3. The latter serves to measure the FIR wavelength and also suppresses unwanted radiation when two or more FIR lines lase simultaneously. The interferometer was constructed at our laboratory using a translation stage with a micrometer screw drive and a machinist's indicator gauge with an accuracy of 0.0001 inches per division to register the distance traveled between maxima. The reflectors are Buckbee–Mears nickel inductive meshes mounted on a holder patterned after one illustrated in the literature (Renk and Genzel, 1962). Two or three sets of meshes with varying spacings are required to

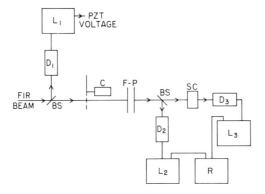

FIG. 3.   Block diagram of the system electronics and FIR optics: BS, beam splitter; L, lock-in amplifiers; D, detectors; C, chopper; F–P, Fabry–Perot interferometer; SC, sample chamber; and R, ratiometer.

maintain a finesse of 15 to 60 over the FIR spectral region. The remainder of the optical and electronic train in Fig. 3 is discussed in the next section.

C.   SYSTEM ELECTRONICS: LASER STABILIZATION AND FIR DETECTION

The block diagram in Fig. 3 illustrates the detectors and electronic apparatus used in our spectrometer. A pyroelectric detector is used to monitor the output of the FIR laser at the first beam splitter before it reaches the sample region. This signal is fed into a commercial phase-sensitive detection system which is used to frequency-stabilize the CO_2 laser. This is accomplished by altering the cavity length with the PZT element to which the output coupler is attached. It is advantageous to use the FIR signal rather than the CO_2 signal for the feedback system since the correct emission frequency to pump the FIR gas may not be at the center of the emission gain profile. Thus, by maximizing the FIR signal, the CO_2 laser can be made to lase slightly off the frequency where it would normally be lasing if its signal were maximized. Changing the voltage on the PZT is also used to assist in bringing the CO_2 emission into coincidence with the gas absorption for the pumping process to occur. Both line widths are approximately 70–80 MHz so a close coincidence is necessary.

Following the first beam splitter the FIR beam is chopped by a commercial chopping wheel. The chopping frequency is set at 11 Hz for detection with Golay cells, but could be increased for use with faster detectors. After passing through the Fabry–Perot interferometer, the beam is again divided at a beam splitter to provide source compensation. Part of the beam goes into a reference channel while the remainder is directed toward the sample. Both of the detectors are Golay cells and their signals are fed into matched lock-in amplifiers. The signals from the lock-ins are combined in a ratiometer and the final ratioed data are recorded either in digital form or on a chart recorder. The ratio method works well because it can compensate for any variations that occur in the amplitude of the FIR laser. It provides excellent results for either transmission or reflection measurements for a broad range of samples.

V.   System Performance

Perhaps the most meaningful analysis of the system's performance is to give examples of its spectroscopic use. Before launching into these applications, however, it may be useful to discuss the basic performance parameters of the source-compensated system.

A.   Power Output

The first consideration in the system operation was the available power level, which was confidently expected to far exceed what was available from a Fourier spectrometer with a blackbody source. Absolute power levels in the submillimeter range are difficult to measure, owing to the absence of a universally accepted calibrated power meter. It is possible, however, to make accurate relative measurements comparing the laser power levels to those available in a commercial Fourier system, a Grubb Parsons Mark II cube interferometer, operating in our laboratory. (For later reference we shall refer to this interferometer as our standard Fourier system). Both the interferometer and the laser system use Golay detectors with the same chopping frequencies and collection geometries, so that differences in the detector signals are directly related to the source intensity levels. The result is that each of the laser lines shown in Table I is from 2 to 50 times stronger than the total integrated intensity available in the Fourier system over the range 10–200 cm^{-1}, giving a power improvement of 50 to 1000 at any given frequency.

Very rough power levels can be given, since the volts-to-watts conversion factor is approximately known for the Golay detector. The result is that the most powerful line, that at 84.2 cm^{-1} from CH_3OH, has a power level of the order of 1 mW. Table I shows the power levels of all the lines relative to this value, indicating that the weakest line observed has a power near 0.1 mW.

B.   Frequency Coverage

The second important parameter in system performance is that of frequency coverage; that is, how much of the submillimeter range is accessible by the system? In a Fourier transform system the blackbody source, in principle, covers the entire FIR range, but the available power drops off rapidly at low frequencies. Further, in a beam splitter type of interferometer, the beam splitter interference fringe shape causes a fall-off in power at low and high frequencies. A typical beam splitter of 12.5-μm mylar will pass useful power between 10 and 220 cm^{-1}, with the half-power points lying at 30 and 160 cm^{-1}. A lamellar grating interferometer provides relatively high amounts of low-frequency power but at the cost of settling for a limited range, perhaps 5 to 50 cm^{-1}. In the beam splitter instrument, the power level in various parts of the submillimeter range can be optimized by selecting beam splitters of appropriate thickness, a procedure which also limits the overall flexibility of the system.

The laser system also has a built-in limitation, namely the Manley–Rowe relation, which severely limits the pumping efficiency at low FIR frequencies. As Table I shows, however, strong and weak lines are fairly evenly distributed over the frequency range, undoubtedly because other factors, such as line-to-line variation in the pump power, mask the decline in efficiency. The net result is that the laser system can cover the range 8–244 cm^{-1} with a power variation of about a factor of 10, with even the weakest lines providing far more power than does a Fourier instrument. Complete coverage does require that the pump media be changed, somewhat analogously to changing beam splitters in a Fourier system.

C. FREQUENCY SPACING

The third obvious system parameter is frequency spacing; that is, how closely can the frequency be varied from value to value? A system with a blackbody source, in principle, has a continuous range of frequencies. In a grating instrument, of course, diffraction at the slits produces an effective bandwidth for the spectrometer. No such strong slit effect exists in a Fourier interferometer (although there is a weak dependence on the size of the aper-tures), but there is another limitation on the frequency spacing. This arises from the digital nature of the data sampling process and subsequent computer analysis. It can be shown that the resulting frequency resolution is given by $\Delta f = 1/L$, where L is the maximum path difference. Nevertheless, it is not possible to approach infinite resolution simply by using a very long path difference. The limitation is that data are meaningful only until the signal-to-noise ratio is one, so the ultimate limit on frequency resolution is the system noise. For a Fourier spectrometer typical values of Δf range between 1 and 10 cm^{-1}, depending on the quality of the detector and the spectral range covered. The laser spectrometer approaches this kind of performance, with an average frequency spacing of 3 cm^{-1} below 100 cm^{-1} and an average spacing of 7 cm^{-1} between 8 and 244 cm^{-1}.

D. SYSTEM STABILITY

Another important question is that of system stability. A blackbody source is capable of providing good long- and short-term stability in the ultraviolet to middle infrared regions. In the submillimeter range, small fluctuations in the total light intensity can have a large effect on the small fraction of power that lies in the region. This is especially likely to happen if the blackbody source is a mercury arc, whose high power output for its small size makes it popular for Fourier spectrometers. One obvious solution to such source fluctuations is to use a double-beam or source-compensated Fourier system, but few such instruments have been attempted. Contributing causes are the difficulty of data analysis for a two-detector Fourier system and the lack of extra power to divert to a secondary beam. Lasers, on the other hand, are liable to develop long- and short-term instabilities. The problem is compounded when one laser pumps a second laser. However, the high power of the laser makes it feasible to use a double-beam or source-compensating arrangement as was done in our system. The inherent stability of some of the lines we used was sufficient to produce excellent results using only source compensation with no additional stabilization. For other lines, the feedback control system was needed to reduce drift in the FIR signal. Data showing the long- and short-term behavior of the laser signal shall be discussed in the next section.

VI. Spectroscopic Applications

The performance of a spectrometer cannot be defined simply in terms of power levels, stability, and frequency coverage. Other characteristics, such as stray light and beam behavior, must also be considered, and often these can be examined only in the course of an actual spectroscopic measurement. This section, therefore, shall describe several different spectroscopic measurements made with the laser system. In all cases the laser results shall be compared with data obtained from more conventional FIR systems and systematic discrepancies and random noise shall be thoroughly discussed. The applications fall into four general categories: transmission measurements in liquid water, transmission measurements in bulk and epitaxial semiconductors, transmission measurements in thin film superconductors in the normal and superconducting states, and reflection and scattering work on highly absorbing FIR coatings.

A. Transmission Measurements in Liquid Water

The absorbing behavior of liquid water is notoriously difficult to measure in the infrared region because the absorption coefficient is very high. The absorption decreases somewhat in the submillimeter range but is still high enough to make its measurement a formidable task. Results reported in the literature reflect these difficulties in their internal random errors and in their lack of agreement among experimenters (Ray, 1972). These problems made water measurements an ideal first application of the laser system (Bean and Perkowitz, 1976).

The measurements used a method developed and applied by Draegert *et al.* (1966). The transmission through a cell filled with water is given by

$$T = Ce^{-\alpha d}, \tag{1}$$

where C is a constant that depends on the optical properties of the cell, α the water absorption coefficient, and d the water path length. From this expression, it can be seen that a plot of $\ln(T)$ versus d gives a straight line whose slope is α, independent of the cell properties.

A sample cell with adjustable pathlength was constructed from crystalline quartz windows, which are excellent submillimeter transmitters. The transmission was measured at three different cell thicknesses for each of six FIR frequencies. The absorption coefficients obtained from these measurements are shown in Fig. 4. For comparison, a curve showing typical results from one set of published measurements is given. Our data agree with this curve within 7%, while the discrepancy among the different published sets of data is of the order of 10%. The error bars appearing in the figure are equal to or less than those obtained in any other reported measurements with grating or Fourier

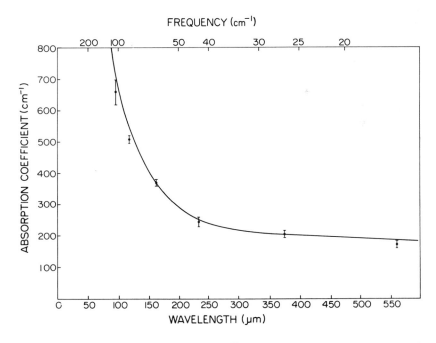

FIG. 4.   Measured values of the absorption coefficient of liquid water at six wavelengths. The solid curve was generated using experimental data from Downing and Williams (1975) (reproduced from Bean and Perkowitz, 1977b).

transform instruments using Golay detectors, although we used water path-lengths of up to 300 μm, several times larger than any used in the past. The penetrating power of the submillimeter laser is clearly shown by these considerations. The use of longer path lengths reduces the relative measurement error in the determination of the path length itself.

B. TRANSMISSION MEASUREMENTS IN BULK AND EPITAXIAL SEMICONDUCTORS

Transmission measurements of bulk semiconductors in the submillimeter range give valuable information about the lattice and the free carrier behavior. Furthermore, for some materials the optical behavior is well understood, so that measurements in such materials provide a good benchmark experiment for the laser spectrometer. In epitaxial systems where a thin film semiconductor lies on a bulk sample, FIR measurements can, in principle, give information about the epitaxial layer, the substrate, and the important interface region. Such measurements are more difficult to analyze, but can still be used to compare laser results with those of a conventional spectrometer.

1. *Bulk Semiconductor: GaAs*

Theoretical considerations suggest that in the submillimeter range the absorption due to the free carriers in a semiconductor is described by the Drude theory as

$$\alpha_{fc} = \alpha_0/(1 + \omega^2\tau^2), \tag{2}$$

where ω is the optical frequency in radians per second, the scattering time τ is the average time between electronic collisions, and α_0 depends on the lattice properties, the scattering time, and the electronic concentration but is independent of ω. This equation holds only if the electronic concentration is not too large (Perkowitz, 1971).

The critical part of Eq. (2) is the quadratic frequency dependence, since other theories give different dependences. To test the validity of this equation, measurements were made with our standard Fourier system on a sample of GaAs (Perkowitz, 1971). The sample was well characterized with a known carrier concentration of 5×10^{15} cm^{-3} and a thickness of 0.054 cm. Even this relatively low concentration gives heavy absorption in the submillimeter region, and the transmission over much of the range was well below 1%.

The Fourier results for this sample are shown in Fig. 5. The system gave good wavelength coverage with more than adequate resolution down to 20 cm^{-1}. The error bars for transmissions above 1% are reasonably small, but the errors become strikingly worse for lower transmissions. Figure 6 shows the value of α_{fc} derived from the transmission data plotted in a form to display

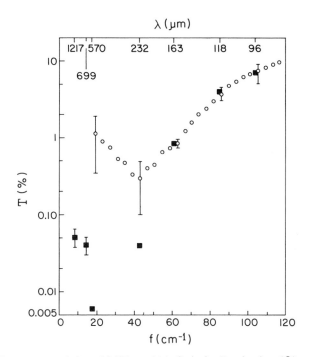

FIG. 5. Percent transmission of 0.054-cm-thick GaAs for Fourier data (○) and laser data (■). The FIR laser wavelengths are marked at the top of the figure (reproduced from Bean and Perkowitz, 1977b).

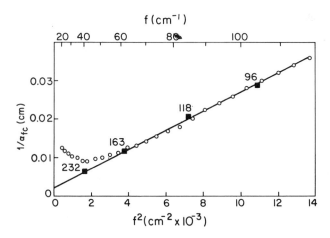

FIG. 6. Plot of the reciprocal of the free carrier absorption for GaAs versus the frequency squared for Fourier (○) and laser (■) data. The solid curve shows the linear nature of the laser data (reproduced from Bean and Perkowitz, 1977b).

the quadratic dependence. The data above 60 cm^{-1} follow the Drude prediction very well, but the divergence below 60 cm^{-1} is unexplained.

To further examine the situation and to compare the laser spectrometer with the standard Fourier system, the same sample was remeasured with the laser system. The sample was mounted in a simple slide holder allowing a sample-in/sample-out measurement of the transmission. The laser results are given in Fig. 5. There are two striking features: first, the laser data have error bars up to ten times smaller than those obtained with the Fourier system; second, the laser values agree extremely well with the Fourier results for higher frequencies and transmissions above about 1 %, but deviate enormously for lower transmissions at low frequencies.

There is strong evidence that the deviant laser values are in fact correct. Figure 6 shows the free carrier absorption calculated from the laser data. Up to 50 cm^{-1} the Fourier and laser data are identical. At 40 cm^{-1}, the absorption value calculated from the laser results lies precisely on the line predicted by the Drude theory, removing the divergence. It would have been valuable to extend the comparison to the frequencies below 20 cm^{-1}, but the necessary information to convert transmission to absorption was not available at the time of the measurement for these very low frequencies.

These results suggest that stray light is playing a role in the Fourier system. The blackbody source provides radiation over a broad range, and this may affect the Fourier measurements. Although it might be possible to remove this erroneous baseline effect by further analysis of the Fourier results, this introduces an additional complexity into the data analysis. The laser system provides directly measured quantities at single laser lines with no broad background radiation entering as a factor.

2. *Epitaxial Semiconductor: InAs on a GaAs Substrate*

InAs/GaAs is of considerable technical interest. Both reflection and transmission measurements should be useful in determining the properties of the entire system. As a comparison of the standard Fourier system and the laser system, the same sample of epitaxial material was examined in both systems, and the results are shown in Fig. 7. The InAs epitaxial layer had a carrier concentration of about 5×10^{15} cm^{-3} and a thickness of about 17 μm. The substrate, composed of high-resistivity GaAs, gave a high transmission even with a thickness of 0.1 cm. Thus the measured transmissions for this system are somewhat higher than those in the earlier GaAs sample.

As in the previous measurement, the Fourier and laser data agree very well over much of the frequency range, with the signal-to-noise ratio for the laser vastly superior to the Fourier results. Below 30 cm^{-1} the two data sets diverge. We have made a detailed theoretical analysis of the FIR optical behavior of the epitaxial/substrate system and find that the low-frequency behavior shown by

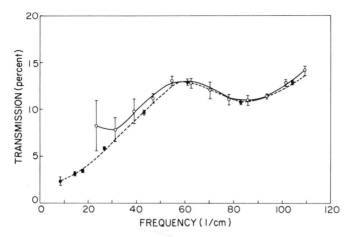

FIG. 7.   Transmission of a 17-μm-thick epitaxial layer of InAs on a high-resistivity GaAs substrate for Fourier (○) and laser (●) data. The superior signal-to-noise ratio for the laser data is evident, especially at very low frequencies (reproduced from Bean and Perkowitz, 1977b).

the laser data is correct. The upturn in the Fourier data below 30 cm^{-1} is anomalous and, again, is most probably indicative of stray light problems in regions of low light intensity.

C.   TRANSMISSION MEASUREMENTS IN THIN-FILM SUPERCONDUCTORS

Superconductors are perhaps the most interesting of solid state materials to examine in the submillimeter range, since their characteristic energy gap falls in this region. The theory of superconducting optical behavior has been developed by Mattis and Bardeen (1958) and has been used to analyze several bulk superconductors. Much current interest lies in thin film superconductors, since several of the materials with the highest superconducting transition temperatures are available in this form. Examples include V_3Si with a transition temperature T_c as high as 17 K and Nb_3Ge with transition temperatures up to 23 K.

In bulk materials, the only measurement that can be made is the determination of the reflectivity. The change in the reflectivity (or the surface impedance) as the sample is cooled from the normal to the superconducting state gives information about the energy gap and other parameters of interest. This change is small, since even in the normal state the superconductor is very highly reflective. For thin films, an alternative method is to measure the film transmission. Since the energy gap leads to an absorption onset, transmission measurements are sensitive to the energy gap and the change in transmission on cooling from the normal to the superconducting state is considerable (Tinkham, 1970). Transmission measurements have been made in elemental superconductors like lead, which can be made in films as thin as a nanometer while retaining good superconducting properties (Palmer and Tinkham, 1968). For the more complex mixed superconductors like V_3Si, film thicknesses of at least tens of nanometers are necessary. In these films the magnitude of the transmission is very small.

The foregoing discussion shows that transmission measurements in thin-film, high-temperature superconductors should provide a demanding test of the capabilities of the laser spectrometer. To test the spectrometer and to obtain information about thin-film V_3Si, two samples were examined at length. They consisted of films of nominal thickness 30 and 60 nm grown on sapphire substrates. At room temperature, the estimated transmission of the thinner sample was about 8%, dropping to 1 to 4% at low temperatures. The second sample, because of its increased thickness and higher electrical conductivity, had a far smaller transmission. The estimated transmission ranged from about 1% at 300 K to 0.01 to 0.06% at low temperature.

The first measurements in these samples were made at room temperature, where the transmission of the film substrate was measured relative to that of the substrate alone. The ratio T(film + substrate)/T(substrate) was formed by using a sample-in/sample-out method, with the reference sapphire nominally of the same thickness as the substrate sapphire. The measurements were made with both the Fourier and the laser systems and the results are shown in Fig. 8. The data can be fitted with a simple expression due to Tinkham (1970):

$$T_{rel} = [1 + \sigma_0 \, dZ_0/(n + 1)]^{-2}, \qquad (3)$$

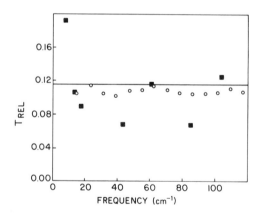

FIG. 8.   Relative transmission of a V_3Si film on a sapphire substrate at 300 K for Fourier (○) and laser (■) data. The solid curve gives the theoretical fit (reproduced from Bean and Perkowitz, 1977b).

where σ_0 is the dc conductivity of the sample, d the film thickness, Z_0 the impedance of free space, and n the refractive index of the substrate. It is apparent that the Fourier results agree with the theory, but the laser results diverge wildly above and below the correct frequency-independent values (McKnight et al., 1977).

This discrepancy pointed out an important difference between the laser and the Fourier systems which had not been appreciated previously. The sapphire substrates used in these samples are of excellent optical quality, with sufficient transmission and parallelism to support interference fringes. In ratioing results including interference fringes from two different pieces of sapphire, the fringes could not be expected to cancel out entirely. However, the fringe effect does not appear in the Fourier data, since the broadband character of the radiation tends to average over many fringe cycles. In the laser system, on the other hand, the frequency width of any of the laser lines is far less than the fringe spacing, and this system is fully capable of following the fringe behavior with the results shown in Fig. 8.

Further measurements were made at temperatures above and below the superconducting transition, namely, 19 and 5.5 K. In these measurements no reference sample was used. Instead, the intensity transmitted through the sample was measured at 19 and 5.5 K, and the ratio $I_{5.5}/I_{19}$ was formed. This could be equated to the ratio of superconducting to normal transmission T_s/T_n, which is a very convenient quantity for comparison with theory. This procedure eliminated fringe effects due to slightly different thicknesses. Golay cells were used to detect the transmitted power. Reflection measurements were also made to obtain the reflectivity R_s/R_n, but those results are not shown here.

Figure 9 shows the raw data for the transmitted intensities at 5.5 and 19 K for the thinner sample, measured at 43.6 cm^{-1}. These data represent the source-compensated signals with no feedback stabilization. The time constant on each of the lock-ins was 1 sec. The relatively large excursion near 200 sec represents a manual readjustment of the pumping frequency. These data show long-term stability of the order of minutes and illustrate the typical random noise with a room-temperature detector and a highly absorbing sample.

The results for T_s/T_n are shown in Fig. 10 for both samples along with theoretical fits (Mattis and Bardeen, 1958). For purposes of assessing the laser spectrometer, the most important feature is the low noise obtained. Even when measuring ratios of two very small intensities, the errors are usually below 2% of the ratio and never exceed 5% of the ratio. This performance is exceptional for a room-temperature detector and is wholly due to the high power of the laser lines. The quality of the data is at least the equal of what can be obtained with a Fourier spectrometer using an excellent cooled detector.

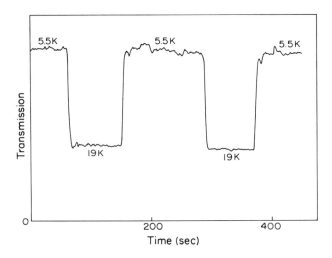

FIG. 9. Transmitted intensities for superconducting and normal state thin film V$_3$Si of 30 mm measured at 43.6 cm^{-1}. This figure shows the long and short term noise in the FIR laser spectrometer.

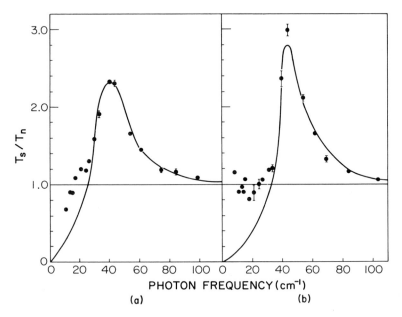

FIG. 10. Measured values of R_s/T_n and theoretical fits (solid lines) for V$_3$Si: (a) thickness of 30 nm and (b) thickness of 60 nm. Error bars not shown are smaller than the symbol size. These plots omit data points beyond 104 cm^{-1}.

The experimental results show some anomalies that may be related to the properties of the laser spectrometer. As the figure shows, the fits do not explain the low-frequency structure seen near 30 cm^{-1} in both samples, or the very low-frequency plateau obtained for the thicker sample. We have conjectured that these deviations arise from a heating effect due to the laser power, but this conjecture seems improbable for several reasons. The samples are in good thermal contact with their supporting copper blocks which have large masses. For several of the laser lines, measurements were made as a function of FIR power with no change in the value of T_s/T_n. Finally, for the thinner sample, runs made on two different Fourier transform systems also gave spectra with the small structure near 30 cm^{-1}. Although the evidence is against the possibility that laser heating is important in this case, such an effect might become significant in other experiments.

D. REFLECTION AND SCATTERING MEASUREMENTS IN ABSORBING FIR COATINGS

Several FIR systems now under development (for instance, astronomical systems) may require knowledge about the properties of supporting devices and materials such as windows, filters, and baffle materials. In a joint project with the Lockheed Missiles and Space Company we have used the laser spectrometer to help characterize the behavior of absorbing FIR coatings.

Several commercial coatings, known as good absorbers in the middle infrared, were examined in the standard Fourier system to determine the normal incidence specular reflectivity between 23 and 220 cm^{-1} with a

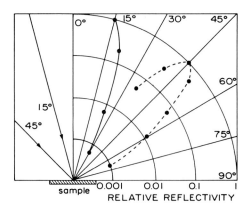

FIG. 11. Plot of the intensity scattered from FIR absorbing materials as a function of angle measured from the specular direction. The solid line gives the relative reflectivity for 103.6 cm^{-1} radiation scattered from a 40 mil thickness of Ecco Sorb CR-110. The dashed line gives the same quantity for 17.5 cm^{-1} radiation scattered from 5 coats of 3M Black Velvet. The angles of incidence are 15° and 45°, respectively.

resolution of 8 cm^{-1}. The laser system was then used to give additional reflection points at 8.2 and 17.5 cm^{-1}, and these matched extrapolations of the Fourier data (Alff *et al.*, 1978).

It was also important to characterize the scattering properties of the coatings by measuring the reflected radiation as a function of angle relative to the specular direction. The Fourier system could have been used to produce a complete frequency spectrum for each angle of interest, but such detailed and time-consuming coverage was not necessary. Instead, the laser was used to spot check the scattering behavior at two characteristic frequencies, 17.5 and 103.6 cm^{-1}. A simple rotating stage was built that made it possible to change the angle of incidence of the laser beam relative to the sample and to change the angle of the Golay detector relative to the specular direction.

Some typical results for these measurements are shown in Fig. 11. They are not corrected for the finite angular resolution of the Golay detector and thus represent upper limits on the scattering behavior. The smallest relative reflectivities on the wings of the curves correspond to absolute reflectivities of about 0.05% or less. Good measurements of these low light levels might have required a cooled detector with the Fourier system, and such a detector, much bulkier than a Golay cell, would have been physically awkward for the off-axis measurements.

VII. Performance Summary and Future Projections

Our laser system has definite strengths and weaknesses relative to other FIR spectrometers. The chief advantage is the high power, which makes it possible to obtain superior results for highly absorbing or barely reflecting samples even with room-temperature FIR detectors. A second advantage, compared to Fourier spectrometers, is the ease of data analysis. No computer is necessary and the determination of noise and baseline levels is straightforward, which is not always the case in Fourier spectroscopy. The greatest disadvantage is the lack of continuous high-resolution coverage over the entire FIR range. The laser spectrometer does provide frequency spacing comparable to what can be achieved with many Fourier systems and sufficient to deal with many spectroscopic problems. The system does not now provide resolutions of the order of 1 cm^{-1} or less, which can be obtained with the very best available Fourier/cooled detector combinations. A second difficulty is that the use of the spectrometer requires a fair amount of laboratory time, since the finding of the desired pumping and FIR lines is not a completely routine process.

Other features of the laser system have both positive and negative aspects. The spectral purity of the laser lines is great enough to form fringes that would not appear in a grating or Fourier spectrometer. These fringes may complicate the measurement in some experiments, but also can serve as a means of measuring the refractive index in transparent materials. Also, the lack of the extraneous and intense ultraviolet and visible light which appears in blackbody FIR sources can be a decided advantage for sensitive samples. As our superconductor results indicate, it may be necessary to at least consider whether the laser power is causing undesirable heating in some cases. On the other hand, the high power may cause new and useful effects, for instance, submillimeter pair breaking in superconductors analogous to the microwave pair breaking already observed (Chi and Langenberg, 1976).

Our experience with the laser system and recent reported work suggest several improvements that could be made. The greatest practical improvement would be to add invar rods to the FIR laser, since the slip joint still allows some thermal drift. Difficulties in finding the desired lines now arise because the FIR cavity length must be adjusted together with the PZT voltage to get both the proper cavity length for lasing and a coincidence between the pump and absorption frequencies. This process could be made far simpler by installing a spectrophone (Busse and Thurmaier, 1977) to detect frequency coincidence. After such coincidence, the cavity would be scanned until lasing was obtained. The spectrophone might also serve as the sensing element in the stabilization loop. Finally, we plan to install a modified output coupler of the type described by Hodges *et al.* (1977). Their investigations indicate that our power could be improved by over an order of magnitude for many of the FIR lines.

Other potential improvements are also under consideration. We are examining the possibility of adding more pumped media to our basic set of four. A recent report by Danielewicz (1978), for instance, shows that CD_3OH is a rich source of FIR laser lines. We expect that the addition of one or two such media would significantly reduce the line spacing without unduly complicating the operation of the spectrometer. In a related effort, we are contemplating the use of a manifold arrangement to allow quick changes from medium to medium. One important study which has not yet been made is to examine the operation of the system with a cooled detector. In the measurements of the thicker superconducting sample described in Section VI.C, there was evidence that the limiting factor on system performance for some of the FIR lines was Golay cell noise. We would like to estimate the relative contributions of detector and laser noise by determining if operation with a cooled detector gives a significant improvement. A high-power laser with a cooled detector could give sensitivity unmatched by any existing submillimeter system.

Even with no further improvements, the pumped laser spectrometer is a functioning instrument which should continue to contribute to FIR spectroscopy. In addition to the areas described here, at least two other major areas could benefit from the laser spectrometer. One is the FIR magneto-spectroscopy of semiconductors and metals. Such experiments have tradi-tionally been done at fixed microwave frequencies while the external magnetic field was varied to produce changes in the sample properties (Palik and Furdyna, 1970). More recently, it became apparent that such measurements were often better done at submillimeter frequencies, and fixed-frequency FIR gas lasers have supplemented the microwave sources. Such lasers do not offer much variety in frequency. Two of the most common ones are the HCN laser, emitting at 29.7 cm^{-1}, and the water vapor laser, with primary emission at 84.3 cm^{-1}. With a tunable pumped laser spectrometer, it would be possible to set the exciting frequency at various values of interest, for instance near the plasma frequency or the important phonon frequencies for a particular material. This flexibility combined with the capability of sweeping the magnetic field and the high power of the laser would produce very rich results for FIR magnetospectroscopy.

The second area is more speculative but may prove of great importance. It is the FIR spectroscopy of water-based biological systems. It is by now well established that many biological molecules including proteins (Beetz and Ascarelli, 1978) and chlorophyll (Boucher et al., 1966, Perkowitz and Bean, 1977) show significant absorptive structure in the FIR. Such measurements and analogous ones in the middle infrared range are usually not made in aqueous solution because of the very heavy absorption of water. This difficulty has contributed to the popularity of Raman scattering as a biological tool, since the Raman effect is insensitive to the presence of water. It would be valuable, however, to make transmission measurements that provide different data from and complement Raman results.

Such measurements are somewhat more possible in the FIR than in the middle infrared, because the water absorption decreases significantly toward lower frequencies, as shown in Fig. 4. Although the absorption is still quite high, the laser spectrometer is capable of penetrating large path lengths of water with a good signal-to-noise ratio. We have attempted to measure the properties of chlorophyll in water with the laser spectrometer. The current power level is not great enough to detect clearly the chlorophyll, but with improved power output and a differential type of measurement the experiment may become possible. Further, the ability of the laser spectrometer to provide good data without cooled detectors points to the possibility of simple in vivo measurements. We have made preliminary studies to show that the laser system can easily be used to determine the FIR transmission through a living leaf.

Our efforts with pumped laser spectroscopy have convinced us that this method is a valuable addition to FIR spectroscopic techniques. The comparisons between the laser system and the well-established Fourier spectrometer show that the former, despite its very recent development and application, is quite capable of giving reliable results of high spectroscopic quality. Our comparisons do not show that laser spectroscopy can replace Fourier spectroscopy. Since the basic submillimeter problem of providing a highpower, continuously tunable source has not yet been fully solved, it is important to be aware of all spectroscopic methods that provide partial solutions. The Fourier and the laser methods can be complementary, each providing the more efficient means of measurement under certain circumstances. Any researcher concerned with submillimeter spectroscopy can only benefit from the use of both Fourier and laser systems in the laboratory.

ACKNOWLEDGMENTS

This work was supported by the Emory University Research Committee, the National Institutes of Health Biomedical Sciences Support Program, the National Science Foundation (Grant No. DMR75-13917-A01), the Office of Naval Research, and the Research Corporation.

REFERENCES

Alff, W. H., Blue, M. D., Grammer, J. R., and Perkowitz, S. (1978). *Proc. Int. Conf. Submillimetre Waves Their Appl. 3rd, March 29–April 1, 1978, Univ. of Surrey* p. 276.
Bean, B. L., and Perkowitz, S. (1976). *Appl. Opt.* **15**, 2617.
Bean, B. L., and Perkowitz, S. (1977a). *Opt. Lett.* **1**, 202.
Bean, B. L., and Perkowitz, S. (1977b). *J. Opt. Soc. Am.* **67**, 911.
Beetz, C. P., and Ascarelli, G. (1978). *Proc. Int. Conf. Submillimetre Waves Their Appl. 3rd, March 29–April 1, 1978, Univ. of Surrey*, pp. 302, 304, 306.
Boucher, L. J., Strain, H. H., and Katz, J. J. (1966). *J. Am. Chem. Soc.* **88**, 1341.
Busse, G., and Thurmaier, R. (1977). *Appl. Phys. Lett.* **31**, 194.
Chang, T. Y. (1974). *IEEE Trans. Microwave Theory Tech.* **MTT-22**, 983.
Chang, T. Y., and Bridges, T. J. (1970). *Opt. Commun.* **1**, 423.
Chi, C. C., and Langenberg, D. N., (1976). *Bull. Am. Phys. Soc.* **21**, 403.
Danielewicz, E. J. (1978). *Proc. Int. Conf. Submillimetre Waves Their Appl. 3rd, March 29–April 1, 1978, Univ. of Surrey*, p. 64.
Danielewicz, E. J., and Coleman, P. D. (1976). *Appl. Opt.* **15**, 761.
Downing, H. D., and Williams, D. (1975). *J. Geophys. Res.* **80**, 1656.
Draegert, D. A., Stone, N. W. B., Curnutte, B., and Williams, D. (1966). *J. Opt. Soc. Am.* **56**, 64.
Gallagher, J. J., Blue, M. D., Bean, B. L., and Perkowitz, S. (1977). *Infrared Phys.* **17**, 43.
Hodges, D. T., and Hartwick, T. S. (1973). *Appl. Phys. Lett.* **23**, 252.
Hodges, D. T., Foote, F. B., and Reel, R. D. (1977). *IEEE J. Quantum. Electron.* **QE-13**, 491.
Mathieu, P., and Izatt, J. R. (1978). *Opt. Commun.* **26**, 86.
Mattis, D. C., and Bardeen, J. (1958). *Phys. Rev.* **111**, 412.
McKnight, S. W., Thorland, R. H., and Perkowitz, S. (1977). *Thin Solid Films* **41**, L61.
Palik, E. D., and Furdyna, J. K. (1970). *Rep. Prog. Phys.* **33**, 1193.
Palmer, L. H., and Tinkham, M. (1968). *Phys. Rev.* **165**, 588.
Perkowitz, S. (1971). *J. Phys. Chem. Solids.* **32**, 2267.
Perkowitz, S., and Bean, B. L. (1977). *J. Chem. Phys.* **66**, 2231.

Ray, P. S. (1972). *Appl. Opt.* **11**, 1836.

Renk, K. F., and Genzel, L. (1962). *Appl. Opt.* **1**, 643.

Rosenbluh, M., Tempkin, R. J., and Button, K. J. (1976). *Appl. Opt.* **15**, 2635.

Tinkham, M. (1970). *In* " Far-Infrared Properties of Solids " (S. S. Mitra and S. Nudelman, eds.), pp. 223–246. Plenum Press, New York.

Yamanaka, M. (1976). *Rev. Laser Eng.* **3**, 57.

CHAPTER 5

Electron Cyclotron Heating of Tokamaks

Wallace M. Manheimer

I. Introduction—Tokamaks, Supplementary Heating, and Electron Cyclotron Resonance Heating

It is now obvious that one of the principal hopes of the controlled fusion program in the United States lies in tokamaks. America is now committed to building no less than five large tokamaks; TFTR, PLT, and PDX in Princeton; Alcator C at MIT; and Doublet III at General Atomic. The tokamak program is worldwide. The Soviet Union and Japan are each building large tokamaks, T20 and JT60, respectively. Finally, the nine Euratom nations have recently agreed to build JET (Joint European Torus) at Culham Laboratory in England. The total worldwide allocation of funds to all of these tokamaks will surely be over one billion dollars. A schematic of PLT is shown in Fig. 1 and an artist's conception of JET is shown in Fig. 2.

Despite this tremendous worldwide commitment, it is far from being a certainty that tokamaks will even be the basis of a controlled fusion power plant. To get from where they are now to where they must be to produce power, tokamaks will have to pass through several as yet uncharted plasma physics regimes. It is of great interest to see whether the use of intense millimeter wave sources will make this job any easier. Probably the greatest potential use of intense millimeter waves in tokamak research and development is in the area of electron cyclotron resonant heating, the subject of this chapter. Before discussing electron cyclotron resonant heating, we present a brief discussion of tokamaks.

A. TOKAMAKS

A tokamak is a toroidal plasma containment device. External coils produce a large toroidal magnetic field B_z. If the vertical magnetic flux through the hole in the torus changes with time, a loop voltage V is induced. This voltage drives a current I through the plasma. There are two important effects of this current. First, it produces a poloidal magnetic field B_θ. Although this poloidal

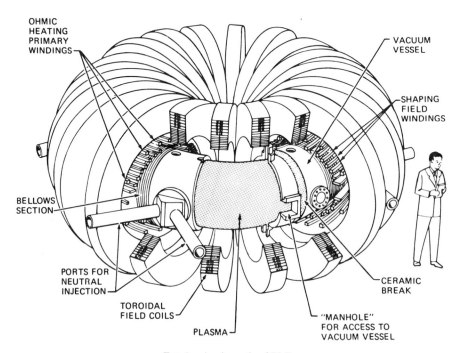

FIG. 1.   A schematic of PLT.

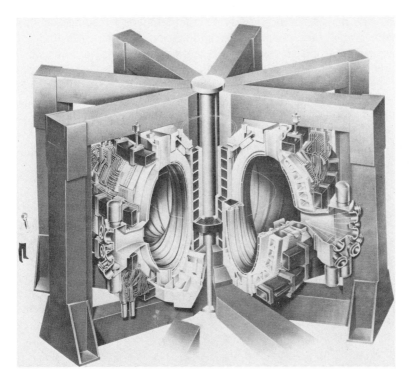

Fɪɢ. 2.   An artist's conception of JET, provided by the JET design team. The principal dimensions are: major diameter of torus, 5.9 m; horizontal plasma diameter, 2.5 m; vertical plasma elongation, 4.2 m; overall diameter, 14.8 m; overall height, 11.5 m.

magnetic field is generally small compared to B_z, the presence of the poloidal field is necessary for a plasma equilibrium to exist. Second, the current heats the plasma by ohmic heating. That is, the total power input is $P = IV$.

Let us define the total energy content of the tokamak plasma as W. Then the energy confinement time τ_E is defined as W/P. Although a tokamak is inherently a pulsed device because it relies on the flux change in the hole to drive the voltage, the pulse time is generally so much longer than the energy confinement time that we regard it as a steady state device here.

In steady state, the power input by the external circuit must be balanced by a power outflow. The most obvious channel of power outflow is via thermal conduction. However, classical thermal conduction is much too small to balance the ohmic power input. There are at least three other potential sources of energy outflow.

First, there is radiation from high Z impurities that get into the plasma from the walls. Second, there is energy loss from magnetohydrodynamic instabilities and the associated large-scale fluid convection. Third, there is anomalous thermal conduction, generally assumed to be caused by micro-instabilities. Various numerical codes attempting to describe and predict tokamak behavior have all attempted to model these effects in various ways. However, since the understanding of the physics of the various loss processes is at best uncertain and at worst totally chaotic, no model has been completely successful in explaining all tokamak data.

Now, consider the condition for a break-even fusion reactor. We envision a deuterium–tritium plasma with number density n and temperature T. (For our purposes here, we ignore the fact that there are at least three different species, deuterons, tritons, and electrons with potentially different densities and temperatures). Whenever a deuteron collides with a triton, there is a certain probability that they will fuse and give off energy E. Denoting the collision frequency for fusion by v_f, we have that the power given off per unit volume is $n v_f E$. The fusion collision frequency, while a complicated function of particle energy, surely depends linearly on the number of target particles, so $v_f = n\alpha$, where α is independent of density. Thus the energy emitted per unit volume is $n^2 \alpha E$. The energy input per unit volume, on the other hand, is $\frac{3}{2} nT/\tau_E$. Thus the condition for more energy output than power input is

$$n\tau_E > 3T/2\alpha E. \tag{1}$$

The quantity on the right-hand side of Eq. (1) is a complicated function of plasma temperature. Typically there is an optimum temperature at which the $n\tau_E$ predicted by Eq. (1) minimizes. For a DT fusion reaction, the generally agreed-upon numbers are

$$n\tau_E \sim 10^{14} \quad \text{cm}^{-3} \text{ sec} \qquad \text{for} \quad T \sim 10 \quad \text{keV}. \tag{2}$$

We shall now survey some recently published data for ohmically heated tokamaks (Dimock *et al.*, 1971; Artsimovich *et al.*, 1972; Bol *et al.*, 1975; Equipe, 1975; Berlizov *et al.*, 1977; Grove *et al.*, 1977; Apgar *et al.*, 1977) to see how close experiments are to the necessary conditions for breakeven as expressed by Eq. (2). This data is summarized in Table I. Much tokamak data is presented in IAEA (International Atomic Energy Agency) meetings which take place now every two years (previously every three years). To give an idea of progress, Table I is divided by spaces into three parts. Above the first space are results from the 1971 IAEA meeting in Wisconsin, above the second space are results from the 1974 meeting in Tokyo, and below this space are results from the 1976 meeting in West Germany.

As is apparent from Table I, the first five years 1971–1976 have witnessed steady progress in increasing $n\tau_E$ but virtually no progress in increasing the temperature. That is, in 1971 the maximum tokamak temperature reported was about 2 keV, as it was also in 1976.

TABLE I

Summary of Recent Tokamak Data

Tokamak	Reference	$\langle n \rangle$ (cm^{-3})	$\langle n \rangle \tau$	T_e ($r = 0$) (keV)
ST	Dimock et al. (1971)	2×10^{13}	2×10^{11}	1.2
T4	Artsimovich et al. (1972)	2×10^{13}	2×10^{11}	2
ATC compressed	Bol et al. (1975)	10^{14}	$\sim 10^{12}$	2
TFR	Equipe (1975)	5×10^{13}	$\lesssim 10^{12}$	2.5
T10	Berlizov et al. (1977)	8.5×10^{13}	5×10^{12}	1.2
PLT	Grove et al. (1977)	$\sim 7 \times 10^{13}$	3×10^{12}	1.8
Alcator	Apgar et al. (1977)	5×10^{14}	10^{13}	0.8

As far as confinement is concerned, tokamak experiments are aided by two lucky results. First, the power losses seem to be diffusive (i.e., like thermal conduction). Therefore the confinement time scales as the system size squared, so that by going to larger systems the confinement time increases. This appears to be the basic factor that helps PLT and T10 in Table I. Second, for some reason, not very well understood, the confinement time seems to increase linearly with density. Thus, by going to higher density, $n\tau_E$ increases as the density squared. This appears to be the basic feature that helps Alcator in Table I.

Physicists involved in tokamak research now seem to be reasonably confident that for ohmically heated tokamaks with $T \sim 1$ keV, $n\tau_E$ can be increased to more than 10^{14} by a combination of higher density and larger machine size. Thus the issue of heating appears to be the next crucial problem.

The problem of heating and the problem of confinement are of course not independent. Both the plasma temperature and confinement time are determined by the overall balance between power input and power outflow. In plasma heating the obvious hope is that by increasing the power input, one can increase the plasma temperature. Hopefully, with supplementary heating, one can also increase, or at least not decrease, the confinement time. The most fundamental requirement of any supplementary heating scheme then is that it deposit power, at least comparable to the ohmic power, into the tokamak plasma. In this chapter we shall discuss one such heating scheme, electron cyclotron resonance heating. Before examining this process we pause briefly to look at other heating schemes under serious consideration.

B.   Supplementary Heating

At this point, far and away the front runner is heating by intense neutral beams. Proton beams are accelerated to tens of kiloelectron volts (possibly up to 100 keV). These beams are then passed through a hydrogen gas cell. By charge exchange, this proton beam emerges as a neutral beam that is then injected across the magnetic field into the tokamak plasma where it is ionized and thereby trapped. The early experiments on beam heating injected beam power smaller than ohmic power (Equipe, 1977). These experiments showed that the ions could be heated to the electron temperature. However, there was very little electron heating, as can be seen in Fig. 3, where electron and ion temperature profiles with and without beam heating are shown for TFR.

More recently, up to four times the ohmic power was injected into PLT (Eubank *et al.*, 1978). Much more impressive heating has been shown. The electron temperature now increases from about 2.5 to 3.5 keV while the ion temperature increases from about 1 to about 5.5 keV. This demonstrates that supplementary heating is possible, at least in principle.

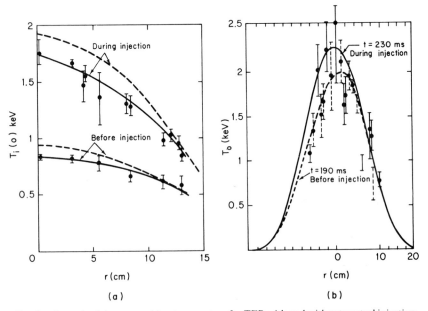

Fig. 3.   A graph of electron and ion temperature for TFR with and without neutral injection.

While neutral beam heating is now in first place, it is not so firmly entrenched that it no longer pays to examine alternatives. There are definite problems with neutral beam heating. Principally neutral beams can only be efficiently produced at energies below about 100 keV. However, larger-sized tokamaks may well require higher energy neutrals in order that the charge-exchange mean free path be comparable to or larger than the minor radius of tokamak. Several schemes to circumvent this difficulty have been proposed, but none have yet been tested. Furthermore, the neutral beam deposits its energy by charge-exchange collisions. The optimum configuration is then that in which the charge-exchange mean free path is roughly equal to the distance from the injection point to the center of the plasma. However, if this can be achieved, a sizable amount of beam energy will be deposited within one charge-exchange mean free path on either side of this point. Thus it is difficult to control the position at which the beam energy is deposited. Also, in a thermonuclear environment, fusion neutrons can travel up the guide tube and in time destroy or make radioactive the beam sources.

The other obvious way to inject energy across the magnetic field and deposit it into the tokamak is electromagnetically (i.e., via radio frequencies, micro-waves, or millimeter waves). We first discuss heating at frequencies below the electron plasma frequency. The problem here is that below this frequency the plasma is generally opaque. Nevertheless, here there are frequency ranges where it is thought that plasma heating may be viable. First, there is the lower hybrid frequency $\omega \sim \omega_{pi}$; second, there is the ion cyclotron frequency $\omega \sim \Omega_{ci}$; third, there are shear Alfven waves with $\omega \ll \Omega_{ci}$.

Lower hybrid and ion cyclotron heating have been attempted on the Wega (Blanc et al., 1977) tokamak and on ST (Adam et al., 1975), respectively. In the Wega experiment, about 80 kW of rf power were injected where the ohmic power was 150 kW. Typically the ion temperature increases by 20 to 30% (from about 140 eV to about 170 eV), while there is no change at all in electron temperature. In the ST experiments, 70 kW of rf power (which is comparable to the ohmic power) at the ion cyclotron harmonic typically double the ion temperature from 100 to 200 eV. No increase in electron temperature was reported.

There are also potential problems with microwave and rf heating schemes. Since the vacuum wavelength is of the order of the system size, it is difficult to control the position at which the energy is deposited if it is not converted into other modes. If it is converted to plasma modes with much shorter wavelength, there are inevitable losses associated with these processes. Even if a long wavelength vacuum mode can be converted to a short wavelength plasma mode with perfect efficiency, the plasma mode often must travel many times around the torus before depositing its energy into the particles. This position of energy deposition will then be difficult to control. Furthermore, the lower the wave frequency, the larger the oscillating electron velocity is at fixed power. However, the threshold for parametric instability generally goes as the oscillating electron velocity not the power. Therefore, as a general rule, the lower the frequency, the easier it is to excite parametric instability. Indeed, experiments on lower hybrid heating do generally see enhanced oscillations at other frequencies, an almost certain indication of parametric instability. The presence of parametric instabilities also makes it much more difficult to control where and how microwave energy will be deposited in the plasma.

There are at least three other particle beam heating schemes proposed for tokamaks (Bottiglione *et al.*, 1976; Sudan, 1976; Benford *et al.*, 1974; Ott and Manheimer, 1977). The first is cluster injection. Solid hydrogen crystals of perhaps 100 to 1000 atoms can be singly charged and accelerated through potentials of about 5 or 10 MV. The larmor radius of a cluster in the tokamak field can be large enough for the cluster to propagate to the center of the torus. However, the energy per particle in the cluster can still be large enough to heat the plasma. Second, it has been proposed that intense relativistic electron beams can be used to heat tokamak plasmas. The obvious problem here is how to inject the beam from an external injection point, across the magnetic field, and into the plasma. Some progress has been made on injecting beams into toroidal chambers with low field and no plasma, but, so far at least, no one has figured out how to inject an electron beam into a tokamak. Third, there have been analogous suggestions to inject an intense ion beam into a tokamak. Here the ion beam can propagate across the tokamak field by dragging along with it an equal number of electrons and so propagate as an intensely energetic plasma beam.

From the foregoing, it is obvious that the problem of supplementary heating of tokamaks is an extremely difficult problem. It is not certain that any of the schemes just mentioned can even clear the first hurdle; that is, can they take power from some external point and deposit it cleanly in the center of the tokamak plasma? The tremendous advantage of electron cyclotron heating is that it offers the potential of doing just this.

The main problem, of course, is that extremely short wavelengths are required. For instance, for PLT the toroidal field will ultimately be about 50 kG, so a wavelength of around 2 mm is required. Furthermore, since this power is to be used for heating, it is of crucial importance that the millimeter wave power be produced efficiently. In this respect the tokamak community is fortunate that just such sources (called *gyrotrons*) are now being actively developed both in the United States and in the Soviet Union. Other chapters describe in great detail the progress along these lines, and that subject will not be pursued here.

To see what sorts of millimeter wave sources are needed, in Table II are listed the magnetic fields of various tokamaks. The magnetic field, of course,

TABLE II

Requirements for Electron Cyclotron Heating of Various Tokamaks

Machine	Major radius (m)	B (kG)	Wavelength (mm)	Minimum power (MW)	Fractional absorption (%)
JET	3.0	30	3.5	5	≈ 100
PLT	1.4	50	2.1	1	≈ 100
Alcator	0.5	75	1.4	0.3	≈ 90
Versator	0.4	8	13.0	0.1	≈ 5
Macrotor (UCLA)	1.0	10	10.0	0.1	≈ 15
TM3	0.6	25	4.2	0.1	30

dictates a wavelength in order that the injected power be at cyclotron resonance. The injected power must also be comparable to or larger than the ohmic
power. Finally, we list the fractional energy deposited by the millimeter
waves on one pass through the resonant region (see Sections II–VI), assuming
an ordinary wave is incident with $(\omega_{pe}/\Omega_c)^2 = \frac{1}{2}$. The assumed electron
temperatures are 1 keV for JET, PLT, and Alcator; 250 eV for Versator and
Macrotor; and 400 eV for TM3 (Richards *et al.*, 1978; Alikaev *et al.*, 1976).
While the power levels indicated by Table II are high, at least this power does
not have to be delivered by a single gyrotron. Since the power can be transmitted through very small waveguides, many gyrotrons can simultaneously
feed power into the tokamak. Depending on things like number and size of
access points, it might be possible to use 10 or perhaps even as many as 100
different sources. Thus the power requirements on a single source are probably
between one and two orders of magnitude lower than indicated in Table II.

While gyrotrons necessary for heating the entire tokamak plasma will not
be developed for some time, gyrotrons sufficient for local heating could be
developed in the near term. There are two basic types of local heating experiments that could be done. First, one could deposit a large amount of
energy at one point in the torus very quickly and watch it diffuse away.
Second, one could continuously heat an interior region of the plasma and
examine the new steady state that is established.

In either case, the power of the gyrotron (or gyrotrons) must be larger than
the energy in the localized region divided by the containment time of that
region. Also, the total energy deposited should be comparable to the energy in
that region of the tokamak. Experiments on PLT show that the electron
confinement time is relatively insensitive to radius and is between about 30
and 50 msec. However, when a region is locally heated, its confinement is
almost sure to drop greatly. Let us arbitrarily consider a confinement time of
1 msec. Then if the plasma temperature is 1 keV, the density is 5×10^{13} cm^{-3},

the major radius is 135 cm (PLT), and the electron energy content of the inner centimeter radius is about 20 J. Therefore, a gyrotron (or gyrotrons) with a power of 20 kW at a wavelength of 2.2 mm and a pulse time $\gtrsim 1$ msec are needed to do local heating experiments on PLT.

Thus for plasma heating one can say (with the confidence typically held before facing hard facts) that electron cyclotron resonance heating looks like an extremely attractive scheme. As we shall see in this chapter, theory indicates that it is a very clean scheme. Virtually nothing seems to go into the plasma except power. Also, it appears to be possible to specify almost any power deposition profile. There is every reason to expect that this power will be deposited into the thermal electrons rather than into a nonthermal tail. As an additional bonus, the theory of electron cyclotron resonance heating in tokamaks is really quite simple. Basically all that is involved is a calculation of electron cyclotron damping along ray trajectories. Since this aspect of the theory is so straightforward, it is even possible to go to the next step and do full wave calculations of reflection coefficients, mode conversion coefficients, and the like (Litvak *et al.*, 1977; Alikaev *et al.*, 1977; Eldridge *et al.*, 1978; Fidone *et al.*, 1977; Antonsen and Manheimer, 1978). These more complicated studies generally confirm the simple theory. A final (and from all past experience in plasma physics, totally unexpected) windfall is that the simple theory seems to work! Experiments on TM3 do, in fact, demonstrate heating of the electrons just when theory indicates they should heat (Alikaev *et al.*, 1976).

In the remainder of this chapter, we shall go through the theory of electron cyclotron heating of tokamak plasmas. Earlier theoretical work in this area has generally regarded the basic interaction as between individual electrons and cavity modes (Kuckes, 1968; Grawe, 1969; Lieferman and Lichtenberg, 1973; Sprott, 1971; Kawaruma *et al.*, 1971; Eldridge, 1972). The reason, of course, is that long wavelength sources and small low-density plasmas were available. In the present work and other recent theoretical work, the emphasis is different in that we are concerned primarily with large, dense plasmas and small wavelength. Thus the primary issues become propagation and deposition of millimeter wave energy. After discussing theory, we shall review some recent experimental work, and close by considering other potentially important theoretical issues.

II. Simple Treatment of Plasma Heating with Millimeter Waves

A. High-Frequency Waves in a Plasma

In this subsection we shall briefly review the properties of high-frequency waves in a collisionless plasma. This is in no way intended as a complete treatment of the subject. A discussion of waves in plasmas can be found in such standard textbooks as Allis *et al.* (1963) or Stix (1962). Generally, the notation used in this chapter is consistent with that used in the former book.

Assuming that the electric field varies as $\mathbf{E}(\mathbf{k}) \exp i(\mathbf{k} \cdot \mathbf{r} - \omega t)$ and $\mathbf{J}(\mathbf{k}, \omega) = \boldsymbol{\sigma}(\mathbf{k}\omega) \cdot \mathbf{E}(\mathbf{k})$, then Maxwell's equations yield the dispersion relation

$$\mathbf{D}(\mathbf{k}, \omega) \cdot \mathbf{E}(\mathbf{k}) = 0, \tag{3}$$

where

$$\mathbf{D} = (c^2/\omega^2)(\mathbf{kk} - k^2\mathbf{I}) + \mathbf{K} \qquad \text{where} \quad \mathbf{K} = \mathbf{I} - (4\pi i/\omega)\boldsymbol{\sigma}, \tag{4}$$

$\mathbf{I}$ being the identity tensor.

We find it most convenient to use rotating coordinates in the x, y plane ($\mathbf{B}$ is parallel to z):

$$\begin{aligned} \sqrt{2}l = x - iy, & \qquad \sqrt{2}E_l = E_x - iE_y, \\ \sqrt{2}r = x + iy, & \qquad \sqrt{2}E_r = E_x + iE_y. \end{aligned} \tag{5}$$

In this coordinate system, the conductivity tensor is diagonal for a cold plasma, and

$$\sigma_r = 1 - [(\omega_p^2/\omega^2)/(1 - \Omega/\omega)], \tag{6a}$$

$$\sigma_l = 1 - [(\omega_p^2/\omega^2)/(1 + \Omega/\omega)], \tag{6b}$$

$$\sigma_\| = 1 - (\omega_p^2/\omega^2), \tag{6c}$$

where $\omega_p^2 = 4\pi n e^2/m$ and $\Omega = eB/(mc) > 0$. Then, in a cold plasma, for wave vector $\mathbf{k}$ perpendicular to the magnetic field, the dispersion relation reduces to

$$(k^2 c^2 - \omega^2 + \omega_p^2)(k^2 c^2[\omega^2 - \omega_p^2 - \Omega^2] - (\omega_p^2 - \omega^2)^2 + \Omega^2\omega^2) = 0. \tag{7}$$

Equation (7) is the product of two terms so that there are two independent modes of propagation, the ordinary wave with dispersion relation

$$k^2 = (\omega^2 - \omega_p^2)/c^2 \tag{8a}$$

and the extraordinary mode

$$k^2 = (1/c^2)[(\omega_p^2 - \omega^2)^2 - \Omega^2\omega^2]/(\omega^2 - \omega_p^2 - \Omega^2). \tag{8b}$$

If k is in the x direction, the electric field of the ordinary mode is linearly polarized in the z direction, while the electric field of the extraordinary mode is elliptically polarized in the x,y plane. For the extraordinary mode, $k = 0$, corresponding to wave reflection, occurs at the cyclotron cutoff

$$(\omega_p^2 - \omega^2)^2 = \Omega^2\omega^2. \tag{9}$$

Defining

$$\alpha^2 = \omega_p^2/\omega^2, \tag{10a}$$

$$\beta^2 = \Omega^2/\omega^2, \tag{10b}$$

the cyclotron cutoff occurs at

$$\beta^2 = (1 - \alpha^2)^2.$$

The point $k = \infty$, corresponding to wave absorption, occurs at

$$\alpha^2 + \beta^2 = 1, \tag{11}$$

the upper hybrid resonance.

The ordinary wave, correspondingly, has a cutoff (reflection) at

$$\alpha^2 = 1 \tag{12}$$

and has no resonance (absorption). It is also essential to note that neither wave interacts strongly with the plasma at the cyclotron frequency. The reason is that at $\omega - \Omega$, one can solve for the polarization of the wave and determine that $E_r = 0$. However, E_r is the component of circularly polarized electric field which rotates in the same sense as the electrons rotate about the magnetic field. On the other hand, E_l is the component of the circularly polarized field that rotates opposite to the direction of electron rotation. Since, at cyclotron resonance ($\omega = \Omega$), the electric field is either parallel to B (ordinary wave) or rotates in the opposite sense to the electrons, there can be no strong interaction at $\omega = \Omega$ in cold plasma theory. However, as shall be seen shortly, a nonzero electron temperature does introduce wave absorption at cyclotron resonance.

One convenient way of visualizing the character of waves in a cold plasma is through a CMA diagram, which plots the positions of cutoffs and resonances in the α^2, β^2 plane. These cutoffs [Eqs. (9) and (12)] and resonances [Eq. (11)] are shown in Fig. 4 as the dotted and solid lines, respectively.

Now envision a wave propagating from some external point into the plasma. As the wave propagates through the plasma, it is useful to trace out the position of the wave front on the CMA diagram. Since this external point is at zero density, it corresponds to a point somewhere on the β^2 axis (i.e., at $\alpha^2 = 0$) of the CMA diagram. As the wave propagates toward increasing density, α^2 increases so the point moves toward the right on the CMA diagram. Once the wave has passed a density maximum, the point reverses direction and begins to move toward the left. If the wave propagates toward increasing magnetic field, the point moves upward on the CMA diagram. If it propagates toward decreasing field, the point moves downward on the CMA diagram. Since the magnetic field in the tokamak is higher on the inside than the outside, a wave launched from the outside (low-field side) traces a curve moving upward on the CMA diagram. Correspondingly, a wave launched from the inside (high-field side) traces a curve moving downward.

According to cold plasma theory, the only way a wave can couple its energy into the plasma is at the upper hybrid resonance point. In the next subsection we shall discuss this.

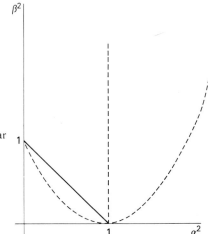

FIG. 4. The CMA diagram for perpendicular propagation. (See text for discussion.)

B. UPPER HYBRID RESONANCE HEATING

The most obvious way (and in cold plasma theory, the only way) to deposit high-frequency millimeter wave energy into the plasma is at the upper hybrid resonance. That is, the ray path must proceed unimpeded from an external source to the upper hybrid resonance point $\omega^2 = \omega_p^2 + \Omega^2$. Specifically, the cyclotron cutoff, that is, the reflection point $(\omega_p^2 - \omega^2)^2 = \Omega^2\omega^2$, cannot occur between the source and upper hybrid resonance. As is apparent from the CMA diagram (Fig. 4), if the wave is launched from the outer wall of the torus (so that the point tracing the position of the wave in the plasma moves upward on the CMA diagram), the wave always meets the cyclotron cutoff and reflects before it can reach the upper hybrid resonance and be absorbed.

Let Ω_0 denote the cyclotron frequency at the center of the torus. If $\omega = \Omega_0$, the positions of the cyclotron cutoff and upper hybrid resonance in the tokamak cross section are shown as the dashed and solid lines, respectively, in Fig. 5. The region between the dotted and solid lines is a region where the extraordinary wave does not propagate, but is evanescent. For a large tokamak, for instance the size of PLT or JET, the wavelength required is 2–3 mm, while the plasma diameter is of order 1 m or larger. Thus the amount of tunneling through the evanescent layer to the upper hybrid resonance should be negligible.

Therefore, in order to rely on heating at the upper hybrid resonance, the millimeter wave energy should be launched from the inside of the torus. This, of course, enormously complicates the technical problem since there is very little space inside the torus, and what space there is is generally fully packed with hardware, for instance the iron core of a transformer. The only

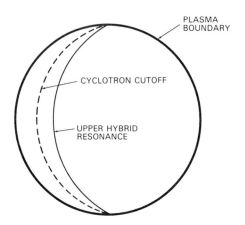

FIG. 5. The position of the upper hybrid resonance and cyclotron cutoff in a tokamak plasma.

other possibility is to launch the wave from the outside of the torus and hope that it reflects often enough from the cyclotron cutoff and metal walls so as to ultimately enter the plasma as if it were launched from the inside and thereby deposit its energy at the upper hybrid resonance.

C. WKB Theory of Absorption at $\omega = \Omega$ and $\omega = 2\Omega$ in a Hot Plasma

The dispersion relations for the ordinary and extraordinary waves obtained from the Vlasov–Maxwell set of equations indicate that these waves will have resonances and cutoffs near each of the cyclotron harmonics for perpendicular propagation. If the gradients in applied magnetic field and density are gentle, then a WKB description is expected to be valid everywhere in the plasma except in the immediate vicinity of resonance or cutoff.

It is instructive to apply WKB theory first to the resonance region to determine whether a wave will suffer significant energy loss in propagating through the resonance. WKB theory, of course, is not completely accurate in that it will not predict reflection or mode conversion, although it is surprisingly accurate in determining the wave energy transmission coefficient.

The WKB treatment of wave absorption at a cyclotron harmonic for arbitrary angle of incidence, i.e., arbitrary k_x, can be formulated in the following way. We wish to solve the matrix equation $\mathbf{D}(\mathbf{k}, x) \cdot \mathbf{E} = 0$ for $k_x(x)$, where $\mathbf{D}(\mathbf{k}, x)$ is the local dielectric tensor $\mathbf{D} = k_x k_x - (\omega^2/c^2)\mathbf{1} + (4\pi i\omega/c^2)\boldsymbol{\sigma}$ and $\boldsymbol{\sigma}$ the conductivity tensor. In the limit $V_e^2/c^2 \ll 1$, $\mathbf{D}$ may be approximated by the cold dielectric tensor $\mathbf{D}_0$ plus a resonant term $\mathbf{D}_1$, which results from finite Larmor radius effects. The tensor $\mathbf{D}_1$ will only be important near a resonance (it contains terms that vary as x^{-1} where $x = 0$ is the location of the resonance), and we will treat $\mathbf{D}_1$ as a perturbation. The solution for the perpendicular wave number $k_x(x)$ and the electric field $E(x)$ will differ slightly from their values predicted by cold theory $k_x = k_0 + k_1$, $\mathbf{E} = \mathbf{E}_0 + \mathbf{E}_1$, where $\mathbf{D}_0(k_0, x) \cdot \mathbf{E}_0 = 0$.

We Taylor-expand $\mathbf{D}$ in k_x about k_0 and multiply the dispersion equation on the left by $\mathbf{E}_0$ and obtain an expression for k_1;

$$k_1(x) = \mathbf{E}_0 \cdot \mathbf{D}_1(k_0, x) \cdot \mathbf{E}_0 (\mathbf{E}_0 \cdot (\partial/\partial k_0) \mathbf{D}_0(k_0) \cdot \mathbf{E}_0)^{-1}.$$

The wave transmission will then be given by $T = \exp(-2\pi\eta)$, where

$$\eta = [1/(\pi i)] \int_{-\infty}^{\infty} dx \, k_1(x).$$

We now specialize to the case of perpendicular propagation.

1. *Ordinary Wave*

The dispersion relation for the ordinary wave propagating perpendicularly to an applied magnetic field, obtained from the Vlasov–Maxwell set of equations, is given by

$$\frac{k_\perp^2 c^2}{\omega^2} = 1 + \sum_n \frac{\omega_p^2}{\omega} \int d^3v \, \frac{J_n^2(k_\perp v_\perp/\Omega) v_z}{\omega - n\Omega} \frac{\partial f_0}{\partial v_z}, \tag{13}$$

where $f_0(v)$ is the velocity distribution function for electrons and J_n the Bessel function.

To determine the total amount of wave energy damping in the resonant region we assume that Ω is a weak function of position $\Omega = \Omega(x)$ and integrate the imaginary part of $k_\perp(x)$ as determined by Eq. (13) through the resonant region. The effect of the resonance will be treated as a perturbation to the cold dispersion relation. For example, near the first cyclotron harmonic we obtain from Eq. (13) by expanding the Bessel functions for small arguments

$$\frac{k_\perp^2 c^2}{\omega^2} = 1 - \frac{\omega_p^2}{\omega^2}\left(1 + \frac{k_\perp^2 v_e^2}{4\Omega^2} \frac{\omega}{\omega - \Omega}\right), \tag{14}$$

where we have assumed a Maxwellian distribution for electrons

$$f_0(\mathbf{v}) = [1/(\pi v_e^2)]^{3/2} \exp(-v^2/v_e^2). \tag{15}$$

Taking $\Omega(x)$ to be $\Omega(x) = \omega[1 - (x/R)]$, where R is the major radius of the tokamak, we find

$$k_{\perp r}^2 c^2/\omega^2 \simeq 1 - (\omega_p^2/\omega^2)$$

and

$$\frac{2k_{\perp r} k_{\perp i}(x) c^2}{\omega^2} \simeq - \mathrm{Im}\left(\frac{\omega_p^2}{\omega^2} \frac{1}{4} \frac{k_{\perp r}^2 v_e^2}{\omega^2} \frac{1}{(x/R) + i\varepsilon}\right),$$

where $k_{\perp r}$ ($k_{\perp i}$) is the real (imaginary) part of $k_\perp$ and ε a positive infinitesimal resulting from the causality condition $\text{Im}(\omega) > 0$. The wave energy transmission coefficient T is given by

$$T = \exp\left\{ \int_{-\infty}^{\infty} dx \, 2k_{\perp i}(x) \right\} = \exp(-2\pi\eta_{01}), \tag{16}$$

where

$$\eta_{01} = \frac{R}{4} \frac{\omega_p^2}{\omega^2} \left(1 - \frac{\omega_p^2}{\omega^2} \right)^{1/2} \frac{\omega}{c} \frac{T_e}{m_e c^2} \tag{17}$$

and $T_e = \frac{1}{2}mv_e^2$. The wave transmission coefficient at first glance would appear to be nearly unity (no wave absorption) owing to the smallness of $T_e/(mc^2)$. However, the quantity $R\omega/c$ can be quite large, indicating small wave transmission and potentially high wave absorption. In a plasma with the following parameters: $R = 130$ cm, $B = 35$ kG, $T_e = 2$ keV, and $n = 0.5 \times 10^{14}$ cm^{-3}, we find $T = 5.3 \times 10^{-3}$, indicating negligible wave transmission.

If the same analysis is applied to the ordinary wave at the second harmonic, we find a similar expression

$$T = \exp(-2\pi\eta_{02}),$$

where

$$\eta_{02} = \frac{R\omega}{c} \left(1 - \frac{\omega_p^2}{\omega^2} \right)^{3/2} \frac{\omega_p^2}{\omega^2} \left(\frac{T_e}{mc^2} \right)^2. \tag{18}$$

The expression for η_{02} in Eq. (18) is smaller than that is Eq. (17) for η_{01} by a factor of the order of T_e/mc^2. Thus the transmission coefficient at the second harmonic is nearly unity, and wave absorption and transformation are not expected to be important in this case.

2. Extraordinary Mode

For purely perpendicular propagation the dispersion equation for the extraordinary mode is

$$\begin{pmatrix} \dfrac{1}{2}\dfrac{k_x^2 c^2}{\omega^2} - K_r & -\dfrac{1}{2}\dfrac{k_x^2 c^2}{\omega^2} - K_{rl} \\[3mm] -\dfrac{1}{2}\dfrac{k_x^2 c^2}{\omega^2} - K_{rl} & \dfrac{1}{2}\dfrac{k_x^2 c^2}{\omega^2} - K_l \end{pmatrix} \begin{pmatrix} E_r \\ E_l \end{pmatrix} = 0, \tag{19}$$

where

$$E_{(r)}_{(l)} = 2^{-1/2}(E_x \pm iE_y).$$

The wave electric field has Cartesian components $E = (E_x, E_y, 0)$, the wave vector has components $k = (k_x, 0, 0)$, and the applied magnetic field is in the z direction. The matrix elements $K_{r,l}$ and K_{rl} contain the response of the electron plasma and are given by

$$K_{\binom{r}{l}} = 1 \mp i \frac{\partial}{\partial \mu} \left\{ \frac{\omega}{\omega \Omega} \mu \int_0^\infty d\eta \exp\left[-i\left(\frac{\omega}{\Omega} \pm 1\right)\eta - \mu \frac{b^2}{2}(1 - \cos \eta) \right] \right\}\bigg|_{\mu = 1},$$

(20a)

$$K_{rl} = i \frac{\partial}{\partial \mu} \left\{ \frac{\omega}{\omega \Omega} \int_0^\infty d\eta \exp\left[-i\frac{\omega}{\Omega}\eta - \eta \frac{b^2}{2}(1 - \cos \eta) \right] \right\}\bigg|_{\mu = 1}, \qquad (20b)$$

where $b = k_x v_e/\Omega$. For our purposes here it is only necessary to have approximations for $K_{r,l}$ and K_{rl} valid for $k^2 v_e^2 \ll \omega^2$. These approximations are found to be

$$K_{\binom{r}{l}} = 1 - \frac{\alpha}{1 \pm \beta}\left(1 + \frac{k_x^2 c^2}{\omega^2} \frac{\delta}{1 \pm 2\beta}\right) - \frac{3}{32}\left(\frac{k_x^4 c^4}{\Omega^4} \frac{\delta^2}{1 \pm \beta}\right)\alpha^2, \quad (21a)$$

$$K_{rl} = -\frac{1}{2}\alpha^2\left(\frac{k_x c}{\omega}\right)^2 \frac{\delta}{1 - \beta^2}, \qquad (21b)$$

where $\alpha^2 = \omega_p^2/\omega^2$, $\beta = \Omega/\omega$, and $\delta = 2T_e/(mc^2)$. The first terms of Eqs. (21a) and (21b) are correct to first order in δ. When the dispersion equation (19) is solved near the first harmonic resonance due to cancellations, it is necessary to retain terms that are second order in δ. Thus we have included the last term in Eq. (21a), which is second order in δ.

The analyses of the preceding section can be carried out in a similar fashion to determine the wave energy transmission coefficient for the extraordinary wave at the first and second electron cyclotron harmonic. We find $T = \exp(-2\pi\eta)$, where, at the first harmonic,

$$\eta_{x1} = \frac{1}{8} \frac{\omega_p^2}{\omega^2}\left(\frac{T_e}{mc^2}\right)^2 \frac{R\omega}{c}\left(2 - \frac{\omega_p^2}{\omega^2}\right)^{3/2} \qquad (22)$$

and, at the second harmonic,

$$\eta_{x2} = \alpha^2 \frac{R\omega}{c}\left(\frac{3 - 2\alpha^2}{3 - 4\alpha^2}\right)^2\left[\frac{4(1 - \alpha^2)^2 - 1}{3 - 4\alpha^2}\right]^{1/2} \frac{T_e}{mc^2}, \qquad (23)$$

where $\alpha^2 = \omega_p^2/\omega^2$. In contrast to the case of the ordinary wave, the wave transmission for perpendicular propagation is nearly unity at the first harmonic and possibly low at the second harmonic.

For the case of oblique propagation, the transmission coefficients have been worked out for both the ordinary and extraordinary mode at the first harmonic. These transmission coefficients (Eldridge *et al.*, 1978) are

$$\eta_{01} = \frac{R\omega}{4c} \frac{\omega_p^2}{\omega^2} \frac{[1 - (\omega_p^2/\omega^2)]^{1/2}}{1 + (k_z^2 c^2/\omega^2)[1 - (\omega_p^2/\omega^2)]} \left(\frac{T_e}{mc^2}\right) \tag{24}$$

and

$$\eta_{x1} = \frac{R\omega}{4c} \left(\frac{k_z c}{\omega}\right)^2 \frac{\omega^2}{\omega_p^2} \left(2 - \frac{\omega_p^2}{\omega^2}\right)^{3/2} \left(1 + \frac{\omega_p^2}{\omega^2}\right)^2 \left(\frac{T_e}{mc^2}\right), \tag{25}$$

where k_z is the component of wave vector along the magnetic field. In the derivation of Eqs. (24) and (25), terms of order $(T_e/mc^2)^2$ were neglected so that Eq. (25) gives $\eta_{x1} = 0$ for $k_z = 0$ instead of $\eta_{x1} \sim (T_e/mc^2)^2$ as given in Eq. (22). The transmission of the ordinary mode is not strongly affected by oblique incidence. However, the transmission coefficient of the extraordinary mode drops sharply as the direction of **k** changes from perpendicular to oblique.

Let us close by discussing whether the wave can be launched from the inside or outside of the torus. The ordinary wave has no cyclotron cutoff, so it can be launched from either the inside or outside wall of the torus. The extraordinary wave at the second cyclotron harmonic also does not have a cyclotron cutoff until much higher densities (to be discussed in Section III). Thus it can also be launched from either the inside or outside wall of the torus. The extraordinary wave at the cyclotron frequency has a cutoff at low density if launched from the outside of the torus. (If $k_\parallel \neq 0$, the reflection point $k_x^2 = 0$ is at even lower density.) Thus, to heat with the extraordinary wave with $k_\parallel \neq 0$, $\omega \sim \Omega$, the wave must be launched from somewhere on the inner wall of the tokamak.

Finally, let us note that a much more extensive treatment of the WKB theory of wave absorption in a tokamak equilibrium has recently been given (Ott *et al.*, 1979).

III. Full Wave Treatment of the Extraordinary Mode at the Cyclotron Harmonic

At perpendicular incidence, WKB theory breaks down for the extraordinary wave at $\omega \approx 2\Omega$. In Antonsen and Manheimer (1978), a full wave treatment of this process was given. There it was found that a wave incident upon a cyclotron harmonic resonance $\omega = 2\Omega$ is partially absorbed, transmitted, reflected, and converted to other types of wave motion, in this case, to a Bernstein mode. The procedure used here was quite complex. It involved first calculating the current density at $\omega \approx 2\Omega$ in an inhomogeneous magnetic field. Then an integral equation was found for the electric field E. This integral equation was solved and an integral expression was then found for the electric field. Taking the asymptotic limits for $x \to \infty$ and $x \to -\infty$ gave transmission, reflection, absorption, and mode conversion coefficients.

We shall not follow that procedure here but rather confine ourselves to general remarks on the mode conversion process. The general method shall be illustrated in Section V where we carry out a full wave theory calculation for the (much simpler) problem of an ordinary wave incident on a cyclotron resonance $\Omega \approx \omega$. Equations (19)–(21) can be combined to give the dispersion relation

$$\frac{\alpha^2\delta}{1-2\beta}\kappa^4 + \left[-1 + \frac{4\alpha^2}{3} + \left(-1 + \frac{2}{3}\alpha^2\right)\frac{2\alpha^2\delta}{1-2\beta}\right]\kappa^2$$
$$+ \left(1 - \frac{8\alpha^2}{3} + \frac{4\alpha^4}{3}\right) = 0, \quad (26)$$

where $\kappa = k_x c/\omega$, and we have assumed $\beta = \frac{1}{2}$ unless it occurs in the combination $1 - 2\beta$. Thus Eq. (27) is the dispersion relation for modes in the vicinity of the cyclotron harmonic resonance.

For small δ, the roots are approximately given by

$$\kappa^2 = (1 - \tfrac{2}{3}\alpha^2)(1 - 2\alpha^2)/(1 - \tfrac{4}{3}\alpha^2) \qquad (27a)$$

and

$$\kappa^2 = [(1 - 2\beta)/(\alpha^2\delta)](1 - \tfrac{4}{3}\alpha^2). \qquad (27b)$$

The value of κ^2 given by Eq. (27a) is the extraordinary wave. As long as $\alpha^2 < \frac{1}{2}$, the value of κ^2 in Eq. (27a) is greater than zero. However, $\alpha^2 < \frac{1}{2}$ simply means that the wave is on the low-density side of the cyclotron cutoff. Thus $\alpha^2 = \frac{1}{2}$ denotes the maximum density to which an extraordinary mode can propagate from a vacuum without being reflected.

Equation (27b) is the dispersion relation for the Bernstein mode. Notice that for the Bernstein mode $\kappa^2 > 0$ only for $1 - 2\beta > 0$; that is, on the low-field side of the cyclotron harmonic resonance. In deriving Eqs. (27a) and (27b), δ was assumed to be small, but $1 - 2\beta$ was not equal to zero. In the limit of $1 - 2\beta \to 0$ for arbitrary δ, the roots of Eq. (26) approach

$$\kappa^2 = 2(1 - \tfrac{2}{3}\alpha^2), \qquad (28a)$$
$$\kappa^2 \doteq [(1 - 2\beta)/(2\alpha^2\delta)](1 - 2\alpha^2). \qquad (28b)$$

Thus Eqs. (28a) and (28b) show that one root approaches zero for $1 = 2\beta \to 0^+$ while the other roots approach a constant value. Thus plotting κ^2 as a function of x (where $1 - 2\beta = x/R$), it is clear that the roots connect as shown in the solid curves of Fig. 6. The distance over which the roots connect is of the order of $\pi R\delta$ as indicated in the figure. Also, in the figure, positive κ

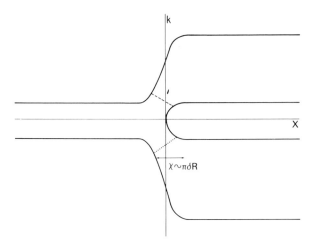

FIG. 6. A plot of wave number as a function of x for perpendicular propagation of the extraordinary $\mathbf{E} \perp \mathbf{B}_0$ wave near $\omega = 2\,\Omega$.

indicates a wave whose phase velocity $(=\omega/k)$ and group velocity are to the right, while negative κ indicates propagation to the left.

Now imagine that a wave is launched from an external source and propagates toward the cyclotron harmonic resonance. To be definite, let us say the wave is launched from the low-field side, so that the κ of the incident wave at large positive x is on the upper of the negative κ lines of Fig. 6. As the wave front propagates toward $x = 0$, it can either remain on the same dispersion curve or else tunnel through an evanescent region and jump to one of the other curves.

The former possibility, that is, the wave front staying totally on the same curve without tunneling through to one of the other curves, indicates total reflection. Tunneling through the evanescent region along the dotted line indicates conversion to a transmitted wave. On the other hand, tunneling through the evanescent region along the dashed line indicates conversion to a reflected, Bernstein mode. Notice particularly that, as indicated by Fig. (6), total reflection does not imply "conversion," in that the wave front remains on the same curve, whereas coupling to a transmitted extraordinary wave does imply "conversion," in that the wave front jumps to another branch of the $\kappa(s)$ curve.

Clearly the measure of whether "conversion" is important is the ratio of the wave number ω/c to the size of the connection region $\pi\delta R$. First, imagine that

$$\pi(\omega R/c)\delta \gg 1. \tag{29}$$

As the dispersion curve $\kappa(x)$ bends around, $\kappa(x)$ does not change much in one wavelength so that the mode follows the curve rather than tunnels through an evanescent layer which is large compared to a wavelength. The amount that tunnels through along the dotted line is measured roughly by the amount the wave attenuates in tunneling through the evanescent region or

$$\ln T_n - \pi(\omega R\delta/c), \tag{30}$$

where T is the transmission coefficient. As long as T is small, most of the energy then proceeds along the $\kappa(x)$ curve and is reflected.

Second, let us imagine that

$$\pi(\omega R/c)\delta \ll 1 \tag{31}$$

so that the $\kappa(x)$ curve bends rapidly over a wavelength. Also, the amount of attenuation in the evanescent region is very small. Hence it is easier for the wave to tunnel through the narrow evanescent region along the dotted line than it is to follow the $\kappa(x)$ curve around sharp corners. Thus if Eq. (31) is satisfied, most of the energy will be transmitted.

Now let us consider a wave incident from the high-field side. This wave propagates to the right so at large negative x the κ of the incident wave is on the positive κ line of Fig. 6. Clearly, if Eq. (29) is satisfied, the natural process is for the wave to mode convert to an electrostatic wave at large positive x. Also, a small amount will tunnel through the evanescent region along the dashed line and a transmitted extraordinary wave will be produced with transmission coefficient given roughly by Eq. (30). However, there is no way to couple to a reflected extraordinary wave (i.e., along the dotted line) without also coupling to an incident extraordinary wave, which is of course forbidden by the boundary conditions. Thus if the wave is incident from the high-field side, we expect no reflection.

The more exact calculation gives the following results for the coefficient of transmitted power T, reflected power R, and mode converted power F.

(a) Wave incident from high-field side:

$$T = \exp(-2\pi\eta_{x2}), \tag{32a}$$

$$F = 1 - T, \tag{32b}$$

$$R = 0. \tag{32c}$$

(b) Wave incident from low-field side:

$$T = \exp(-2\pi\eta_{x2}), \tag{33a}$$

$$R = (1 - T)^2, \tag{33b}$$

$$F = T(1 - T), \tag{33c}$$

where η_{2x} is given in Eq. (23).

In Eqs. (32) and (33) transmitted, reflected, and mode converted are used in the conventional sense. (For instance, an extraordinary wave for, say $x < 0$, gives a *transmitted* extraordinary wave for $x > 0$.) It is easy to verify from Eqs. (32) and (33) that

$$T + R + F = 1. \tag{34}$$

As long as $2\pi\eta_{2x} \gtrsim 1$, Eqs. (32) show that an extraordinary mode is converted to a Bernstein mode if it is incident from the high-field side. Since a Bernstein mode is electrostatic and cannot propagate out of the plasma, the incident wave energy is effectively deposited in the plasma. However, if the extraordinary mode is incident from the low-field side, it is reflected so that the incident energy is not deposited in the plasma. We shall now show that a very small amount of k_z serves to eliminate this reflection.

To do this, let us assume that the electron distribution parallel to the magnetic field is given by a Lorentzian rather than a Maxwellian;

$$f(v_z) = v_e/[\pi(v_x^2 + v_e^2)].$$

Then it is possible to show with a calculation like that in Antonsen and Manheimer (1978) that the dispersion relation is as given in Eq. (26) except that $1 - 2\beta$ is replaced with $1 - 2\beta + (i|k_z|v_e/\omega)$. While this has no effect on the extraordinary mode for large x, in the region of $x = 0$ ($\omega = \Omega$), the effect of a nonzero k_z is to introduce (cyclotron) damping as is apparent from Eq. (28b). Thus, as the wave front follows the inner curve in Fig. 6, before it can bend around and become a reflected wave, it experiences a region of large damping near $x = 0$. Thus the reflected wave amplitude will be less than the incident wave amplitude, and energy will be deposited near $x = 0$.

For a Lorentzian distribution function, it is very easy to calculate the reduction of the reflection coefficient. The only difference in the calculation of reflection coefficient R is that the origin of the complex x plane is vertically translated to

$$x \rightarrow x + (i|k_z|v_e R/\omega). \tag{35}$$

Therefore, as k_z increases from zero, the sum of the incident and reflected wave transforms from

$$\underbrace{\exp(-ik_x x)}_{\text{Incident}} + \underbrace{\exp(ik_x x)}_{\text{Reflected}} \tag{36a}$$

to

$$\underbrace{\exp(-ik_x(x + (i|k_z|v_e R/\omega)))}_{\text{Incident}} + \underbrace{\exp(ik_x(x + (i|k_z|v_e R/\omega)))}_{\text{Reflected}}. \tag{36b}$$

Hence the coefficient of reflected power is reduced by

$$R(k_x)/R(k_z = 0) = \exp(-4k_x|k_z|v_e R/\omega). \tag{37}$$

Thus a very small amount of k_z, $k_z \approx c/(v_e R)$, is sufficient to eliminate the reflected wave if power is injected from the outside (low-field side) of the tokamak. This amount of k_z is unavoidable if only by virtue of the size of the waveguide feeding in the power. Specifically, the amount of k_z needed to eliminate the reflected wave is very much less than the amount of k_z needed to generate absorption of the extraordinary wave at the first harmonic, as can be seen by comparing Eq. (37) with Eq. (25).

It is also worth pointing out that this nonzero k_z gives rise to cyclotron damping of the Bernstein wave in the vicinity of $x = 0$. Thus, with a very small amount of k_z, the energy from the millimeter wave source is deposited near the position of cyclotron harmonic resonance $\omega = 2\Omega$.

Finally, we shall mention here (and discuss more fully in the next section) that because the wave energy is converted to electron thermal energy at a position where all electrons are resonant, we expect heating of the body of the distribution function not acceleration of a nonthermal tail.

IV. Full Wave Treatment of the Extraordinary Mode at the Upper Hybrid Resonance

The dispersion relation for the extraordinary mode is given in Eqs. (19) and (21). Combining them, we find that the dispersion relation for the extraordinary wave reduces to

$$\frac{3\alpha^2\delta}{2(1-4\beta^2)}\kappa^4 + \left[(\alpha^2+\beta^2-1)\left(1-\frac{2\alpha^2\delta}{1-4\beta^2}\right) - \frac{2\alpha^2\delta\beta^2}{1-4\beta^2}\right]\kappa^2$$
$$+ (1-\alpha^2)^2 - \beta^2 = 0, \tag{38}$$

where $\kappa = k_\perp c/\omega$, and we have neglected terms of order δ^2. Far from the upper hybrid resonance there are two separate modes, the extraordinary modes with

$$\kappa^2 = [(1-\alpha^2)^2 - \beta^2]/(1-\alpha^2-\beta^2) \tag{39a}$$

and the Bernstein modes with

$$\kappa^2 = 2(1-4\beta^2)(1-\alpha^2-\beta^2)/3\alpha^2\delta. \tag{39b}$$

As we discussed in Section II, the extraordinary wave propagates on the high-density side of the upper hybrid resonance and on the low-density side of the cyclotron cutoff. Assuming that at the upper hybrid resonance $1 - 4\beta^2 < 0$ (which means $\omega_p^2 < 3\Omega^2$), then the Bernstein mode also propagates on the high-density side of the cyclotron resonance. The Bernstein mode is a backward wave in this density regime. This can easily be seen by calculating the group velocity from Eq. (39b):

$$V_g^{-1} = dk/d\omega = (4\omega/3\omega_p^2\delta c^2 k)[(\omega^2 - 4\Omega^2) + (\omega^2 - \omega_p^2 - \Omega^2)]. \quad (40)$$

For $1 - 4\beta^2 < 1$ and $\alpha^2 > 1 - \beta^2$, the group velocity is in the opposite direction to the phase velocity.

The values of κ as a function of density are shown in Fig. 7 for the four modes. For densities below the electron cyclotron cutoff, there are only two extraordinary modes, the incident and outgoing, and they connect at $\kappa = 0$ at the cyclotron cutoff. For $\beta^2 < 1 - \alpha^2 < \beta$, there is an evanescent region where no modes exist. Finally, on the high-density side of the upper hybrid resonance, all four modes exist and connect as shown.

For a large tokamak, the size of the evanescent region is very long compared to a free space wavelength. Therefore, if energy is propagated from the low-field side (from the left in Fig. 5), it will be reflected.

Now imagine the energy being propagated in from the high-field side (from the right in Fig. 5). The wave front starts out along the lower positive κ curve in Fig. 7. As the wave front approaches the upper hybrid resonant point, its wave number increases. Clearly the natural tendency is for the wave to stay on the same dispersion curve, that in the second quadrant of Fig. 7. Therefore, the incident extraordinary mode energy will be converted into a Bernstein mode near the upper hybrid resonance point. The group velocity of

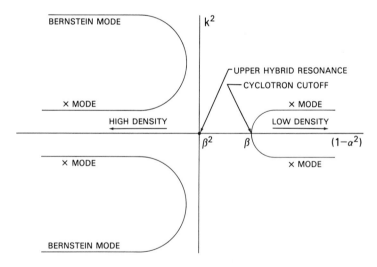

FIG. 7.   A plot of wave number as a function of x for perpendicular propagation of the extraordinary mode near the cyclotron cutoff and upper hybrid resonance.

this Bernstein mode is in the direction of increasing density, even though its phase velocity remains in the direction of decreasing density. Thus the most natural process, for evanescent length large compared to wavelength, is mode conversion to a reflected Bernstein wave. Hence energy propagates toward the upper hybrid resonant point as an extraordinary wave. Once it reaches this point, it reflects and propagates away, toward the high-density region, as a Bernstein wave. Since this latter wave is electrostatic, it cannot propagate out again, so it is effectively trapped and ultimately heats the plasma.

Full wave calculations of mode conversion at the upper hybrid resonant point have been carried out (Stix, 1965; Kuehl, 1967; Horton, 1966; Gorman, 1966; Longren *et al.*, 1966; Golant and Piliya, 1972). It is shown that if the extraordinary wave is incident from the high-density side, all of its energy is in fact converted into a reflected Bernstein mode and none is transmitted or reflected (as extraordinary waves) no matter what the wavelength is, relative to the size of evanescent region. On the other hand, if the wave is incident from the low-density side, most of the energy is reflected or transmitted, but some can be mode-converted into a Bernstein wave. The conversion efficiency maximizes at about 40% if the density gradient scale length is of order c/ω. However, for the larger density gradient scale lengths expected for a tokamak, the conversion efficiencies are vanishingly small.

In order to see how the millimeter wave energy is ultimately deposited into the plasma electrons, it is necessary to see how the Bernstein mode propagates and damps in the tokamak plasma. A numerical study of this problem has recently been carried out (Kochetkov, 1976).

The Bernstein mode is produced at the upper hybrid resonance point with frequency $(\omega_p^2 + \Omega^2)^{1/2} > \omega_c$. It is created somewhere along the solid line in Fig. 5. Once produced, it propagates toward the higher-density, higher-field region. Thus it propagates toward the cyclotron resonance. As the wave front approaches cyclotron resonance, the cyclotron and collisional damping increase, and the Bernstein wave energy is ultimately deposited into plasma electrons.

For high-temperature plasmas, cyclotron damping generally dominates collisional damping. A particle can resonate with the wave and give rise to cyclotron damping if its parallel velocity V_z satisfies

$$V_z = (\omega - \Omega)/k_z. \tag{41}$$

It is not clear what should be taken for k_z. There are at least two effects which give rise to nonzero k_z. First, even for perpendicular injection, there is some k_z present simply because of the geometry of the injection scheme (for instance, finite size of the waveguide parallel to the field). Second, even if $k_z = 0$ at the upper hybrid resonance point, it is not clear that $k_z = 0$ at each point along the ray path of the Bernstein mode. For instance, the poloidal magnetic field B_θ changes throughout the tokamak cross section, so that the direction of the ambient field changes along the ray path. Thus a $k_z \approx B_\theta/B_z k_\perp$ could be induced simply by the changing direction of the fields in the tokamak, as assumed in Grove *et al.* (1977).

We shall now discuss the nature of the energy deposition into the electrons. That is, is the energy deposited into the thermal electrons or into a non-thermal runaway tail on the electron distribution function? This question is very difficult to answer in any quantitative way. However, the theory does give some qualitative indications into which electrons the incident wave energy is deposited.

Let us assume that $(\omega - \Omega)/k_z \gg V_z$ at the hybrid resonance. Then initially there will be negligible damping of the Bernstein mode. As it propagates toward cyclotron resonance, $(\omega - \Omega)/k_z$ gradually decreases. Therefore, the first particles encountered by the wave, which can interact strongly with it, have very high parallel energy. As the wave continues along its ray path, it interacts with lower and lower velocity particles until the wave amplitude finally damps down to zero.

If the wave damps significantly before it reaches the position where $(\omega - \Omega)k_z \sim V_e$, as can be the case according to Grove *et al.* (1977), the wave energy is then deposited into electrons with high parallel phase velocity. Therefore, the natural tendency may well be to deposit the millimeter wave energy into a nonthermal tail of energetic electrons rather than into the thermal part of the electron distribution function.

Let us contrast this with the mode conversion to a Bernstein mode at $\omega \approx 2\Omega$, as discussed in the previous section. There the Bernstein mode was produced at the resonant point, and a very small k_z brings virtually all of the particles into resonance. As the wave propagates away from the cyclotron harmonic resonance, it damps to zero amplitude very quickly because virtually all electrons of the plasma resonate with it. Thus, we do not expect the wave to single out high parallel velocity electrons for energy deposition. The simplest theory indicates then that for the extraordinary wave at $\omega \approx 2\Omega$, energy is probably deposited into the thermal part of the electron distribution function, whereas for upper hybrid resonance energy may well be deposited into a runaway tail on the electron distribution function.

V.   Full Wave Treatment of the Ordinary Mode
at Cyclotron Resonance

To demonstrate the techniques used in full wave calculations of absorption, transmission, and mode conversion near a cyclotron resonance, in this section we shall perform the simplest such full wave calculation, that of the ordinary wave incident upon a cyclotron resonance. The dispersion relation for the ordinary mode propagating purely perpendicular to an applied magnetic field is given by Eq. (14). Solving for $k_\perp^2 c^2/\omega^2$, we find

$$\frac{k_\perp^2 c^2}{\omega^2} = \left(1 - \frac{\omega_p^2}{\Omega^2}\right)\left[1 + \frac{V_e^2}{4c^2}\frac{\omega_p^2}{\Omega^2}\frac{\omega}{\omega - \Omega}\right]^{-1}. \tag{42}$$

Let us now assume that $k_\perp^2$ can be replaced with $-d^2/dx^2$. [In Dubois and Goldman (1967) it is shown that this prescription is valid for the ordinary mode where the polarization forbids any coupling to an electrostatic wave. In general, if k depends strongly on magnetic field, i.e., as for an electrostatic mode, such a replacement is invalid and the integral expression for the perturbed current density must be more carefully worked out.]

If $k_\perp^2$ is replaced with $-d^2/dx^2$ and $\omega/(\omega - \Omega)$ is replaced with R/x, the equation for the z component of electric field can be written out as

$$(d^2/dx^2)E_z + k_0^2\left\{\frac{(x/R)}{[(x/R) + \delta + i\varepsilon]}\right\}E_z = 0, \tag{43}$$

where

$$\delta = \frac{V_e^2}{4c^2}\frac{\omega_p^2}{\omega^2} \quad \text{and} \quad k_0^2 = \frac{\omega^2}{c^2}\left(1 - \frac{\omega_p^2}{\omega^2}\right).$$

The quantity $i\varepsilon$ is included because ω is taken to have a positive imaginary part because of the causality condition.

Equation (43) has a reflection point at $x = 0$ and a resonance at $x = -R\delta$. This then is the familiar problem of Budden tunneling through a back-to-back cutoff and resonance (Budden, 1961). A plot of

$$k(x) = k_0\left[\frac{x/R}{(x/R) + \delta}\right]^{1/2}$$

is given in Fig. 8. An ordinary wave incident from the right (the low-field side) first sees a point of reflection ($x = 0$) and some energy will be reflected. However, some energy also tunnels through the evanescent region and is absorbed at the resonant point $x = -R\delta$. What is left emerges as a transmitted ordinary wave which continues to propagate to the left. On the other hand, a wave incident from the left first encounters a resonance so that wave energy is deposited in the plasma. Additional wave energy tunnels through the evanescent region and emerges on the right as a transmitted wave. However, we do not expect a reflected wave in this case because the wave energy would have to tunnel twice through the evanescent region.

As we shall see shortly, a full wave treatment does in fact show that this qualitative picture is correct. We now proceed to solve Eq. (43) by Fourier

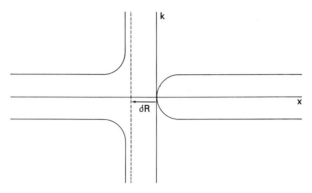

FIG. 8.   A plot of wave number as a function of x for perpendicular propagation of the ordinary wave near $\omega = \Omega$.

transformation. Let us assume that $E_z(x)$ can be written as

$$E_z(x) = \int_c E(k) \exp ikx \, dk, \tag{44}$$

where the contour c will be specified later. Multiplying Eq. (43) by $(x/R) + \delta$, we find that the equation reduces to

$$0 = \int_c dk \left\{ \frac{(-i)}{R} \left[k_0^2 - k^2 \right] E(k) \frac{d}{dk} - \delta k^2 \, E(k) \right\} \exp ikx \, dk. \tag{45}$$

If Eq. (45) is integrated by parts in k, it reduces to

$$\int_c \left\{ \frac{d}{dk} \left[(k_0^2 - k^2) E(k) \right] + \frac{iR\delta k^2}{(k_0^2 - k^2)} \left[(k_0^2 - k^2) E(k) \right] \right\} \exp ikx \, dk = 0. \tag{46}$$

In writing Eq. (46), we have assumed that the contributions from the end points of the partial integration vanish. This then restricts the choice of contour c to those on which the integrand vanishes at the end points.

The only way to satisfy Eq. (46), for arbitrary contour, is to choose $E(k)$ so that the expression in the curly braces itself vanishes. Then $E(k)$ can then be found by solving a simple first-order differential equation in k space. The result is

$$(k_0^2 - k^2)E(k) = \exp\left(-iR\delta \int^k \frac{k'^2}{k_0^2 - k'^2} \, dk' \right) \tag{47a}$$

or

$$E(x) = \int_c \frac{(k - k_0)^{-1 + i\eta}}{(k + k_0)^{1 + i\eta}} \exp ik(x + R\delta) \, dk, \tag{47b}$$

where $\eta = k_0 R\delta/2$. Without loss of generality, the zero on the x axis can be shifted to $-R\delta$. Thus we will shift the axis but will not change notation, so that the exponential in what follows will be written as exp ikx.

In order to determine coefficients of reflection, transmission, and mode conversion, it is necessary to asymptotically evaluate the integral in Eq. (47) for $x \rightarrow +\infty$ and also for $x \rightarrow -\infty$. Let us say that the solution is known at, for instance, $x \rightarrow -\infty$. The question is how to analytically continue the solution to $x = +\infty$. This is a somewhat subtle problem. For instance, one way to analytically continue to $+\infty$ is to rotate the argument of x forward by π. However, rotating backward by π is also a way to get from $-x$ to $+x$. Since there are two branch points at $k = \pm k_0$, rotating x by $\pm\pi$ will give two different answers. Hence the very first thing is to specify the proper way to do the analytic continuation from $x \rightarrow +\infty$ to $x \rightarrow -\infty$.

The clue to this arises from the causality condition. Equation (43) has a pole in the lower half x plane since $\varepsilon > 0$. Thus the solution to Eq. (43) must be analytic in the upper half x plane, but not necessarily in the lower half x plane. Therefore to analytically continue from large positive x to large negative x (or vice versa), one rotates the argument of x in the *upper* half x plane. For instance, if arg x for $x \rightarrow \infty$ is chosen to be zero, then arg x for $x \rightarrow -\infty$ must be π (and not $-\pi$!).

As we discussed earlier in this section, we expect that as $x \rightarrow -\infty$ the two linearly independent solutions consist of either only an incident wave or only a transmitted wave but no reflected wave. Thus we shall first examine the asymptotic expansion of Eq. (47) for $x = -\infty$ (arg $x = \pi$) and then analytically continue to $x = \infty$ (arg $x = 0$). For arg $x = \pi$, the contour of integration in Eq. (47) must close in the lower half k plane. There are two branch points, one at k_0 and the other at $-k_0$. Contours for the two linearly independent solutions are shown as C_1 and C_2 in Figs. 9a and 9b. In Fig. 9, the contours of integration are shown as solid lines, while the branch cuts are shown as cross-hatched lines. As we shall see shortly in the limit of $x \rightarrow -\infty$, the solution from contour C_1 represents an incident wave (i.e., $E_z \approx e^{ik_0x}$). On the other hand, the solution from C_2 represents a transmitted or reflected wave (i.e., $E_z \approx e^{-ik_0x}$).

Now let us consider how these solutions can be analytically continued to $x \rightarrow +\infty$. For positive x, the contour of integration must close in the upper half plane. In general, if the argument of x is θ ($0 \lesssim \theta \lesssim \pi$), the contours in the k plane must close for Im $kx < 0$ or

$$0 < \arg k + \arg x < \pi. \tag{48}$$

Thus as arg x rotates clockwise from π to zero, the argument of the region in the k plane where closure is possible rotates counterclockwise.

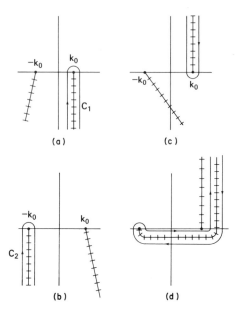

FIG. 9. Contours of integration for the solutions of Eq. (43). Contours in a and b are for the two linearly independent solutions for $x \rightarrow -\infty$. (See text for discussion.)

Let us first consider the solution along contour C_1 of Fig. 9a. As x approaches positive infinity, the contour simply rotates counterclockwise through the k plane to the contour in Fig. 9c. As we shall see shortly, a contour integral like C_1 in either Fig. 9a or 9c represents a wave propagating to the right ($e^{ik_0 x}$). Thus the integral in Fig. 9a represents the incident wave, and the integral in Fig. 9c represents the transmitted wave. Comparing the two solutions will then give the transmission coefficients.

The situation in Fig. 9b is slightly more complex. As the contour rotates counterclockwise 180°, it encounters the other branch point. Since a contour obviously cannot cross a branch line and remain an analytic continuation, the branch line originating from $k = k_0$ must also rotate in the k plane to keep out of the way of the original rotating contour. Thus for $x \rightarrow +\infty$ the contour in Fig. 9b rotates to one like that shown in Fig. 9d.

The contour in Fig. 9d may be deformed into that shown in Fig. 10. Thus as $x \rightarrow +\infty$, there is one piece of the solution like $e^{ik_0 x}$ (the transmitted wave) and another piece going like $e^{-ik_0 x}$ (the reflected wave). The horizontal part of the contour in the upper half k plane does not contribute to the integral in Eq. (47b) because e^{ikx} is vanishingly small. However, it is included in the figure so as to keep clear which branch on the left contour connects with which branch on the right.

To summarize, one can show only by manipulating the contours of integration of Eq. (47) but without working out the integrals that a wave incident from $x = -\infty$ (Figs. 9a and 9c) is transmitted but not reflected. On the other hand, a wave incident from $x = +\infty$ (Figs. 9b, 9d), and 10 is both reflected and transmitted just as speculated earlier in this section.

We now proceed to work out the reflection and transmission coefficients. We start with the simpler calculation that of the integral along the contour C_1 in Fig. 9a. The integral is most conveniently done by transforming the independent variable to

$$-u = i(k - k_0)|x|e^{i\theta}, \tag{49}$$

where θ is the argument of x, that is, $\theta = \pi$ for the integral along contour C_1 of Fig. 9a. The integral in the u plane is shown in Fig. 11. Thus

$$E(x \to -\infty) = [i/(|x|e^{i\theta})]^{i\eta}(2k_0)^{-i\eta-1}e^{ik_0x} \int_{C_1} du\, u^{i\eta-1} \exp -u. \tag{50}$$

Let us say that arg $u = 0$ along the lower part of the contour in Fig. 11. Then the argument of u along the upper part of the contour is -2π since the path

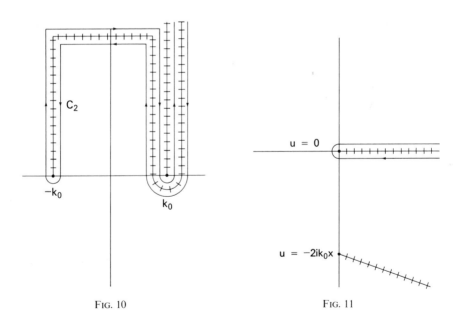

FIG. 10

FIG. 11

FIG. 10. A deformation of the contour of Fig. 9d.

FIG. 11. The contour for integral of Fig. 9a in the u plane.

of integration goes around the pole in the negative direction. Using the fact that

$$\Gamma(t) = \int_0^\infty z^{t-1} e^{-z}\, dz, \tag{51}$$

we find that for $\theta = \pi$,

$$E(x \to -\infty) = [i/(|x|e^{i\pi})]^{i\eta}(2k_0)^{-i\eta-1}(e^{2\pi\eta} - 1)\Gamma(i\eta)e^{ik_0 x}. \tag{52}$$

As $x \to +\infty$, $\theta = \arg x$ rotates from π to zero but the u integration remains unchanged. Thus

$$E(x \to +\infty) = \exp(-\pi\eta)E(x \to -\infty) \tag{53a}$$

and

$$|E^2(x \to \infty)|/|E^2(x \to -\infty)| = e^{-2\pi\eta} \tag{53b}$$

so that the transmission coefficient is

$$T = e^{-2\pi\eta}, \tag{54}$$

where, once again, $\eta = k_0 R\delta/2$. The transmission coefficient in Eq. (54), is exactly that given in Eqs. (16) and (17). Therefore, the WKB treatment of wave transmission gives exactly the same result as the full wave theory for an ordinary mode incident upon the cyclotron resonance from the high-field side.

We now turn to the problem of transmission and reflection of a wave incident from the low-field side. A completely analogous argument gives

$$E(x \to -\infty) = [i/(|x|e^{i\theta})]^{-i\eta}e^{-ik_0 x}$$

$$\times \int_{c_1} V^{-i\eta-1}\{2k_0 - [iV/(|x|e^{i\theta})]\}^{i\eta-1}e^{-V}\, dV, \tag{55}$$

where now

$$V = i(k + k_0)|x|e^{i\theta}$$

and for $x \to -\infty$, $\theta = \arg x = \pi$. Hence

$$E(x \to -\infty) = [i/(|x|e^{i\pi})]^{-i\eta}(2k_0)^{i\eta-1}(e^{2\pi\eta} - 1)\Gamma(-i\eta)e^{-ik_0 x}. \tag{56}$$

As $x \to +\infty$ ($\theta = \arg x = 0$), the contour of integration in the V plane deforms to that shown in Fig. 12. The integral along the top contour of Fig. 12 goes through exactly as in the case of the wave incident from the high-field side so that the transmission coefficient is still given by Eq. (54).

The integral along the bottom contour of Fig. 12, which gives rise to the reflected wave, is slightly more complicated for two reasons. First, the contour integral must be done twice since the path of integration goes twice

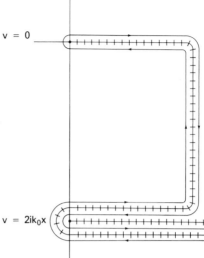

$v = 0$

$v = 2ik_0x$

FIG. 12.   The contour for the integral of Fig. 9d in the V plane.

around the lower pole. Second, care must be taken that the lower contours connect to the upper contours as shown in Fig. 12.

The lower part of the upper contour in Fig. 12 connects with the outer of the contours going around the lower branch point; the upper part of the upper contour connects with the inner contour going around the lower branch point. By our convention, on the lower part of the upper contour, the argument of V is zero. Therefore, on the upper part, it is -2π. Thus the integrand at each point on the upper branch of the upper contour is simply exp $-2\pi\eta$ times the integrant at the analogous point on the lower branch of the upper contour. Hence the integral around the lower branch point is simply $-\exp -2\pi\eta$ times the outer integral around the lower branch point. The overall minus sign, of course, accounts for the fact that the inner and outer contours go around the branch in opposite directions.

Hence we now have only to do the integral around the outer contour around the lower branch point. Before doing this, it is necessary to see how this outer contour connects to the lower part of the upper contour. Particularly, we have to examine what the factor in the integrand of Eq. (14)

$$V^{-i\eta-1}\{2k_0 - [iV/(|x|e^{i\theta})]\}^{i\eta-1}$$

becomes. By convention, arg $V = $ arg $k_0 = 0$ on the lower branch of the upper contour. To get from the lower branch of the upper contour to the upper branch of the lower contour, the argument of V changes by $-\pi/2$ and the magnitude of V becomes approximately $2k_0 x$. Thus in going from the lower branch of the upper contour to the upper branch of the lower,

$$V^{-i\eta} \rightarrow (2k_0 x)^{-i\eta} e^{-\eta\pi/2}. \tag{57a}$$

A similar argument shows that

$$[2k_0 - (iV/|x|)]^{i\eta - 1} \rightarrow u^{i\eta - 1} e^{\pi/2(\eta + i)}, \tag{57b}$$

where u is defined in Eq. (49). Converting the integral over the lower contour to a u integral, one can find an expression for the reflected wave. The coefficient of reflected power turns out to be

$$R = (1 - T)^2. \tag{58}$$

To summarize, for a wave incident from the high-field side (inside) of the torus, there is no reflection, and the transmission coefficient is given by Eq. (53b). If the wave is incident from the low-field side, the transmission coefficient is still given by Eq. (53b) and the reflection coefficient by Eq. (58). If the transmission coefficient is small, Eq. (58) indicates that for purely perpendicular incidence, nearly all of the wave energy is reflected if the wave is incident from the low-field side. Actually, however, if a small by finite k_z is modeled as in Eq. (35), a completely analogous argument shows that the reflection coefficient is reduced as in Eq. (37).

Although the ordinary wave is characterized by $E_z \neq 0$, but $E_x = E_y = 0$, the wave energy is nevertheless deposited into the perpendicular motion of electrons if $k_z \approx 0$. This can be seen by noting that the electromagnetic fields of the ordinary wave can be generated by the single vector potential $\mathbf{A} = A(x, t)\mathbf{i}_z$. Since there is no z dependence, the z component of angular momentum is conserved, or

$$mV_z - eA(x, t)/c = \text{const.} \tag{59}$$

Since $A_z(x, t)$ is a bound, oscillating function, V_z is also so the electrons do not gain parallel energy. Thus the ordinary wave energy is deposited into perpendicular electron energy. Since for $k_z \approx 0$ the energy is deposited into electrons right at the cyclotron resonance, where all electrons are resonant with the wave, we expect that the energy will go into the thermal electrons rather than into an energetic tail of runaway electrons.

VI. Summary of the Properties of the Different Absorption Mechanisms

In this section we summarize the properties of the absorption mechanisms discussed in the previous four sections. The mechanisms are absorption at the upper hybrid resonance, absorption of an ordinary wave at the cyclotron resonance, absorption of an extraordinary wave at the cyclotron harmonic, and absorption of an extraordinary wave with $k_z \approx k_x$ at the cyclotron resonance. The absorption coefficient for the first process is unity. The transmission coefficients for the next three processes are given by Eqs. (16) and (17), (23) and (24), and Eq. (25), respectively.

We shall now discuss the accessibility of the cyclotron and upper hybrid resonances. This condition, that the resonance be accessible from an external launch point, essentially imposes a density limit on the tokamak plasma. This can most easily be visualized by examining the ray paths, in parameter space, in the Allis diagram. The Allis diagram and the ray paths are shown in Fig. 13. For the ordinary mode at the cyclotron frequency and the extra-ordinary mode at the cyclotron harmonic, injection is possible either from the outside (low-field side) or inside (high-field side) of the torus. Thus two possible ray paths are shown for each of these possible mechanisms.

For the ordinary wave at $\omega \approx \Omega_c$, the condition that it reach the cyclotron resonance is that it not be reflected between the point of injection and the point of absorption. Since the ordinary wave is reflected at the critical density, the accessibility condition (i.e., density limitation) reduces to $\omega_p^2 < \Omega^2$.

Now consider the extraordinary wave at either the cyclotron frequency or upper hybrid frequency. In either case the wave must be launched from the inside of the torus. The condition that the wave reach the (cyclotron or upper hybrid) resonance is that it not be reflected at the high-density cyclotron cutoff between the launching point and absorption point. The cyclotron cutoff is given by $\beta^2 = (1 - \alpha^2)^2$, so for $\beta \approx 1$, the high-density cyclotron

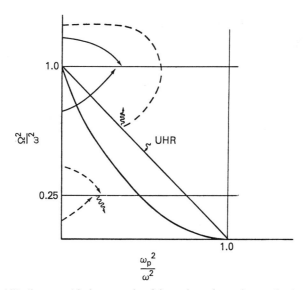

FIG. 13. Allis diagram with the ray paths of the various absorption mechanisms. Shown are ordinary (solid curves), extraordinary (dashed curves), and electrostatic (wavy curves) waves.

cutoff is at $\alpha^2 = 2$. Thus for upper hybrid waves near the cyclotron frequency, the accessibility condition (i.e., density limitation) is $\omega_p^2 < 2\Omega^2$. Thus by using the extraordinary wave rather than the ordinary wave, one can heat plasmas at double the density. However, one must pay the price of launching the wave from the inside of the torus rather than the outside.

Finally, we consider the extraordinary wave at the cyclotron harmonic. In order for the resonant point to be accessible, the wave cannot reflect from the cyclotron cutoff between the launching and absorption points. For $\beta^2 \approx \frac{1}{4}$, the reflection point is at $\alpha^2 \approx \frac{1}{2}$, or $\omega_p^2 \approx 2\Omega^2$, since $\omega \approx 2\Omega$. Thus the density limitation is $\omega_p^2 < 2\Omega^2$. Again, by using the extraordinary wave at the cyclotron harmonic rather than the ordinary wave at the cyclotron frequency, one can heat plasmas at double the density. However, now one must pay the price of operating at twice the frequency.

We now discuss how well the energy deposition process can be controlled. A fundamental control one would like to have is control of the position in the plasma at which the millimeter wave energy is deposited. The energy is deposited ultimately by cyclotron damping at roughly the position where

$$[\omega - n\Omega(x)]/k_z \approx V_e. \tag{60}$$

As was shown in Sections III and V, for the ordinary wave at the cyclotron frequency and the extraordinary wave at the cyclotron harmonic, k_z can be very small, $k_z \approx c/(V_e R)$. Replacing $\omega - \Omega(x)$ with $\Omega(x/R)$, Eq. (60) shows that energy deposition takes place in a region of width of roughly c/Ω around the cyclotron resonance. That is, energy is deposited in a region of order of the free space wavelength. Clearly, for these two processes, excellent control of the position of energy deposition is possible, if k_z can be reduced to $\approx c/(V_e R)$.

Now consider the extraordinary wave at the cyclotron resonance with $k_z \approx k_x \approx \Omega/c$. Equation (60) now indicates that the energy will be deposited in a region of size roughly $x/R \approx V_e/c$ about the cyclotron resonance. For an electron temperature of about a kilovolt, the energy is deposited over the entire cross section. Thus the position of energy deposition cannot be controlled very well. The same is true for absorption at the upper hybrid resonance. At the upper hybrid resonance, the extraordinary mode is converted to a Bernstein mode which propagates toward the cyclotron resonance and is absorbed. If $\Omega^2 \approx \omega_p^2$, the upper hybrid and cyclotron resonances are well separated, and the Bernstein mode must propagate over a sizable portion of the plasma cross section before it deposits its energy in the electrons. Once again, the position of energy deposition cannot be controlled very well.

Finally we turn to the question of just how the wave energy is deposited into the plasma electrons. That is, is the energy deposited into the thermal electrons or into a nonthermal tail on the electron distribution function?

This question is difficult to answer accurately, and the discussion here is not meant to be the last word, rather it is a speculation guided to some extent by the theories presented here and to some extent by experiment. More accurate theories and future experiments may give different results.

We first discuss the ordinary wave at the cyclotron frequency. Since the polarization of the incident wave prevents coupling to other wave types, the energy is deposited directly into plasma electrons at the cyclotron resonance. Since k_z is nearly zero, virtually all electrons are simultaneously resonant with the wave, and we expect the wave energy to be deposited into the thermal electrons rather than into a nonthermal tail. For the extraordinary wave at the cyclotron harmonic, the situation is quite similar. If the wave is incident from the low-field side with a very small but nonzero k_z, energy is once again deposited directly into the plasma electrons near the cyclotron harmonic resonance. Thus all electrons are simultaneously resonant, and we expect the energy to be deposited into thermal electrons rather than into a nonthermal tail.

The situation would appear to be quite different for the other two absorption mechanisms, absorption at the upper hybrid resonance and absorption of an extraordinary wave with $k_z \approx k_x$ at cyclotron resonance. Consider first the former process. At upper hybrid resonance, a Bernstein mode is produced and it propagates from a region of virtually no damping toward the cyclotron

TABLE III

CHARACTERISTICS OF THE FOUR ABSORPTION MECHANISMS

Mechanism	Fractional absorption	Density limit	Direction of energy incident	Spatial deposition control	Deposition
0 wave at $\omega \approx \Omega$, $k_z \ll k_x$	$1 - \exp(-2\pi\eta_{01})$ η_{01} from Eq. (17)	$\omega_p^2 < \Omega^2$	Outside or inside of torus	Good	Probably thermal electrons
x wave at $\omega \approx 2\Omega$, $k_z \ll k_x$	$1 - \exp(-2\pi\eta_{x2})$ η_{x2} from Eq. (23)	$\omega_p^2 < 2\Omega^2$	Outside or inside of torus	Good	Probably thermal electrons
Upper hybrid resonance	100%	$\omega_p^2 < 2\Omega^2$	Inside of torus	Poor	Probably a runaway tail
x wave at $\omega \approx \omega_{ce}$, $k_z \approx k_\perp$	$1 - \exp(-2\pi\eta_{x1})$ η_{x1} from Eq. (25)	$\omega_p^2 < 2\Omega^2$	Inside of torus	Poor	Probably a runaway tail

resonance and strong damping. Then, as discussed in Section IV, the wave first interacts with high parallel velocity particles and may damp all of its energy before ever interacting with thermal electrons. Thus, there is a tendency in this case to deposit the wave energy preferentially into an energetic tail on the electron distribution function.

The situation is similar for an extraordinary mode with $k_x \approx k_z$ incident on the cyclotron resonance. Here again, far from resonance, the wave is undamped. As the wave front approaches the resonance point, it first encounters resonant particles with high-phase velocity. Thus again one would expect the millimeter wave energy to be deposited preferentially into an energetic tail of runaway electrons.

To summarize, the basic properties of these four absorption mechanisms are enumerated in Table III.

VII. Experimental Results on Electron Cyclotron Resonance Heating

There are a great number of experimental results, obtained over a period of many years on electron cyclotron heating of plasmas. However, because plasma dimensions have generally been small, and wavelengths available have been long, there are relatively few experiments on electron cyclotron resonance heating in which a plasma with size much larger than a wavelength has been heated. For this reason, most theory on electron cyclotron heating has emphasized the stochastic acceleration of electrons in, for instance, specified cavity modes, rather than the propagation of energy from an external launch point to a point of deposition in the plasma. There has been some recent experimental work in tokamaks and other toroidal devices for which the theory developed here should be a reasonable approximation to the experiment. Also there is other experimental work on mirrors which we shall touch on very briefly here.

For our purposes here, the most interesting series of experiments is that of Alikaev *et al.* (1975, 1976) on electron heating in TM3. Sources are now available that deliver up to 60 kW of millimeter wave power at 4-mm wavelength (resonant field of 25 kG) for up to 1 msec. If the power is delivered at the second harmonic, the central plasma electron temperature rises by 200 eV and is apparently still increasing at the end of the pulse. In Fig. 14 the electron temperature profile is shown before and during the heating pulse for the case where the power is injected at the fundamental cyclotron resonance. Comparison of laser scattering measurements and diamagnetic measurements of temperature indicate that the energy is deposited into the thermal electrons and not into a runaway tail.

These experimental results seem to be reasonably consistent with the theory detailed in Sections III and V, namely, the heating at cyclotron

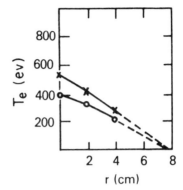

FIG. 14.   Electron temperature before (○) and after (×) the millimeter wave pulse in the experiments of Alikaev *et al.*

resonance arose from the ordinary wave, while the heating at the cyclotron harmonic arose from the extraordinary wave. Alikaev mentions other experimental evidence that supports this interpretation. For the interaction at the second harmonic, the upper hybrid resonance did not occur in the plasma, so the heating could not have been from this process. Furthermore, the absorbed microwave power was independent of power input, thereby indicating a linear absorption process. Also, by selecting the frequency of the incident wave, they could control whether the energy was deposited into the thermal electrons or a runaway tail. If the frequency is sufficiently below the cyclotron frequency ($\Omega/\omega = 1.3$), the wave propagates freely into the plasma from the outside. It does not interact with thermal electrons. However, the wave energy was deposited into a tail of runaway electrons of high parallel velocity. This feature also seems to be consistent with the discussions in the previous section. Finally, another series of experiments used microwave power with $\Omega < \omega < 2\Omega$ so that neither cyclotron nor cyclotron harmonic resonance could occur, but the upper hybrid resonance was in the plasma. In this case, no heating was observed. Whether the absence of heating was due to deposition in the peripheral regions, where energy is not well confined; or due to reflection from the plasma and deposition in the metal walls was not clear. To summarize, the main features of Alikaev's experiment appear to be consistent with a plasma–ordinary wave interaction at the cyclotron resonance and a plasma–extraordinary wave resonance at the cyclotron harmonic.

Perhaps as interesting as the electron heating are the experimental results on energy confinement. It turns out that the electron energy confinement time increases as a function of temperature for this heating mechanism. Figure 15 shows the measured electron energy confinement time in microseconds as a function of central electron temperature.

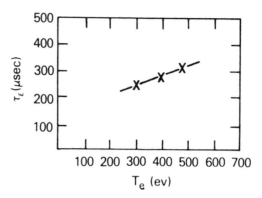

FIG. 15.   The electron confinement time as a function of temperature.

We now briefly discuss three other electron cyclotron heating experiments in toroidal plasmas: the Wisconsin toroidal octopole (Barter *et al.*, 1974); the Princeton FM spherator (Okabayashi *et al.*, 1973); and the Culham Levitron (Riviere *et al.*, 1978). The electron temperature in all three devices is much lower than in a tokamak, typically $T_e \approx 10$ eV. We first discuss the octopole experiment since it is the simplest to interpret in the sense that low power microwaves are absorbed in a preformed plasma. This experiment regards the containment devices as a resonant cavity. The microwave absorption is then found by a measurement of the Q of the cavity with and without plasma.

One object of this experiment was to test the single particle theories of wave absorption. Generally, these theories assume a wave spectrum in the cavity and integrate particle orbits in the electromagnetic fields. When the particle is on that part of its orbit which is in cyclotron resonance with the microwave field, it gains energy irreversibly, and Barter *et al.* (1974), Dandl *et al.* (1964, 1971), Eldridge (1972), Lieberman and Lichtenberg (1973), and Sprott (1971) calculate this energy gain in various geometries.

Actually, it is rather amazing that such a theory could be expected to work at all since cold plasma theory predicts that the plasma is not resonant at the electron cyclotron frequency. That is, in these references, it was always E_r, the component of circular polarization that rotates in the same sense as the electrons, that gives rise to this wave–particle energy exchange. However, cold plasma theory shows that E_r is zero at cyclotron resonance. For instance, the standard cold plasma expression for wave polarization is (Allis *et al.*, 1963)

$$\frac{E_r}{E_l} = \left(\frac{c^2 k^2}{\omega^2} - 1 + \frac{\alpha^2}{1 + \beta}\right) \Big/ \left(\frac{c^2 k^2}{\omega^2} - 1 + \frac{\alpha^2}{1 - \beta}\right), \qquad (61)$$

where, in Eq. (61), $\mathbf{k}$ is no longer assumed to be perpendicular to $\mathbf{B}$. Clearly $E_r = 0$ at $\beta = 1$.

However, despite this uncertainty, Barter *et al.* (1974) show that the single-particle theory does work reasonably well for low density. Figure 16 shows a plot of measured absorption versus theoretically predicted absorption as a function of plasma density. One possible interpretation is the following. Equation 61 shows that far enough away from cyclotron resonance that $\alpha^2/(1 - \beta) \approx 1$, the polarization is not strongly affected by the presence of the plasma. For small α, this position can be very close to the position of resonance. However, if the distance between $\beta = 1$ and $\beta = 1 - \alpha^2$ is small compared to a wavelength, the cavity mode tunnels easily through the narrow evanescent region and E_r at $\beta = 1$ should not be significantly different from E_r far away. Since the vacuum wavelength was about 3 cm, while the transverse dimension was about 18 cm, the vacuum theory could be valid up to large values of α as shown in Fig. 16. However, it is worth pointing out that such a vacuum theory should not be valid for a tokamak where $\alpha \approx 1$ and the wavelength is small compared to the size of the evanescent region. Barter *et al.* (1974) also show that electron heating rate is independent of both electron-neutral collision frequency and also of microwave power. This is further evidence that the absorption is a collisionless, linear mechanism.

Electron cyclotron heating has also been tried in the Princeton FM-1 Spherator (Okabayashi *et al.*, 1973). This experiment also sees strong electron heating. However, the heating has a threshold-like behavior, i.e., it abruptly increases above a critical input power level. This is interpreted as the onset

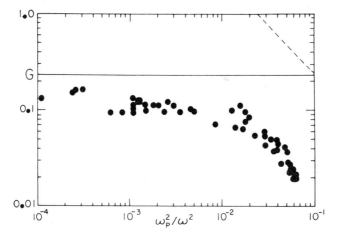

FIG. 16. Absorption as a function of density in the experiments of Barter *et al.* The solid line indicates the theoretical and the dashed line indicates G_{max}.

of a parametric instability. The measured fluctuation spectrum in the plasma does in fact show that above this threshold power there is a low-frequency spectrum at roughly the ion plasma frequency. This is characteristic of a parametric instability.

Another recent experiment on electron cyclotron heating in a toroidal confinement device is that recently reported in the Culham Levitron (Riviere et al., 1978). In this experiment, microwave power of 3-cm wavelength is incident upon the plasma from the outside. The density gradient scale length is roughly 1 cm, so that the distance between cyclotron cutoff, cyclotron resonance, and upper hybrid resonance is less than one vacuum wavelength. This experiment finds good absorption. The absorption is interpreted as mode conversion at the upper hybrid resonance followed by cyclotron damping of the electrostatic wave somewhere between the upper hybrid resonance and cyclotron resonance.

Finally, we very briefly mention studies of electron cyclotron heating in mirror machines (Dandl et al., 1964, 1971; Bernhardi et al., 1976). These experiments generally show absorption in agreement with the single-particle theories. However, the absorption is usually by a superthermal tail of energetic electrons which reaches typically megavolt energy. Hence, the plasmas produced in these mirror experiments appear to be quite different from those produced in tokamaks or toroidal devices.

VIII.   Parametric Instabilities

The absorption mechanisms discussed in Sections II–VI were all linear. One question is whether the linear picture presented here is modified by the presence of a very intense millimeter wave source. In other words, can parametric instabilities play a role? There is a vast literature on parametric instabilities in plasmas. For instance, more than half of Volume 6 of *Advances in Plasma Physics* is devoted to a series of reviews of this area. Clearly it is far beyond the scope of this chapter to do any more than scratch the surface.

For someone interested in the effect of parametric instabilities on electron cyclotron heating of tokamaks, the first and most fundamental question is: Are parametric instabilities beneficial or harmful? They may be beneficial because they give rise to absorption where it does not otherwise exist. For instance, if $\omega < \Omega$, it is apparent from the Allis diagram in Fig. 4 that millimeter wave energy propagates freely into the tokamak from the low-field side as long as the density is less than the critical density for the ordinary mode or the high-density cyclotron cutoff for the extraordinary mode. Of course linear theory would predict no absorption by the plasma. However, if a parametric instability could give rise to absorption, it would obviously be a very desirable effect. The basic process is the decay of the transverse wave into two or more electrostatic waves which are then trapped in the plasma. In general, the electrostatic waves propagate parallel to the electric field of the original millimeter wave (the so-called pump wave). A diagram of this basic process is shown in Figs. 17a and 17b. There are two basic types of decay process (Silin, 1965; Dubois and Goldman, 1967). First, the transverse wave at $\omega = \omega_0'$ $k_0 = 0$ can decay into two propagating electron and ion waves (Fig. 17a). The waves produced satisfy the standard selection rules

$$\omega_0 = \omega_e + \omega_i, \qquad 0 = k_e + k_i. \tag{62}$$

This process is usually called the *decay instability*. Second, the transverse wave can decay into a standing electron wave and a purely growing ion wave

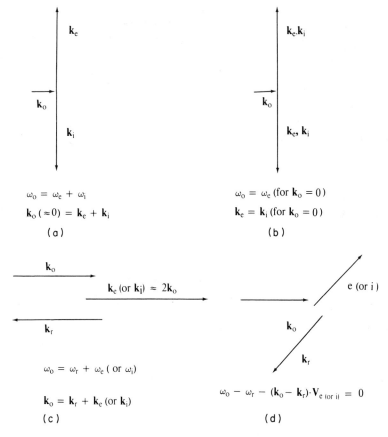

FIG. 17. Diagram of four basic parametric instabilities: (a) the decay instability; (b) the oscillating two-stream instability; (c) Raman or Brillouin backscatter; (d) induced electron or ion Compton scattering.

shown in Fig. 17b. These waves do not satisfy the selection rules, and the decay products do not propagate in the undriven system. The process can be viewed as a longitudinal self-focusing. As electric field energy concentrates in one region of space, the enhanced electrostatic pressure forces plasma out of this region. However, when plasma is forced out, the index of refraction increases, and electrostatic energy is further refracted into the region of higher field strength. This instability is usually called the *oscillating two-stream instability*.

On the other hand, parametric instabilities could be harmful in a number of ways. For instance, rather than giving rise to anomalous absorption, they could give rise to anomalous reflection. The basic process here is pump decay into an electrostatic wave and a reflected wave, shown in Fig. 17c. If the electrostatic wave is an electron wave, the process is usually called *Raman scattering*; if it is an ion wave, *Brillouin scattering*. Also, the incident wave can scatter off an electron or ion, as shown in Fig. 17d. This is usually called *induced Compton scattering*. Even if the parametric instability gives rise to absorption, the process might still be harmful, as simulations of parametric instabilities have shown energy is likely to be deposited into an energetic tail rather than into thermal electrons. For instance, the heated electron distribution function (Thompson *et al.*, 1974) resulting from a particle simulation of a parametric instability is shown in Fig. 18. It is not the purpose of this chapter to conclude whether or not parametric instabilities are beneficial or not. Indeed, such a conclusion is impossible at this time because thresholds and growth rates of some of the relevant backscatter instabilities have not yet been worked out. However, the personal opinion of this author (and this opinion may in time be proven incorrect) is that parametric instabilities are harmful and should be avoided if possible.

This opinion is based largely on experience in laser fusion. Here laser light propagates through the underdense plasma to the critical surface where it is (hopefully) absorbed. The problem is, of course, that the critical density is a reflection point, not an absorption point. It was originally assumed that at the critical density, the laser light would excite a parametric instability and decay into electrostatic electron and ion waves. It was, of course, also realized that Brillouin or Raman scattering in the underdense plasma could prevent the laser light from ever reaching the critical surface.

Opinion in the laser fusion community now seems to be shifting. It was found that if the laser intensity is large enough to generate parametric instability at the critical surface, its momentum density there is also high enough to cause a density discontinuity over a free space wavelength. This steep density gradient then stabilizes the decay instability at the critical surface. Thus opinion now is that parametric instabilities do not play a

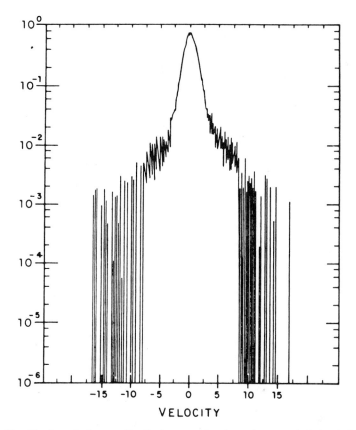

FIG. 18. The heated electron distribution resulting from the numerical simulation of a parametric instability (taken from Thomson *et al.*, 1974).

significant role in absorption, but that resonant absorption (Friedberg *et al.*, 1972; Estabrook *et al.*, 1975) and perhaps other processes do (Manheimer *et al.*, 1977).

However, the gradient scale length in the underdense plasma can still be long, so that Raman and Brillouin scattering can (Forslund *et al.*, 1973; Kruer *et al.*, 1973) and do (Ripin *et al.*, 1974; Kruer, 1978) take place in laser-produced plasmas. Thus, while the beneficial effects of parametric instabilities are deemphasized, their harmful effects persist. Most physicists involved in laser fusion probably now feel their jobs would be easier if there were no such thing as a parametric instability.

For electron cyclotron heating of tokamaks, the same conclusion might well apply. As shown in Sections II–VI, there is no shortage of linear absorption mechanisms, so anomalous absorption mechanisms are certainly not necessary. However, there is still a potential for anomalous reflection, heating of the tail rather than the body of the electron distribution function, or other mischief.

We now very briefly discuss what a parametric three-wave decay instability is, since this instability can be responsible for either anomalous absorption or anomalous scattering. Denoting the amplitude of the pump, electron, and ion waves, respectively, by a_0, a_e, and a_i, generally, one finds that if Eq. (62) is satisfied, the equations for a_e and a_i are

$$(da_e/dt) + v_e a_e = Ma_0 a_i^*, \tag{63a}$$

$$(da_i^*/dt) + v_i a_i^* = M^* a_0^* a_e, \tag{63b}$$

where M is the so-called matrix element describing the strength of the interaction and v_e and v_i the linear damping rates of the waves. Assuming temporal dependence $\exp \gamma t$, one can easily combine Eqs. (63) to obtain

$$\gamma = \tfrac{1}{2}\{-v_e - v_i \pm [(v_e - v_i)^2 + 4|Ma_0|^2]^{1/2}\}. \tag{64}$$

For $v_e = v_i = 0$ the growth rate is given by $\gamma = |Ma_0|$. For nonzero linear damping the mode is unstable only if

$$|Ma_0|^2 > v_e v_i, \tag{65}$$

so that both modes must have nonzero damping in order to stabilize the system. Thus one way to eliminate parametric instabilities is simply to operate at low enough pump wave strength that the system is stable.

Even if the system is unstable in homogeneous media, the presence of inhomogenieties can still have a stabilizing effect. In inhomogeneous media, the k matching condition can only be satisfied in a limited region of space. If density inhomogenieties had been included in Eq. (63), the coupling term on the right-hand side of, for instance, Eq. (63a) would be multiplied by (Rosenbluth, 1973)

$$\exp i \int [k_{0x}(x) - k_{ex}(x) - k_{ix}(x)] \, dx,$$

where now we have generalized to allow $k_0 \neq 0$. There can only be strong coupling in the region of space where the waves are in phase. The size of this region L_p is given roughly by

$$\int_0^{L_p} [k_{0x}(x) - k_{ex}(x) - k_{ix}(x)] \, dx \approx 1, \tag{66}$$

where $x = 0$ is the position of perfect wave number matching. Thus the interaction takes place in a spatial region of dimension of order L_p. However the electron and ion waves only spend a finite time in this region. The condition for instability is that the time the waves spend in the region is long enough to allow significant growth.

The simplest case is where the waves convect out of the unstable region at their group velocities V_{ge} and V_{gi}. In this case the condition for instability is

$$2\pi |Ma_0|^2 L_p^2/(V_{ge} V_{gi}) > 1. \tag{67}$$

If the wave is produced in a region where its group velocity is zero, the calculation is slightly more complex, but the idea is the same. Instead of convecting out of the unstable region, one or both of the decay waves refract out (Klein *et al.*, 1973). The time each wave spends in an unstable region can then be calculated and an analogous threshold condition can be worked out.

If the effect of linear damping and inhomogeneity are still not sufficient to stabilize the system, there is still one additional possibility. As we have seen by considering the case of inhomogeneity, a parametric decay instability requires a particular phase relation between all three waves involved. Another way to destroy this phase relation and thereby help stabilize the system is to send in a pump wave with a frequency bandwidth (Thompson, 1975). In a homogeneous system, a wide-band pump cannot completely stabilize the system if $v_e = v_i = 0$, but the growth rate is reduced roughly to $\gamma^2/\Delta\omega$, which may be important if the frequency band width $\Delta\omega$ is significantly larger than the growth rate $\gamma = |Ma_0|$. If v_e and v_i are not zero, then the threshold for instability is correspondingly raised. Thus if anomalous scattering proves to be a significant problem in electron cyclotron resonance heating of tokamak plasmas, one possible solution might be to use a relatively wide-band millimeter wave source.

Analogous calculations also show that the oscillating, two-stream instability can be stabilized by linear damping of the electron wave alone. Since the threshold does not depend on the damping rates of both decay waves but only of one, the threshold field strength for instability is higher than for the decay instability. Since the oscillating, two-stream instability also depends on a particular phase relation between the pump and decay waves, analogous calculations show that both inhomogeniety and finite bandwidth can be important stabilizing effects (Perkins and Flick, 1978).

The induced Compton scattering instabilities, on the other hand, do not depend on any particular phase relation between the pump and decay wave. Thus damping of the decay wave can provide stabilization, but inhomogeneity and finite bandwidth cannot. These instabilities, however, generally have much lower growth rates than decay and oscillating, two-stream instabilities.

We conclude by very briefly discussing calculations of parametric instabilities relevant to electron cyclotron heating of tokamaks. As far as anomalous scattering is concerned, it appears that the relevant calculations have not yet been done. Calculations of parametric instabilities relevant to scattering have been done for an unmagnetized or weakly magnetized ($\Omega \ll \omega_p$) plasma (Manheimer and Ott, 1974), or else for the case where all three waves propagate parallel to the magnetic field (Kaw, 1976). The basic configuration in electron cyclotron resonance heating, of course, is that where all waves propagate perpendicular or nearly perpendicular to the field in a strongly magnetized plasma.

Calculations of threshold and growth rate for the decay of a dipole ($k_0 = 0$) pump wave into electrostatic waves in a strongly magnetized plasma have been done (Porkolab, 1972). The dispersion relation [Eq. (1) from Porkolab (1972)] is quite complicated and numerical calculations have been made only at a few points in parameter space. A *very* rough summary of the results is the following:

(a) Far above threshold, the growth rate maximizes at about the ion plasma frequency.

(b) The threshold is given by $\mu \gtrsim 0.1$.

The quantity μ [Eq. (2a) of Porkolab (1972)] is a fairly complicated quantity involving the electric field, frequency, and polarization of the pump wave. If $\omega_0 \approx \Omega$, we may regard it approximately as

$$\mu \approx [eE_0/(m\omega_0 c)](\Omega/(|\omega_0 - \Omega|)), \tag{68}$$

where the pump wave is $\mathbf{E} \cos \omega_0 t$. The incident irradiance I for threshold, as given by Eq. (68), then is

$$I \approx 2 \times 10^9 (B/30 \text{ kG})^2 [(\omega_0 - \Omega)/\Omega]^2 \quad \text{W/cm}^2. \tag{69}$$

Thus the threshold for decay instability appears to be quite high.

IX. Speculations on the Effect of Electron Cyclotron Heating on Energy Confinement

We now turn to the most speculative part of this chapter, the effect of electron cyclotron heating on electron energy confinement. As discussed in the first section, there are at least three ways in which energy might be lost from the center of tokamak plasmas. The first is radiation by high Z impurities that get to the center of the plasma. The second is by excitation of MHD instabilities. These instabilities are generally excited in the center of the plasma if the current density there is sufficiently high that the inverse rotational transform $q = B_z r/(B_\theta R)$ (B_z is the toroidal field, B_θ the poloidal field, r the radial position in the plasma, and R the major radius of the torus) is less than unity. Finally, electron energy can be lost by electron thermal conduction. This thermal conduction can be either neoclassical or anomalous. If it is neoclassical, any process that enhances the collision frequency of the electrons is deleterious to confinement. If it is anomalous, the situation is less clear cut. Recent theories of relevant plasma instabilities seem to show that as the collision frequency is reduced, these instabilities are more strongly excited so that increasing collisionality may be beneficial. In any case, we shall show here that millimeter wave power perturbs the collisionality only very slightly.

We now turn to a discussion of the possible effects of electron cyclotron resonance heating on these three loss mechanisms. At first sight it may seem obvious that electron cyclotron heating cannot increase the impurity content of the plasma. However, experiments on neutral beam heating of tokamaks have generally shown that the impurity radiation does increase with injected power. In some experiments the increase in impurity content was so great that the central electron temperature decreased rather than increased with injected power (Equipe, 1977). Experiments on lower hybrid heating have also shown an increase in impurity radiation with injected power.

The effect of input power on impurity influx is an extremely complicated problem about which this author knows very little, so this discussion shall be confined to a few very qualitative considerations. There are two possible sources of impurities in the plasma, the inner wall of the vacuum chamber and the limiter. The limiter is a ring-shaped flange inside the vacuum chamber. It has an inner hole that defines the radius of the plasma. Since the tokamak is run in steady state (at least for times long compared to an energy confinement time), all input power is ultimately deposited on either the chamber wall or else on the limiter.

As an example, consider a tokamak like TFR, where the input power is roughly $\frac{1}{2}$MW, the major radius is 100 cm, and the minor radius is 20 cm. The maximum power flux on the chamber wall is then less than 1 W/cm^2 roughly comparable to the power flux on the glass wall of a 100-W light. However, since the limiter only touches the plasma on area of perhaps 10 cm^2, the maximum potential power flux on it is 5×10^4 W/cm^2, which is about 100 times larger than the power flux on the inner cylinder wall of an automobile engine. This then suggests that the limiter is a potentially greater source of impurities.

The way to minimize impurities then is to minimize the power flux to the limiter and have the remaining power strike the chamber wall as "gently" as possible. For instance, Alcator has greatly reduced impurity content by surrounding the plasma with relatively dense neutral gas. Presumably the incident power is deposited on the chamber walls by ordinary thermal conduction through this neutral gas, thereby cushioning the chamber wall. That is, energy is deposited by warm gas, not by x-rays, energetic particles, or anything else that can sputter the wall.

Therefore, the outer part of the plasma should be kept as cool as possible, and the hot plasma should be confined to the center. Anything that heats the edge of the plasma will most likely increase the impurity content and therefore decrease the energy confinement time. In this respect, electron cyclotron heating appears to have a unique advantage in that it is the only mechanism for which theory indicates that the position of energy deposition can be precisely controlled. This especially appears to be an advantage over schemes that rely on injected neutral or charged particle beams. There, not only is there a possibility of heating the plasma edge if the deposition length is too small, but there is also a possibility of some beam propagation to the opposite chamber wall if the deposition length is too long. Thus at least these simple considerations show that because one can control the position of energy deposition in electron cyclotron resonance heating, it is about the most attractive scheme as far as minimizing impurities is concerned.

We now discuss whether electron cyclotron heating can affect MHD instabilities. The most obvious potential effect would be to cause a narrow heated region in the plasma center. Since the plasma electrical conductivity is proportional to $T^{3/2}$, the electric current tends to channel into this heated region and reduce q. If it were to fall below unity, MHD instabilities could be excited, and anomalous loss would result.

If one could model this loss process by an anomalous electron thermal conduction which increased whenever q fell below unity, then the effect of these MHD instabilities would be to spread the injected energy over the central region of the plasma so that q is everywhere greater than one. Thus these instabilities would not reduce the overall electron energy confinement time; they would only prevent the energy from concentrating too much in the center.

However, even if these instabilities were absolutely catastrophic, electron cyclotron heating is still a viable mechanism. By using several different millimeter wave sources at several nearby frequencies, one could tailor the energy deposition profile with virtually no constraint. Thus there does not appear to be any reason to suppose that electron cyclotron resonance heating will lower the confinement time by exciting MHD instabilities.

We finally turn to the effect of electron cyclotron heating upon the electron collisionality. To simplify the problem we specialize to the case of ordinary waves near Ω incident at perpendicular incidence. The quasilinear equation for the electron distribution function is

$$\frac{\partial f}{\partial t} = \pi \left(\frac{e}{m\omega}\right)^2 \sum_\omega \frac{1}{V_\perp} \frac{\partial}{\partial V_\perp} \Omega^2 \left| V_z E_z(\omega) J_1 \left(\frac{kV_\perp}{\Omega}\right) \right|^2 \delta(\omega - \Omega) \frac{1}{V_\perp} \frac{\partial f}{\partial V_\perp}$$

$$= \pi \left(\frac{e}{m\omega}\right)^2 \sum_\omega \frac{1}{V \sin \theta} \left\{ \sin \theta \frac{\partial}{\partial V} + \frac{\cos \theta}{V} \frac{\partial}{\partial \theta} \right\}$$

$$\times \Omega^2 \left| V \cos \theta\, E_z(\omega) J_1 \left(\frac{kV \cos \theta}{\omega}\right) \right|^2 \frac{1}{\sin \theta} \left\{ \sin \theta \frac{\partial}{\partial V} + \frac{\cos \theta}{V} \frac{\partial}{\partial \theta} \right\} f \tag{70}$$

where $V_\perp$, V_z denote cylindrical coordinates and V, θ denote spherical coordinates in velocity space (Davidson, 1972). The summation over ω accounts for the fact that there may be more than a single incident frequency present. If one defines the collision frequency for pitch angle scattering induced by the millimeter waves $\nu_M(\theta)$ as the coefficient of the $\partial^2 f / \partial \theta^2$ term on the right-hand side of Eq. (70), one finds

$$\nu_M(\theta) = \pi (e/2m)^2 \sum_\omega \cos^4 \theta (k/\omega)^2 E_z^2(\omega) \delta(\omega - \Omega), \tag{71}$$

where we have assumed that the argument of the Bessel function is much less than one. If the electron distribution function is Maxwellian, the average value of $\cos^4 \theta$ is $\frac{1}{5}$. Defining ν_M as the collision frequency averaged over θ,

$$\nu_M = \frac{1}{5} \pi (e/2m)^2 \sum_\omega |E_z(\omega)(k/\omega)|^2 \delta(\omega - \Omega). \tag{72}$$

By calculating the heating rate of the electrons, it is possible to relate this collision frequency to the millimeter wave power input. For $k_z = 0$, the only heating is in the perpendicular direction, so the millimeter wave power input to the electrons per unit volume is

$$P_m = \int \frac{1}{2} m V_\perp^2\, \partial f / \partial t\, d^3 V. \tag{73}$$

Using Eq. (70) in cylindrical velocity space coordinates, we find

$$P_m = 10 n T_e \nu_M. \tag{74}$$

From Eq. (74) it is possible to relate the millimeter wave energy to the confinement time τ_e, which is defined as the energy per unit volume $\frac{3}{2} n(T_e + T_i)$ divided by the total power input per unit volume $P = P_0$ (ohmic power) + P_m. The result is

$$\nu_M = [3/(20\tau_e)][1 + (T_i/T_e)] - [P_0/(10nT_e)]. \tag{75}$$

Thus the induced collision frequency for pitch angle scattering ν_M is considerably less than the reciprocal of the energy confinement time τ_e^{-1}.

However, the actual electron pitch angle collision frequency is roughly the time for an electron to move one larmor radius across the magnetic field. Clearly it takes many such collisions for an electron to get from the center of the tokamak to the edge. That is, the actual electron pitch angle scatter collision frequency is much greater than reciprocal of the energy containment time. Therefore the millimeter, wave-induced, collision frequency ν_M is much less than the (classical or anomalous) collision frequency characteristic of the electrons in the absence of the millimeter waves. The effect of the millimeter waves on the (classical or anomalous) electron thermal conduction should then be very small.

To summarize, let us first reemphasize that the entire problem of the energy loss in tokamaks is very speculative, since the loss mechanisms themselves have not even been positively identified. Thus no one can really say what the effect of electron cyclotron resonance heating on energy confinement will be. However, the simple theoretical ideas presented here all indicate that there is not now any reason to expect that electron cyclotron heating will in itself reduce the electron energy confinement time.

REFERENCES

Adam, J. *et al.* (1975). "Plasma Physics and Controlled Nuclear Fusion Research 1974," Vol. 1, p. 65. IAEA, Vienna.

Alikaev, V. V., Bobrovskii, G. A., Poznyak, V. I., Razumova, K. A., and Sokolov, Yu. A. (1975). *Nucl. Fusion Suppl.* 33.

Alikaev, V. V. *et al.* (1976). *Sov. J. Plasma Phys.* **2**, 212.

Alikaev, V. V., Dnestrovskii, Yu. N., Parail, V. V., and Perevezev, G. V. (1977). *Sov. J. Plasma Phys.* **3**, 127.

Allis, W. P., Buchsbaum, S. J., and Bers, A. (1963). "Waves in Anisotropic Plasmas," Chapter 4. MIT Press, Cambridge, Massachusetts.

Antonsen, T. M., and Manheimer, W. M. (1978). *Phys. Fluids* (to be published); also see NRL Memo 3583, Naval Research Laboratory, Washington, D.C. 20375.

Apgar, E., Coppi, B., Condhalekar, A., Helava, H., Komm, D., Martin, F., Montgomery, B., Pappas, D., Parker, R., Overskei, D. (1977). "Plasma Physics and Controlled Thermonuclear Research 1976," Vol. 1, p. 247, IAEA, Vienna.

Artsimovich, L., Vershkov, V., Glukhov, A., Gorbunov, E., Zaveryaev, V., Lysenko, S., Mirnov, S., Semenov, I., Strelkov, V. (1972). *Nucl. Fusion Suppl.* 41.

Barter, J. D., Sprott, J. C., and Wong, K. L. (1974). *Phys. Fluids* **17**, 810.

Benford, J., Ecker, B., and Bailey, V. (1974). *Phys. Rev. Lett.* **33**, 574.

Berlizov, A., Bobrovskii, G., Bagdasarov, A., Vasain, N., Vertiporokh, A., Vinogradov, V., Vinogradova, N., Gegechkory, N., Gorbunov, E., Dnestrovskii, Yu. N., Zaveryaev, V., Izvozchikov, A., Lukyanov, S., Lysenko, S., Maksimov, Yu., Notkin, G., Petrov, M., Papkov, G., Razumova, K., Strelkov, V., Schcheglov, D. (1977). "Plasma Physics and Controlled Nuclear Fusion Reseach 1976," Vol. 1, p. 3. IAEA, Vienna.

Bernhardi, K., Fuchs, G., Goldman, M. A., Hebert, H. C., Obermann, D., Walcher, W., Wiesemann, K. (1976). *Plasma Phys.* **18**, 77.

Berry, L., Bush, C., Callen, J., Colchin, R., Dunlap, J., Edmonds, P., England, A., Foster, C., Harris, J., Howe, H., Isler, R., Jahns, G., Ketterer, H., King, P., Lyon, J., Mihalczo, J., Murakami, M., Neidigh, R., Nielson, G., Pare, V., Shaeffer, D., Swain, D., Wilgen, J., Wing, W., Zweben, S. (1977). "Plasma Physics and Controlled Nuclear Fusion Research 1976," Vol. 1, p. 49. IAEA, Vienna.

Blanc, P., Hess, W., Ichtchenko, G., Javal, P., Lallia, P. Mahn, C., Nguyen, T. K., Ohlendorf, W., Pacher, G. W., Pacher, H. D., Takamura, S., Tonon, G., Wegrowe J. G. (1977). "Plasma Physics and Controlled Fusion Research 1976," Vol. 3, p. 49. IAEA, Vienna.

Bol, K., Cecchi, J., Daughney, C., Ellis, R., Eubank, H., Furth, H., Goldston, R., Hsuan, H., Mazzucato, E., Smith, R., Atott, P. (1975). "Plasma Physics and Controlled Nuclear Fusion Research 1974," Vol. 1, p. 83, IAEA, Vienna.

Bottiglioni, R., Coutant, J., and Fois, M. (1976). *Int. Meeting Theoret. Exp. Aspects Heating Toroidal Plasmas, 3rd, Grenoble.*

Budden, K. G. (1961) "Radio Waves in the Ionosphere," Chapter 21.15. Cambridge Univ. Press, London and New York.

Dandl, R. A., England, A. C., Ard, W. B., Eason, H. O., Becker, M. D., and Haas, G. M. (1964). *Nucl. Fusion* **4**, 344.

Dandl, R. A., Eason, H. O., Edmunds, P. H., and England, A. C. (1971). *Nucl. Fusion* **11**, 411.

Davidson, R. C. (1972). "Methods of Nonlinear Plasma Theory," Chapter 8. Academic Press, New York.

Dimock, D., Eckhartt, D., Eubank, H., Hinnov, E., Johnson, L., Meservey, E., Tolnas, E., Grove, D. (1971). "Plasma Physics and Controlled Fusion Research 1971," Vol. 1, p. 451. IAEA, Vienna.

Dubois, D. F., and Goldman, M. V. (1967). *Phys. Rev.* **164**, 207.

Eldridge, O. (1972). *Phys. Fluids* **11**, 676.

Eldridge, O., Namkung, W., and England, A. C. (1978) *Phys. Fluids* (to be published); also see ORNL/TM-6052, Oak Ridge National Laboratory, Oak Ridge, Tennessee.

Equipe, T. F. R. (1975). "Plasma Physics and Controlled Fusion Research 1974," Vol. 1, p. 135, IAEA, Vienna.

Equipe, T. F. R. (1977). "Plasma Physics and Controlled Fusion Research 1976," Vol. 1, p. 69, IAEA, Vienna.

Estabrook, K. G., Valeo, E. J., and Kruer, W. L., (1975). *Phys. Fluids* **18**, 1151.

Eubank, H., Goldston, R., Arunasalam, V., Bitter, M., Bol, K., Boyd, D., Bretz, N., Bussac, J.-P., Cohen, S., Colestock, P., Davis, S., Dimock, D., Dylla, H., Efthimion, P., Grisham, L., Hawryluk, R., Hill, K., Hinnov, E., Hosea, J., Hsuan, H., Johnson, D., Martin, G., Medley, S., Meservey, E., Sauthoff, N., Schilling, G., Schivell, J., Schmidt, G., Stauffer, P., Stewart, L., Stodiek, W., Stooksberry, R., Strachan, J., Suckewer, S., Tait, G., Ulrichson, M., von Goeler, S., and Yamada, M. (1978). "Plasma Physics and Controlled Thermonuclear Fusion 1978," paper IAEA-CN-37-C-3. IAEA, Vienna.

Fidone, I., Granata, G., Meyer, R., and Ramponi, G. (1977). Eur-CEA-FC912, Euratom Rep. Association Euratom, 92260, Fontenay-anx-Roses, France.

Forslund, D. W., Kindell, J. M., and Lindman, E. L. (1973). *Phys. Rev. Lett.* **30**, 739.

Friedberg, J. P., Mitchell, R. W., Morse, R. L., and Rudsinki, L. J. (1972). *Phys. Rev. Lett.* **28**, 795.

Golant, V. E., and Piliya, A. D. (1972). *Sov. Phys. Usp.* **14**, 413.

Gorman, P. (1966). *Phys. Fluids* **9**, 1262.

Grawe, H., (1969). *Plasma Phys.* **11**, 151.

Grove, D., Arunasalam, V., Bol, K., Boyd, D., Bretz, N., Brusati, M., Cohen, S., Dimock, D., Dylla, F., Eames, D., Eubank, H., Fraenkel, B., Girard, J., Hawryluk, R., Hinnov, E., Horton, R., Hosea, J., Hsuan, H., Ignat, D., Jobes, F., Johnson, D., Mazzucato, E., Merservey, E.,

Sauthoff, N., Schivell, J., Schmidt, G., Smith, R., Stauffer, F., Stodiek, W., Strachan, J., Suckewer, S., von Goeler, S., Young, K. (1977). "Plasma Physics and Controlled Nuclear Fusion Research 1976," Vol. 1, p. 21. IAEA, Vienna.

Horton, C. W. (1966). *Phys. Fluids* **9**, 815.

Kaw, P. (1976). *Adv. Plasma Phys.* **6**, 207, references therein.

Kawamura, T., Momota, H., Namba, C., and Tershima, Y. (1971). *Nucl. Fusion* **11**, 339.

Klein, H. H., Manheimer, W. M., and Ott, E. (1973). *Phys. Rev. Lett.* **31**, 1187.

Kochetkov, V. M. (1976). *Sov. Phys.-Tech. Phys.* **21**, 280.

Kruer, W. L. (1978). *Comments Plasma Phys.* **4**, 13.

Kruer, W. L., Estabrook, K. G., and Sinz, K. N. (1973). *Nucl. Fusion* **13**, 952.

Kuches, A. F. (1968). *Plasma Phys.* **10**. 367.

Kuehl, H. H. (1967). *Phys. Rev.* **154**, 124.

Lieberman, M. A., and Lichtenberg, A. J. (1973). *Plasma Phys.* **15**, 125.

Litvak, A. G., Permitin, G. V., Scworov, E. V., and Frajmin, A. A. (1977). *Nucl. Fusion* **17**, 659.

Longren, K. E., Sjolund, A., and Weissglass, P. (1966). *Plasma Phys.* **8**, 657.

Manheimer, W. M., and Ott, E. (1974). *Phys. Fluids* **17**, 1413,

Manheimer, W. M., Colombant, D. G., and Ripin, B. H. (1977). *Phys. Rev. Lett.* **38**, 1135.

Okabayashi, M., Chen, K., and Porkolab, M. (1973). *Phys. Rev. Lett.* **31**, 1113.

Ott, E., and Manheimer, W. M. (1977). *Nucl. Fusion* **17**, 1057.

Ott, E., Hui, B., and Chu, K. R. (1979). To be published.

Perkins, F. W., and Flick, J. (1971). *Phys. Fluids* **14**, 2012.

Porkolab, M. (1972). *Nucl. Fusion* **12**, 329.

Richards, B., Stone, D. S., Fisher, A. S., and Bekefi, G. (1978). *Comments Plasma Phys. Controlled Fusion* **3**, 117 (1978).

Ripin, B. H., McMahon, J. M., McLean, E. A., Manheimer, W. M., and Stamper, J. A. (1974). *Phys. Rev. Lett.* **33**, 634.

Riviere, A. C., Alcock, M. W., and Todd, T. N. (1978). *Topical Conf. Radio Frequency Heating of Plasmas, 3rd, Pasadena, California.*

Rosenbluth, M. N., White, R. B., and Liu, C. S. (1973). *Phys. Rev. Lett.* **31**, 1190.

Silin, V. P. (1965). *Sov. Phys.* **21**, 1127.

Sprott, G. (1971). *Phys. Fluids* **14**, 1795.

Stix, T. H. (1963). "A Theory of Plasma Waves." McGraw–Hill, New York.

Stix, T. H. (1965). *Phys. Rev. Lett.* **15**, 878.

Sudan, R. (1976). *Int. Meeting Theoret. Exp. Aspects Heating Toroidal Plasmas, 3rd, Grenoble.*

Thomson, J. J. (1975). *Nucl. Fusion* **15**, 2012.

Thomson, J. J., Faehl, R. J., Kruer, W. L., and Bodner, S. B. (1974). *Phys. Fluids* **17**, 973.

INDEX

353